SCIENCE AND EMPIRE IN EASTERN EUROPE

Imperial Russia and the Habsburg Monarchy
in the 19th Century

Bad Wiesseer Tagungen des Collegium Carolinum

Band 38

Herausgegeben vom
Vorstand des Collegium Carolinum
Forschungsinstitut für die Geschichte Tschechiens
und der Slowakei

Science and Empire in Eastern Europe

Imperial Russia and the Habsburg Monarchy in the 19th Century

Proceedings of the Annual Conference
of Collegium Carolinum
Bad Wiessee, 5–8 November 2015

Edited by
Jan Arend

Vandenhoeck & Ruprecht

Bibliografische Information der Deutschen Nationalbibliothek

Die Deutsche Nationalbibliothek verzeichnet diese Publikation in der Deutschen Nationalbibliografie; detaillierte bibliografische Daten sind im Internet über <http://dnb.ddb.de> abrufbar.

Bibliographic information published by the Deutsche Nationalbibliothek

The Deutsche Nationalbibliothek lists this publication in the Deutsche Nationalbibliografie; detailed bibliographic data available online: <http://dnb.ddb.de>.

ISBN 978-3-525-31074-8

© 2020 Collegium Carolinum, 81669 München
(www.collegium-carolinum.de)

Verlag: Vandenhoeck & Ruprecht GmbH & Co. KG, Göttingen
(www.vandenhoeck-ruprecht-verlage.com)

Das Werk einschließlich aller Abbildungen ist urheberrechtlich geschützt. Jede Verwertung außerhalb der Grenzen des Urheberrechtsgesetzes ist ohne Zustimmung des Collegium Carolinum unzulässig und strafbar. Das gilt insbesondere für Vervielfältigungen, Übersetzungen, Mikroverfilmungen und die Einspeicherung und Bearbeitung in elektronischen Systemen.

All rights reserved. No part of this book may be reproduced or translated in any form, by print, photoprint, microfilm or any other means without written permission from the Collegium Carolinum. Violations of the above may result in criminal prosecution or civil damage awards.

Für Form und Inhalt trägt der/die jeweilige Verfasser/in die Verantwortung.

Redaktion: Collegium Carolinum, München

Satz: Collegium Carolinum, München

Layout des Einbands: SchwabScantechnik, Göttingen (www.schwabscantechnik.de)

Druck und Einband: Verlagsdruckerei Michael Laßleben, Kallmünz
(www.oberpfalzverlag-lassleben.de)

Gedruckt auf säurefreiem, alterungsbeständigem und chlorfrei gebleichtem Papier.

CONTENTS

Jan Arend: Science and Empire in the European Continental Empires: An Introduction 1

I. Learned Societies and Academies

Martin Franc: An Academy of Sciences in Prague?: Purkyně's Akademia and Josef Hlávka's Emperor Franz Josef Czech Academy for Sciences, Literature, and Arts 23

Maciej Janowski: General Scientific Associations in Partitioned Poland: Socializing the Intelligentsia, Preserving the Nation, and Cultivating the Sciences 43

II. Universities

Jan Surman: Habsburg Higher Education in the Eyes of Foreigners: Ludwik Tęgoborski and William Wilde 55

Mark Hengerer/Sabrina Rospert: Did the Universities of the Late Habsburg Monarchy Impart "Imperial Knowledge"?: A Survey of the Course Catalogues of the Universities of Budapest and Vienna (Summer Semesters 1866, 1886, 1906) 77

Andrej Andreev: Between Science and Empire: Superintendents in Russian Universities during the First Half of the Nineteenth Century 97

III. Science on the Imperial Periphery

Daniel Baric: Classical Archeology and Science Policy in Habsburg Bosnia and Herzegovina: Carl Patsch in Sarajevo, 1891–1918 119

Matthias Golbeck: Letters from the Edge of Empire: Self-Descriptions of the Imperial Official and Scientific Amateur N. F. Petrovskii from Turkestan, 1870–1895 135

IV. Oriental Studies

Johannes Feichtinger: "Orientalistik" in the Habsburg Monarchy
between Imperial Pragmatism and "Pure" Scholarship 151

Arpine Maniero: Oriental Studies in the Russian Empire in the
Context of Imperial Politics and Regional Discourse 169

V. Political Geography and "Landeskunde"

Borbála Zsuzsanna Török: Exploring the k. u. k. Province:
"Landeskunde" and "Honismeret" in Nineteenth-Century
Transylvania 187

Peter Haslinger: National Geopolitics in Habsburg Central Europe:
Imperial and Post-imperial Perspectives on Hungary and Poland,
1890–1930 205

Guido Hausmann: Between Complicity in Imperial Politics and
Exclusion: Politico-Geographical Sciences in the Tsarist Empire
and the Early Soviet Union 227

VI. Criminology

Volker Zimmermann: Criminology and Imperial Diversity: The
German Empire, the Habsburg Monarchy, and Tsarist Russia in
Comparison 251

VII. The Study of Natural Resources

Marianne Klemun: Interwoven Functionalities between Empire and
Science in the Habsburg Monarchy: A Comparison of
Nineteenth-Century Geology and Botany 275

Jan Arend: From Tsarism to Nazism: Evolution and Transfer of
Agro-Colonial Knowledge 289

David Moon: Scientific Innovation in the Russian Empire: The Case
of Genetic Soil Science 305

List of Abbreviations and Acronyms 321

Index of Personal Names 323

Contributors 333

Jan Arend

SCIENCE AND EMPIRE IN THE EUROPEAN CONTINENTAL EMPIRES: AN INTRODUCTION

If we wish to dig channels to drain mires that previously made large areas infertile and exude harmful vapours—if we wish to make the water of these mires usable for shipping and irrigation and turn the flood zone into fertile land—then we must explore not only the moor itself, but the hydrography of the entire region, the natural veins and effluences, the torrents, the evaporation and seepage conditions, etc.[1]

It may come as a surprise that the author of the above lines, a staff member of the Royal Hungarian Statistical Office during the final decade of the nineteenth century, was not actually referring to the exploration of swamps. In fact, he was using the reclamation of foul-smelling moors as a metaphor for a different kind of "reclamation:" the "gypsies" in the Austro-Hungarian Empire, he wrote, were to be enlightened and civilized so as to release them from their "dubious existence and suspicious activity" and lead them to a "humane life." To do so, the "numeric and societal circumstances" of the Roma would have to be studied, just as a bog had to be explored to make it usable. Only then could they be transformed from a "burden" on society into a productive and loyal group of subjects.[2]

Hidden behind this call for comprehensive investigation of the "gypsy question" was an additional topic: the Habsburg Monarchy as an empire. The entire report was riddled with metaphorical images alluding to Austria-Hungary's nature as a large empire. For one thing, it referred to the notion of an imperial mission of civilization aimed at the group of the Roma through the path of enlightenment and education expressed within it. Secondly, it referred to the issue of the multiethnicity of empires—specifically to the Roma as a group of subjects in a culturally heterogeneous, federated imperial realm. Finally, the description of the development of rivers and arable land bespoke an expansive relationship to space that also included the idea of internal and external colonization.

---

[1] József Jekelfalussy and Antal Herrmann, *A Magyarországban 1893 január 31-én végrehajtott Czigányösszeirás Eredményei* (Budapest, 1895), 5. Cited in Wolfgang Göderle, *Zensus und Ethnizität: Zur Herstellung von Wissen über soziale Wirklichkeiten im Habsburgerreich zwischen 1848 und 1910* (Göttingen: Wallstein, 2016), 246.

[2] Jekelfalussy and Herrmann, *A Magyarországban 1893 január 31-én végrehajtott Czigányösszeirás Eredményei*, 5. Cited in Göderle, *Zensus und Ethnizität*, 246.

The Hungarian official's report thus raises the question of how its call for scientific enquiry related to the notions of imperial statehood resonating throughout the text—in fact, how "imperium" and scientific empiricism related to each other within the Habsburg Monarchy in general.

Jürgen Osterhammel defined an empire as a

> large-scale, hierarchically ordered federated dominion with a polyethnic and multireligious character, whose coherence is safeguarded by means of the threat of violence, administration, indigenous collaboration and the universalistic programmes and symbolisms of an imperial elite (usually with a monarchic head), not by means of societal and political homogenization and the idea of general citizenship.[3]

The question of the interrelationship between an imperial order interpreted along these lines and the practice of science suggests itself not least because science and politics can be considered "resources for each other" in the modern era.[4] And while this topic can be viewed as relatively well researched in regard to the European overseas empires of Great Britain or France, much less is known about the connections between science and empire in the east of Europe.[5] In publications dedicated to the subject, the land empires in Eastern

---

[3] Jürgen Osterhammel, "Europamodelle und imperiale Kontexte," *Journal of Modern European History* 2, no. 2 (2004): 157–182, here 172.

[4] Mitchell Ash, "Wissenschaft und Politik als Ressourcen füreinander," in *Wissenschaften und Wissenschaftspolitik: Bestandaufnahmen zu Formationen, Brüchen und Kontinuitäten im Deutschland des 20. Jahrhunderts*, ed. Rüdiger vom Bruch and Brigitte Kaderas (Stuttgart: Steiner, 2002), 32–51.

[5] This topic has been addressed continually within the framework of the imperial history of Western European empires since the 1980s. The greatest progress in this regard has been made in research on the British case, which has been the subject of many specific case studies as well as serving as a model for the development of systematic considerations on the interwoven history of science and empire. On the British case, cf. the comprehensive state-of-research overview in Mark Harrison, "Science and the British Empire," *Isis* 96, no. 1 (2005): 56–63. For newer studies investigating individual aspects of science and empire, cf. e.g., Helen Tilley, *Africa as a Living Laboratory: Empire, Development, and the Problem of Scientific Knowledge, 1870–1950* (Chicago: University of Chicago Press, 2011); Peter Gottschalk, *Religion, Science, and Empire: Classifying Hinduism and Islam in British India* (Oxford: Oxford University Press, 2013). Roy MacLeod has used the example of the British Empire to develop important systematic suggestions for research on *imperium* and science. Cf. Roy Macleod, "On Visiting the 'Moving Metropolis': Reflections on the Architecture of Imperial Science," in *Scientific Colonialism: A Cross-Cultural Comparison*, ed. Nathan Rheingold and Marc Rothenberg (Washington DC: Smithsonian Institution Press, 1987), 217–249; Roy Macleod, "Passages in Imperial Science: From Empire to Commonwealth," *Journal of World History* 4, no. 1 (1993): 117–150. Systematic observations can also be found in: Bernard S. Cohn, *Colonialism and Its Forms of Knowledge: The British in India* (Princeton, NJ: Princeton University Press, 1996). For an introduction to the research literature on additional European colonial empires, cf. the following research overviews: Michael A. Osborne, "Science and the French Empire," *Isis* 96, no. 1 (2005): 80–87; Jorge Cañizares-Esguerra, "Iberian Colonial Science," *Isis* 96, no. 1 (2005): 64–70. For further important contribu-

Europe—the Russian Empire, the Habsburg Empire, and the Ottoman Empire—have regularly remained unconsidered. One symptomatic example is a relevant multi-authored volume from 2005 that refers to the "European Empires" in its title, but discusses none of the three above-mentioned Eastern European examples.[6] Likewise symptomatic is the fact that one of the key contributions to the topic, the book "Nature and Empire: Science and the Colonial Enterprise" edited by Roy MacLeod in the year 2000, suggests expanding the comparative spectrum to include countries like Spain, Portugal, and the Netherlands without even considering that European continental empires might likewise offer useful comparisons.[7] The following volume addresses this lacuna by examining the subject of science and empire using the example of two land empires in eastern Europe—the Russian Empire and Austria-Hungary—in the nineteenth century.

The term "land empire" itself is indicative of one of the reasons for the lack of consideration of Eastern European realms in existing research. Older historiography on imperial history in particular is dominated by the differentiation between "land treaders" and "sea foamers"[8]—with the land-based empires often described as backward. This view was frequently linked to a diagnosis of the continental empires like tsarist Russia or the Habsburg Monarchy as being in decline during the nineteenth century—a decline that ended (with a perceived inevitability) with the fracturing of these empires following the

---

tions to research on science in overseas colonies, cf. e.g., Rebekka Habermas and Alexandra Przyrembel, eds., *Von Käfern, Märkten und Menschen: Kolonialismus und Wissen in der Moderne* (Göttingen: Vandenhoeck & Ruprecht, 2013).

[6] Benedikt Stuchtey, ed., *Science across the European Empires, 1800–1950* (Oxford: Oxford University Press, 2005). The Russian Empire and the Habsburg Monarchy were likewise not taken into consideration in Catherine Delmas, Christine Vandamme, and Donna S. Andréolle, eds., *Science and Empire in the Nineteenth Century: A Journey of Imperial Conquest and Scientific Progress* (Newcastle upon Tyne: Cambridge Scholars Publishing, 2010); Patrick Petitjean, Catherine Jami, and Anne M. Moulin, eds., *Science and Empires: Historical Studies about Scientific Development and European Expansion* (Dordrecht: Kluwer, 1992); Roy Macleod, ed., *Nature and Empire: Science and the Colonial Enterprise* (Chicago: University of Chicago Press, 2000). A conference held in Budapest in 2016 was dedicated to the topic with a view to exploring the situation in the eighteenth century, which is beyond the scope of this book. See the conference convened by the Institute for Advanced Study, Central European University Budapest, "Intertwined Enlightenments? Studies of Science and Empire in the Habsburg, Ottoman and Russian Realms during the Eighteenth Century," https://ias.ceu.edu/intertwined-enlightenments (accessed 17 April 2017).

[7] Roy Macleod, "Introduction: Nature and Empire. Science and the Colonial Enterprise," in *Nature and Empire: Science and the Colonial Enterprise*, ed. Roy Macleod (Chicago: University of Chicago Press, 2000), 1–13, here 6.

[8] The original German terms *Landtreter* and *Seeschäumer* go back to: Carl Schmitt, *Land und Meer: Eine weltgeschichtliche Betrachtung* (Leipzig: Verlag von Philipp Reclam jun., 1942).

First World War. The consequence was the establishment of a largely separate treatment of sea and land empires in various historical areas of research.[9]

The paradigm of backwardness may also have contributed to the fact that the extent of the constitutive role of modern science for the Eastern European land empires has hitherto hardly been examined systematically.[10] The notion that the continental empires were ponderous hulks with little potential for modernization could not easily be aligned with the question of whether modern science played an essential role within them. By illuminating the interrelationship between empire and science in the Russian Empire and the Habsburg Monarchy, this volume links up with newer research that neither denies land empires the potential for modernization nor insists on investigating them independently of the colonial empires.[11]

Empire and Science: Perspectives of a Mutual Relatedness

Was science a "tool" of imperial rule? Older research has regularly taken this view, often conceptualizing the relationship between empire and science as a unidirectional function: science in the service of the empire.[12] And indeed, part of the deeply interwoven history of science and empire can be interpret-

---

[9] Cf. Ricarda Vulpius, Kerstin S. Jobst, and Julia Obertreis, "Neue Imperiumsforschung in der Osteuropäischen Geschichte: Die Habsburgermonarchie, das Russländische Reich und die Sowjetunion," *Comparativ* 18, no. 2 (2008): 27–56, here 30–33.

[10] This is the case although a considerable number of individual studies on specific branches of science exist. Cf. e.g., Christian Marchetti, "Von hybriden Pflügen und kultureller Neugestaltung: Volkskunde und Kolonialismus im Habsburgerreich," in *Wissenschaft und Kolonialismus*, ed. Marianne Klemun (Innsbruck: StudienVerlag, 2009), 98–118; Yvonne Kleinmann, "Wissenschaft imperial—Wissenschaft national: Entwurf einer Geschichte der Ethnographie im Russländischen Reich," in *Imperienvergleich: Beispiele und Ansätze aus osteuropäischer Perspektive*, ed. Guido Hausmann and Angela Rustemeyer (Wiesbaden: Harrassowitz, 2009), 77–104; Marina Mogilner, *Homo Imperii: A History of Physical Anthropology in Russia* (Lincoln: University of Nebraska Press, 2013); Deborah R. Coen, *Climate in Motion: Science, Empire, and the Problem of Scale* (Chicago: University of Chicago Press, 2018). For synthesized approaches, cf. Jan Surman, "Imperial Knowledge? Die Wissenschaften in der späten Habsburgermonarchie zwischen Kolonialismus, Nationalismus und Imperialismus," in *Wissenschaft und Kolonialismus*, ed. Klemun, 119–133; Jonathan Oldfield, Julia Lajus, and Denis J. B. Shaw, "Conceptualizing and Utilizing the Natural Environment: Critical Reflections from Imperial and Soviet Russia," *Slavonic and East European Review* 93, no. 1 (2015): 1–15.

[11] Cf. Vulpius, Jobst, and Obertreis, "Neue Imperiumsforschung," 30–33; Pieter M. Judson, *Habsburg Empire: A New History* (London: Belknap Harvard, 2018).

[12] Cf. e.g., John Gascoigne, *Science in the Service of Empire: Joseph Banks, the British State and the Uses of Science in the Age of Revolution* (Cambridge: Cambridge University Press, 1998). Bernhard Cohn's portrayal is likewise dominated by the perspective of science as a functional element of imperial rule. Cf. Cohn, *Colonialism and Its Forms of Knowledge*.

ed in this way—for science was frequently instrumental in imperial rule, and occasionally also instrumentalized politically.

Nevertheless, anyone wishing to examine connections between science and empire should certainly not exclusively consider the political functionalization of science. Rather, science should be viewed as a complex and multilayered social and cultural practice that frequently follows its own logic and cannot be reduced to the mere function of serving politics.[13] Seen from this point of view, reciprocal influences within the relationship of scientific practice and imperial order come into focus, and we must therefore ask questions in both directions: How is science originated and shaped by the imperial order, and how does scientific practice influence imperial order in return? If we regard science and imperial order as two asymmetrically overlapping social and cultural systems, then we must ask to what extent each of them plays a role in generating and supporting elements of the other—and on the other hand, to what extent they limit and disturb each other's functional logic.[14]

In keeping with the goal of shedding light on this entanglement, the following section entitled "Science in Empires" will discuss the significance of imperial state structures for scientific practice, while the subsequent section entitled "The Scientific Empire" will deal with the effects of science on the imperial state order. The final section will provide a summary of the reciprocal relationship between the two aspects. Meanwhile, the superordinate intention of this introduction is to highlight the still considerable research lacunae and show how the scholarly texts in this volume contribute to filling them.

The studies collected in this book elucidate aspects of the reciprocal relationship between empire and science by addressing a broad spectrum of topics: from the plans to establish academies to the "university question" and the activity of individual researchers in the imperial provinces. The majority of the texts, however, deal with individual disciplines and scientific fields in the imperial context.

Our conference call elicited numerous replies with suggestions for case studies on the (imperial) history of individual scientific disciplines. We immediately noticed, however, that the variety of disciplines observable in the scientific landscape of the nineteenth century was not uniformly represented in the proposals received. Studies on areas of natural history like botany or geology were suggested frequently, as were those on topics from cultural and social sciences, for example, political geography, history, ethnography, and Orientalism. On the other hand, papers on laboratory-based (chemistry) and

---

[13] Cf. the criticism of the view of science as a "tool" of imperial rule in: Suman Seth, "Putting Knowledge in Its Place: Science, Colonialism, and the Postcolonial," *Postcolonial Studies* 12, no. 4 (2009): 373–388, here 375.

[14] Banu Subramaniam, "Science and Postcolonialism," in *A Companion to the History of American Science*, ed. Georgina M. Montgomery (Chichester: John Wiley & Sons, 2015), 491–501, here 493–494.

cognitive-systematic sciences (mathematics, philosophy) were rarely offered. This may be due to various reasons, and even to some degree to chance. But it also appears to confirm a result that research on the Western European empires with overseas colonies has likewise arrived at: namely, that when asking about the interaction between empires and science, increased consideration must be given to "spatially oriented" disciplines—that is, those areas of learning that investigate their objects of examination (such as animals, plants, humans, customs, languages, and so on) in their respective referentiality to cultural and natural spaces and are thus closely linked to questions of territorial expansion and the cognitive harnessing of dominions.[15]

Science in Empires: State Structures and Scientific Practice

The empires of the nineteenth century formed specific structures, differing from other types of statehood in the form of their government, their legal order, and their territorial organization. When comparing the ideal type of the nineteenth-century empire with the nation states being established at the same time, it is clear that empires were not restricted to an ancestral homeland but instead expanded by way of conquest and economic activity. In doing so, they often incorporated territories that were spatially and culturally far removed from the metropolitan centres, a circumstance entailing considerable challenges relating to the exertion of rule. And unlike the legal systems of nation states, which were geared toward equality among their citizens, empires featured complex, historically grown legal orders in which hierarchical relations between ethnically, religiously, and culturally defined groups of subjects were embodied. Over the course of the nineteenth century, many of these groups developed aspirations to autonomy or independence—often related to an emerging national consciousness.[16] The Russian and Habsburg Empires alike thus saw themselves challenged by the competing principle of the nation state, which was also becoming more attractive to the members of the culturally dominant Russian and German/Hungarian population groups respectively.

---

[15] Iris Schröder, "Disziplinen: Zum Wandel der Wissensordnungen im 19. Jahrhundert," in *Von Käfern, Märkten und Menschen*, ed. Habermas and Przyrembel, 147–160. Cf. also the disciplines discussed in: Stuchtey, *Science across the European Empires, 1800–1950*; Delmas, Vandamme, and Andréolle, *Science and Empire in the Nineteenth Century*; Macleod, *Nature and Empire*. One exception with considerable significance, especially for the overseas empires, was laboratory-based tropical medicine.

[16] Jane Burbank and Frederick Cooper, *Empires in World History: Power and the Politics of Difference* (Princeton, NJ: Princeton University Press, 2010), 287–368; Jörn Leonhard and Ulrike v. Hirschhausen, *Empires und Nationalstaaten im 19. Jahrhundert* (Göttingen: Vandenhoeck & Ruprecht, 2009); Dominic Lieven, *Empire: The Russian Empire and Its Rivals* (London: Murray, 2000).

On the basis of these premises, we may ask in which ways science was structurally impacted in the imperial context. On the one hand, we can look at the empire as a "facilitating space" for research, that is, the scientifically productive effects of imperial state structures: Which forms of scientific thinking, practice, and organization did empires stimulate, and how?

On the other hand, structures always possess not only an enabling, but also a limiting or obstructing effect, and empires are no exception in this regard. They naturally must also have curtailed options for thought and action in the field of science, and we therefore also have to ask to what extent empires discouraged certain forms of scientific practice: Which scientific paths of thinking and acting were structurally restricted in the imperial context?

The following paragraphs will highlight three aspects of how science was shaped by the structural conditions of the imperial state: Firstly, specifically imperial forms of "mobility of the elites" affected scientific practice. Secondly, the same applies to the "tension between the imperial state order and national aspirations" apparent in both empires. And thirdly, the "territorial structure of the extensive land empires" naturally suggested investigating natural and cultural phenomena in their spatial variability, focussing scientific attention on different zones and the transitions between them. While this list of characteristics of the structure of imperial states affecting the opportunities and boundaries of scientific activity should by no means be considered exhaustive, the three aspects described are the ones most frequently encountered in the contributions to this volume.

Mobility of the Elites

Nikolai Petrovskii was a young Russian financial clerk serving in the state regulatory authority in St. Petersburg in the 1860s. He developed scientific ambitions on the side and became a member of one of the many learned societies being founded in the Russian Empire during the second half of the nineteenth century, following the European example. In 1870, Petrovskii was dispatched as imperial representative to Russian Turkestan, where he was to collect information on local trade on behalf of the Ministry of Finance. Upon arriving in Tashkent, Petrovskii began to develop into an "amateur specialist" engaging intensively in the study of the geography, history, and archaeology of the area besides his activity for the government. He continued his research activity even after being appointed to the office of consul to the Central Asian city of Kashgar in 1882, using his contacts with members of the local elite to establish a valuable and scientifically important archaeological collection on the early history of the region.

The example of Petrovskii—analysed in detail in this volume by Matthias Golbeck—is indicative of a specifically imperial form of mobility of the elites and the forms of scientific practice it enabled. Representatives of the political, administrative, or cultural elites of the Russian and Habsburg Empires moved actively between the urban centres of the respective realms and their peripheries, and individual imperial biographies often included multiple sojourns in various outlying regions. This pattern of mobility was characteristic not only of persons like Petrovskii who served the imperial state directly, but also of people from various different fields and professions who represented imperial interests in an indirect sense.[17]

During the late nineteenth century, Turkestan was an example of an imperial frontier region in which the Russian state was quasi still in the process of nascence, a peripheral region of the empire at a great geographic and cultural distance from its heart in St. Petersburg.[18] From the perspective of imperial functionaries like Petrovskii, faraway regions like Turkestan or certain areas of Habsburg Galicia must have appeared as largely uncharted areas, as disorderly "nature" to be transformed into "culture." Practices of knowledge accumulation—which were always also linked to categorization and classification—were central to this programme of the ordering and cultivation of supposedly orderless spaces.[19] And it therefore came naturally to representatives of imperial interests on the periphery to not restrict themselves to the fulfilment of their sometimes very specifically defined political assignments, but to also act as imperial "scouts" seeking broader or deeper knowledge of local circumstances.

The contacts and authority enjoyed by imperial functionaries could also often be exploited for such semi-amateurish knowledge generation—and vice versa: imperial officials were able to establish informal contacts in the course of their scientific activities that could be used to further imperial political interests. Science and politics could thus become resources for one another in a very specific fashion on the periphery of empires.

When examining the patterns of mobility of agents like Petrovskii, one begins to follow the flows of knowledge between different locations and spaces of its production. On the one hand, there were the capitals and other large

---

[17] On the "imperial biographies" of imperial elites and their patterns of mobility, cf. Malte Rolf and Tim Buchen, eds., *Elites and Empire: Imperial Biographies in Russia and Austria-Hungary, 1850–1918* (Berlin: De Gruyter Oldenbourg, 2015); Martin Aust and Frithjof B. Schenk, eds., *Imperial Subjects: Autobiographische Praxis in den Vielvölkerreichen der Romanovs, Habsburger und Osmanen im 19. und frühen 20. Jahrhundert* (Cologne: Böhlau, 2015); Malte Rolf, "Einführung: Imperiale Biographien. Lebenswege imperialer Akteure in Groß- und Kolonialreichen, 1850–1918," *Geschichte und Gesellschaft* 40, no. 1 (2014): 5–21, here 13–15; David Lambert and Alan Lester, "Introduction: Imperial Spaces, Imperial Subjects," in *Colonial Lives Across the British Empire: Imperial Careering in the Long Nineteenth Century*, ed. David Lambert and Alan Lester (Cambridge: Cambridge University Press, 2006), 1–31.

[18] Jeff Sahadeo, *Russian Colonial Society in Tashkent, 1865–1923* (Bloomington: Indiana University Press, 2010).

[19] Cf. e.g., Michael Adas, *Machines as the Measure of Men: Science, Technology, and Ideologies of Western Dominance* (Ithaca, NY: Cornell University Press, 1989).

cities, like Vienna, Budapest, St. Petersburg, and Moscow, which were the seats of academies and scientific societies. Besides these, there existed regional centres of learned activity like Prague, Cracow, and Kazan. Lastly, we see the peripheral territories like the Central Asian Kashgar, Petrovskii's place of activity since the 1880s.[20]

The Habsburg Monarchy and the Russian Empire differed significantly in regard to the structural preconditions for the mobility of their respective elites. In contrast to Russia, Austria-Hungary was a polycentric realm. While Vienna and (after 1867) Budapest as metropoles enjoyed a preeminent status, numerous other urban centres of considerable political, cultural, and scientific relevance existed as well (e.g., Prague, L'viv, and Cracow, to name but a few). Compared to this structure, which was rich in major nodes, the position of St. Petersburg and Moscow in tsarist Russia was unique, and this meant that the distance to the nearest urban centre of learning was generally greater than in the Habsburg Empire.

One focal point of existing research on knowledge production in empires was the relationship between these urban centres and the peripheral territories. Older research, which was largely oriented around the British case, assumed a relatively clear division of labour between the core of an empire and its periphery, that is, the colonies. Science in the colonies was thought to be restricted to the collection of locally relevant information and data subsequently transmitted to the researchers in the (capital) cities, who were responsible for synthesis, analysis, and theory formation. A related argumentation emphasized the primarily applied character of research in the colonies, and thus also their remoteness from theory.[21] From this perspective, the colonies or imperial peripheries were merely spaces of secondary scientific character: they were either considered quarries of sorts for empiric resources, the raw materials for the achievement of insight, or as places for testing and application of knowledge produced elsewhere.[22]

Newer research on science and empire has called this notion of an asymmetric division of labour between centres and peripheries into question, how-

---

[20] Hans-Christian Maner, "Zentrum und Grenzregionen in der Habsburgermonarchie im 18. und 19. Jahrhundert: Eine Einführung," in *Grenzregionen der Habsburgermonarchie im 18. und 19. Jahrhundert: Ihre Bedeutung und Funktion aus der Perspektive Wiens*, ed. Hans-Christian Maner (Münster: LIT-Verlag, 2005), 9–24.

[21] Donald Fleming, "Science in Australia, Canada, and the United States: Some Comparative Remarks," in *Proceedings of the Tenth International Congress of the History of Science: Ithaca 1962*, ed. Hermann Barsdorf, 2 vols. (Paris: Hermann, 1964), 179–196, here 182; Michael Worboys, "Science and British Colonial Imperialism, 1895–1940," (PhD diss, University of Sussex, 1980). Similarly see Macleod, "Passages in Imperial Science," 125.

[22] For a critique of this approach, cf. Harrison, "Science and the British Empire," 56 f.

ever.[23] As recent studies have been able to show, scientists from the central cities neither arrived at the peripheries with preconceived paradigms nor restricted themselves to the collection of data while there. Rather, peripheries like Turkestan were spaces of comprehensive knowledge production in their own right, and they also delivered theoretical impulses for science in the urban centres.[24] Even though research in the colonies presumptively often dealt with questions of applied science, new theoretical paradigms could emerge from these application contexts as well.[25]

Peripheries were scientifically productive not least because they were spaces of encounter. Whenever imperial officials, regionally active scientists, and indigenous knowledge holders met, the cultures of knowledge of the centres came into contact with those of the peripheries. Naturally, such relationships often had a colonial "flavour" to them: scientists from the imperial hubs frequently acted out a metropolitan sense of superiority, viewing the regional knowledge holders either as mere suppliers of information or as addressees of their own scientific mission of civilization. Examples of this can be found in the contribution by Arpine Maniero, which describes the attitudes of metropolitan Orientalists in the eastern and southern peripheries of the Russian Empire.

Not all encounters followed this pattern, however; indeed, sometimes they unleashed considerable productive energy. Scientists from the imperial centres regularly displayed willingness to learn and openness to local scientific traditions.[26] A concrete example of this is discussed in Daniel Baric's contribution, which examines the cooperation between Habsburg archaeologist Carl Patsch and his local partners in Bosnia-Herzegovina, for example, Cath-

---

[23] David W. Chambers and Richard Gillespie, "Locality in the History of Science: Colonial Science, Technoscience, and Indigenous Knowledge," *Osiris* 15 (2000): 221–240.

[24] Jakob Vogel, "Public-private Partnership: Das koloniale Wissen und seine Ressourcen im langen 19. Jahrhundert," in *Von Käfern, Märkten und Menschen*, ed. Habermas and Przyrembel, 261–284, here 284. Here an insight that advanced imperial history in general was reflected in the field of the history of science: the centre was not exclusively a "sender" and the peripheries were not purely "receivers"—rather, it is essential for a proper understanding of imperial states to also ask in which ways the peripheries influenced the centre. Cf. e.g., Kathleen Wilson, "Introduction: Histories, Empires, Modernities," in *A New Imperial History: Culture, Identity, and Modernity in Britain and the Empire, 1660–1840*, ed. Kathleen Wilson (Cambridge: Cambridge University Press, 2006), 1–28.

[25] Cf. Désirée Schauz, "What is Basic Research? Insights from Historical Semantics," *Minerva* 52, no. 3 (2014): 273–328.

[26] For examples concerning the British Empire, cf. Mark Harrison, "Medicine and Orientalism: Perspective's on Europe's Encounter with Indian Medical Systems," in *Health, Medicine, and Empire: Perspectives on Colonial India*, ed. Biswamoy Pati and Mark Harrison (Hyderabad: Orient Longman, 2001), 37–87; Christopher A. Bayly, *Empire and Information: Intelligence Gathering and Social Communication in India, 1780–1870* (Cambridge: Cambridge University Press, 1997).

olic priests with knowledge of regional architectural relics. In such scenarios, Baric argues, research became a "collective undertaking."

In addition, studies on the European colonial empires have shown that regional and indigenous elites likewise often exhibited openness to scientific impulses from the imperial centres. Research on the Indian case, for example, suggests that the local elites in imperially ruled territories often absorbed influences from Western science selectively and pragmatically.[27] In her contribution to this volume, Arpine Maniero arrives at similar results by demonstrating a comparable openness among the local elites on the southern and eastern peripheries of the Russian Empire—in this case to the knowledge of Orientalists in St. Petersburg. In contrast to the state of affairs implied by George Basalla's long-influential diffusionist model of knowledge spillover from the West, there existed in some cases a dynamic of knowledge exchange in which the roles of "senders" and "recipients" were not clearly assigned.[28]

Loyalties: Between Nation, Empire, and International Science

Besides their patterns of elite mobility, the Habsburg Monarchy and the Russian Empire shared a further structural attribute: during the nineteenth century, nationalist and imperialist forms of political logic were superimposed in both empires. Nationalist integration ideologies became virulent in manifold ways in Russia and Austria-Hungary alike, encompassing the culturally dominant ethnicities—of the Germans and Hungarians, and Russians respectively—as well as numerous subdominant ethnic groups. This resulted in considerable tension in both imperial realms that also affected science.[29]

Among office-holders in the Russian Empire, the policy-shaping notion developed that the "Russians"—a very fuzzily defined group—had an entitlement to primacy within the realm. These nationalist politics were also directed at Russia's "own" rural East Slavic population with the intent of awakening their national consciousness and pride. In regard to the non-Russian ethnicities, tsarist nationality policy fluctuated between tolerance and efforts at "Russification," with the ratio between the two approaches varying significantly, depending on a number of factors ranging from the inclinations of

---

[27] Ahsan J. Qaisar, *The Indian Response to European Technology and Culture, A.D. 1498–1707* (Delhi: Oxford University Press, 1998); Harrison, "Science and the British Empire," 59–60.

[28] George Basalla, "The Spread of Western Science," *Science* 156, no. 3775 (1967): 611–622. On criticism of Basalla's model, cf., e.g., Subramaniam, "Science and Postcolonialism" in *A Companion to the History of American Science*, ed. Montgomery, 494; Roy Macleod, "Introduction" in *Nature and Empire*, 2–5.

[29] Geoffrey Hosking, *Russia: People and Empire* (Cambridge, MA: Harvard University Press, 1997); Endre Kiss and Justin Stagl, eds., *Nation und Nationenbildung in Österreich-Ungarn, 1848–1938: Prinzipien und Methoden* (Münster: LIT-Verlag, 2006).

individual tsars and their political entourages to the standing of particular ethnic groups within the changing cultural and political hierarchies of the empire.[30] In many regions of the Russian Empire, imperial authority was challenged by non-Russian ethnic groups with a desire for autonomy or independence. Polish resistance, for instance, flared up early on and with vehemence, as embodied in the national uprisings of 1830 and 1864. In the course of the Russian revolutions of 1905 and 1917, the empire would eventually be confronted with an abundance of national aspirations that could not be quelled.[31]

In Austria-Hungary, the Germans as well as the Hungarians held claims to primacy in the empire during the nineteenth century. Following the "Compromise" of 1867, the development of structural elements of a nation state that conflicted with the imperial order began in Transleithania.[32] What is more, as was the case in tsarist Russia, representatives of the Habsburg Empire viewed the aspirations of subdominant ethnicities—whose demands included a spectrum of language rights and autonomous administration all the way up to full independence as nation states—as threats to the empire.[33]

This complex and dynamic entanglement of imperial pretensions of order, increasing articulation of national ambitions, and the beginning development of nation-state structures generated a problem that affected many different areas of society—and science in particular: the question of loyalty to the respective empire and/or nation.[34]

Over the course of the nineteenth century, scientists and scientific institutions increasingly saw themselves confronted with the question whether their research and teaching activities served the empire as a whole or constituted an expression of specific national loyalties. Members of the culturally dominant titular nations—like the Russian financial official, diplomat, and amateur scientist Petrovskii—were occasionally able to escape this compulsion to "take sides," since loyalty to the titular nation and to the entire empire were

---

[30] Andreas Kappeler, *Russland als Vielvölkerreich: Entstehung, Geschichte, Zerfall* (Munich: Beck, 2001), 177–229.
[31] Ibid., 268–299.
[32] Cf. also the contribution by Peter Haslinger in this volume.
[33] Pieter M. Judson and Marsha L. Rozenblit, eds., *Constructing Nationalities in East Central Europe* (New York: Berghahn, 2006); Jonathan Kwan, "Nationalism and All That: Reassessing the Habsburg Monarchy and its Legacy," *European History Quarterly* 41, no. 1 (2011): 88–108.
[34] Martin Schulze Wessel, "'Loyalität' als geschichtlicher Grundbegriff und Forschungskonzept: Zur Einleitung," in *Loyalitäten in der Tschechoslowakischen Republik, 1918–1938: Politische, nationale und kulturelle Zugehörigkeiten*, ed. Martin Schulze Wessel (Munich: Oldenbourg, 2004), 1–22; Jana Osterkamp and Martin Schulze Wessel, eds., *Exploring Loyalty* (Göttingen: Vandenhoeck & Ruprecht, 2017); Jana Osterkamp and Martin Schulze Wessel, "Texturen von Loyalität: Überlegungen zu einem analytischen Begriff," *Geschichte und Gesellschaft* 42, no. 4 (2016): 553–573.

sometimes compatible. This was not the case for members of subdominant elites, however, and as a result they frequently had to decide between exclusively national and imperial loyalties.[35]

The previously mentioned example of the Habsburg archaeologist Patsch analysed by Daniel Baric shows that these loyalties could be situational and unsteady. During a lengthy sojourn in the Bosnian-Herzegovinian periphery (during which he also started a family), Patsch developed an "unconditional feeling of belonging" regarding his provincial surroundings that increasingly estranged him from the "distanced Viennese perspective." Significantly, it was precisely the imperial pattern of mobility described in the previous section that caused the shift in loyalty in this case.

In the realm of science, a third reference point for loyalties besides empire and nation began to establish itself in the second half of the nineteenth century as well: the international scientific community. The notion of science as a universalist endeavour restricted neither by national nor cultural boundaries was in part fuelled by early modern ideas of a *république des lettres*. These were joined in the second half of the 1900s by a new scientific internationalism which—in contrast to the older concept of the republic of letters—already took for granted the nation as a premise or sphere of reference.[36] According to the internationalist credo, which many scientists had begun to internalize, especially by the late nineteenth century, researchers could escape potential conflict between their loyalty to their respective nation and their connectedness with the worldwide scientific community: for if one's achievements found recognition on the international stage, that fame would be reflected on one's own nation as well.[37]

Numerous studies on the question of clashes and compatibility between the national and international loyalties of scientists have already been undertaken. Among other things, they have shown that reality often differed considerably from the internationalist ideal, and that many learned men and

---

[35] Theodore R. Weeks, "Jan Badouin de Courtenay: The Linguist as Anti-Nationalist and Imperial Citizen," in *Elites and Empire: Imperial Biographies in Russia and Austria-Hungary, 1850–1918*, ed. Malte Rolf and Tim Buchen (Berlin: De Gruyter Oldenbourg, 2015), 338–353; Mitchell Ash and Jan Surman, "The Nationalization of Scientific Knowledge in Nineteenth-Century Central Europe: An Introduction," in *The Nationalization of Scientific Knowledge in the Habsburg Empire, 1848–1918*, ed. Mitchell Ash and Jan Surman (Basingstoke: Palgrave Macmillan, 2012), 1–29.

[36] Mitchell Ash, "Internationalisierung und Entinternationalisierung der Wissenschaften im 19. und 20. Jahrhundert: Thesen," in *zeitgeschichte.at: Österreichischer Zeithistorikertag 1999*, ed. Manfred Lechner and Dietmar Seiler (Innsbruck: StudienVerlag, 2000), 4–12; Brigitte Schroeder-Gudehus, "Nationalism and Internationalism," in *Companion to the History of Modern Science*, ed. Robert C. Olby et al. (London: Routledge, 1990), 909–919.

[37] Paul Forman, "Scientific Internationalism and the Weimar Physicists: The Ideology and Its Manipulation in Germany after World War I," *Isis* 64, no. 2 (1973): 153–156.

women did indeed become involved in various forms of loyalty conflicts.[38] The example of the land empires of the Habsburgs and the Romanovs, however, allows investigation of the hitherto insufficiently answered question of how the trilateral loyalty structure of empire—nation—international community shaped scientific practice. This structure is the topic of the contribution by Jan Arend, which uses the example of an encounter between a German and a Russian soil scientist in the first half of the twentieth century to examine the relationship between national, international, and imperial affiliations.

The significance of the specific logics of individual fields of science also becomes visible: some disciplines were affected by the loyalty question far more than others—and here, too, it was the spatially and culturally oriented sciences that were the most heavily contested. The genre of scientific cartography, for example—be it for the purpose of political geography or that of botany and soil science—was particularly relevant in terms of loyalties. Maps not only touched upon the issue of the territoriality of existing states, but also upon the hopes for future establishment or re-establishment of countries—an aspect that continually gained in importance over the course of the nineteenth century. This is documented in Peter Haslinger's contribution, which examines the territorial designs of political geographers in Austria-Hungary and points out that nationally oriented efforts in this context delineated "horizons of encouragement" concerning territorial units to be established. Meanwhile, Marianne Klemun's contribution on Austria-Hungary discusses how the creation of botanical maps was linked to questions of loyalty in the Dual Monarchy: botanists occasionally attempted to make natural and (existing as well as desired) political borders coincide in their work.

The loyalty problem obviously also featured in the history of scientific institutions. At a very general level, the long nineteenth century represented an intensive phase of the establishment and expansion of academies, universities, research facilities, and learned societies in Russia and Austria-Hungary. All of the activities in this field occurred within the described precarious framework of loyalties spanning imperial pretensions of order, nationalist aspirations and feelings of membership in the international scientific community.

The contribution by Martin Franc investigates how this was exemplified in the plans for a Czech academy of sciences in the Dual Monarchy. His comparison of the differing visions of Jan Evangelista Purkyně and Josef Hlávka (with the latter being implemented in 1890) shows unambiguously that sci-

---

[38] Soňa Štrbáňová, "Patriotism, Nationalism and Internationalism in Czech Science: Chemists in the Czech National Revival," in *Nationalization of Scientific Knowledge*, ed. Ash and Surman, 138–155; Mitchell Ash and Jan Surman, "The Nationalization of Scientific Knowledge in Nineteenth-Century Central Europe" in *Nationalization of Scientific Knowledge*, ed. Ash and Surman, 6–13.

ence was considered an important element of a national high culture in this context. The deliberations about separate Czech institutions also expressed a desire for internationally visible national-cultural independence and individuality—which could only be realized with certain restrictions in the Habsburg context, however.

Maciej Janowski describes the activities of Polish learned societies within the three partitioned territories and analyses them in the context of "organic work" for the Polish nation, a political programme stipulating continuous efforts to establish the educational and economic prerequisites for future independence rather than armed resistance. In this context, the cultivation of scientific companionability was an expression of national loyalty in that it was understood as part of an endeavour aimed at ensuring the long-term survival of the Polish nation. Finally, the contribution by Jan Surman investigating the educational policies of the Habsburg Monarchy through the eyes of the Russian diplomat Ludwik Tęgoborski shows that questions of loyalty could easily become acute precisely in the context of educational institutions.

"Vast Land:" Zones and Transitions as Focal Points of Scientific Attention

In 1942, the botanist Heinrich Walter (1898–1989) from Stuttgart recalled his youth spent as the child of ethnic German parents in pre-revolutionary Russia. During this time, he travelled the European part of tsarist Russia "from the Black Sea to the Gulf of Finland" on multiple occasions.[39] On his journeys, he noticed how the vegetation changed with the climatic conditions from south to north: steppes passed into mixed and fir forests, which ultimately gave way to tundra. Walter's travels through the Russian Empire of the early twentieth century contributed to "the orderly sequence of vegetation zones becoming deeply imprinted" on his memory, as he wrote.[40]

Like Walter, other scientists in the Russian Empire and the Habsburg Monarchy also recognized geographic regularities and fluent transitions in the spaces they observed. The vast territories of the continental empires invited researchers to investigate natural and cultural phenomena in the horizontal plane. Whether they analysed vegetation, soil composition, settlement patterns, or the distribution of dialects or religious practices, scientists found themselves inspired by the large, contiguous study areas in Russia and Austria-Hungary to ask questions pertaining to spatial variance and its causes.[41]

---

39 Heinrich Walter, *Die Vegetation des europäischen Russlands: Unter Berücksichtigung von Klima, Boden und wirtschaftlicher Nutzung* (Berlin: P. Parey, 1942), preface, unpaginated.
40 Ibid. Cf. also the contribution by Marianne Klemun in this volume (specifically, the remarks on botanist Wilhelm Daniel Koch and his attention to "gradual transitions").
41 Coen, *Climate in Motion*, 10–13; Nailya Tagirova, "Mapping the Empire's Economic Regions from the Nineteenth to the Early Twentieth Century," in *Russian Empire: Space, People, Power, 1700–1930*, ed. Jane Burbank, Mark von Hagen, and Anatolyi Remnev

The Austrian meteorologist and geophysicist Heinrich von Ficker, who travelled both the Russian and the Habsburg Empire, developed a productive interest in "climatic contrasts" between regions.[42] In his contribution on soil science in tsarist Russia, David Moon shows how the specific geography of this expansive continental empire, which he compares to that of North America, became a laboratory for the contemplation of soil zones. In this context, an article by Yuri Slezkine focussing on ethnography and anthropology and published in 1994 analysed the phenomenon of scientific attention to the spatial dimension of cultural diversity.[43]

The increasing focus on gradual transitions, zones, and regional variance was in part connected to the manner and infrastructure of travel itself. During the nineteenth century, scientists journeyed through the European land empires by foot and on horseback, in coaches and on riverboats, and increasingly by train as well. Individually, these means of transportation naturally differed in terms of their significance for the practice of science. In a more general sense, however, their slowness and linearity of movement raised the traveller's awareness of spatial transition phenomena.[44] Recent studies on the history of science dealing with the connections between transportation infrastructures and scientific practice have shown that means of transport can also be viewed as "instruments of research" since they place the travelling researcher in a specific relation to the investigated space, thereby shaping the conditions of scientific observation and realization.[45] This is certainly an area in which research on the Eastern European continental empires could be extended beyond this volume.

The development of imperial spaces by way of transportation infrastructures observable in the nineteenth century—for example, the expansion and densification of road and rail networks for military and mail purposes—thus had far-reaching effects on the practice of science.[46] Further investigation of this topic will have to take the differing degrees of development of the various regions in the Russian and Habsburg Empires into consideration. Archaeologist Patsch, for example, was able to use the narrow-gauge railway system maintained by the military as well as a network of stagecoaches during his

(Bloomington: Indiana University Press, 2010), 125–138; Marina Loskutova, "Regionalization, Imperial Legacy and the Soviet Geographical Tradition," in *Decentering the Soviet Empire: New Spatial Histories of Russia*, ed. Sanna Turoma and Maxim Waldstein (Farnham: Ashgate, 2013), 135–158.

[42] Coen, *Climate in Motion*.
[43] Yuri Slezkine, "Naturalists Versus Nations: Eighteenth-Century Russian Scholars Confront Ethnic Diversity," *Representations* 47 (1994): 170–195.
[44] Coen, *Climate in Motion*, 78–80.
[45] Jeremy Vetter, "Science along the Railroad: Expanding Field Work in the U.S. Central West," *Annals of Science* 61, no. 2 (2004): 187–211.
[46] Jörg Ganzenmüller and Tatjana Tönsmeyer, eds., *Vom Vorrücken des Staates in die Fläche: Ein europäisches Phänomen des langen 19. Jahrhunderts* (Cologne: Böhlau, 2016).

epigraphic exploration of Bosnia in 1891 even though the region was among the less developed in Austria-Hungary. Meteorologist von Ficker, on the other hand, frequently had to travel on horseback in early twentieth-century Central Asia.[47]

Travelling scientists from Paris or London had to cross oceans in order to investigate their respective colonies. To them, the sea was a palpable sign of the cultural distance between their home and the colonial outlands. What they saw and researched overseas seemed far removed from their normal living environment, and as a result their view of the nature and culture they encountered there was generally "exotifying" instead of focussing on the familiar. In tsarist Russia and Austria-Hungary, however, there was no vast ocean separating the centre from the peripheries—and the transitions were thus less obvious.[48] This leads us to ask how this simple fact influenced the practice of science in the imperial space, and the observation that researchers in the Habsburg and Russian Empires developed a sensitivity to gradual spatial transitions seems relevant in this context: it suggests the hypothesis that the differences between centres and peripheries—and hence between what was considered "own" and "foreign"—may have appeared more gradual and less absolute to scientists in the context of continental empires than in that of overseas empires. In this vein, Johannes Feichtinger's contribution on Habsburg Orientalism shows that the influential idea of an "inner Orient" in the Habsburg Empire was linked to a specific awareness of the transitions and overlaps between "Orient" and "Occident." As Arpine Maniero argues with reference to the work of Vera Tolz, the situation was similar for Orientalism in the Russian Empire.

The Scientific Empire: Knowledge as a Constitutive Element of Imperial Rule

While the previous section dealt with the ways in which empire shaped the practice of science, we will now shift our point of view and examine the effects of science on the imperial order. How did science contribute to creating imperial statehood, and to what extent was it a constitutive part of imperial rule? Conversely, how could science destabilize imperial rule or disturb its functional logic?

Imperial rule has various dimensions to be examined in terms of their science-related constitutive aspects. The primary focus of the essays in this volume is on administrative and governmental action, the legitimation of rule, and the representation of statehood.

As happened in other states, science in the Russian Empire and the Habsburg Monarchy produced "useful" knowledge that informed administrative

---

[47] Coen, *Climate in Motion*, 336.
[48] Ibid., 9.

and governmental activities in manifold ways. Practically all the disciplines and fields of learning discussed in this volume fulfilled this function to some extent. In this context, the disciplines emerging from the cameralistic and political science traditions of the eighteenth century, whose tasks included inventorying the natural and cultural as well as the human resources of the state, were particularly important.[49]

Borbála Zsuzsanna Török's contribution uses the example of Transylvania to describe how this area of science was transformed and diversified over the course of the nineteenth century. Her analysis of scientific societies for local and regional studies (Landeskunde) examines tendencies in the development of the interest in "the land and its people" in the context of an ethnically and culturally highly complex province. The previously mentioned essay by Marianne Klemun discusses the work of the k. k. Geologische Reichsanstalt founded in Austria-Hungary in 1849, whose tasks included the inventorying of natural resources within the empire's territory. Klemun's comparison with the activity of the k. k. Zoologisch-botanische Gesellschaft reveals different forms of functional entanglement between empire and science aimed at the investigation of natural resources.

Disciplines like ethnography, Orientalism, and political geography engaged in a form of social sorting—a categorizing subdivision of the subject population into socially and culturally specific groups that the imperial nationality policy and social policy could latch on to. Volker Zimmermann describes a case of scientific social sorting with particular relevance for imperial legal practice in his contribution focussing on criminology in the German, Russian, and Habsburg Empires. His text shows how multiethnicity in empires became a central reference point for considerations on conditions for the development of criminality and the possibilities of crime prevention.

Scientists in the Habsburg Monarchy and tsarist Russia also produced knowledge suited for legitimizing imperial rule. In contrast to nation states, which used the concept of the nation as a community of equals for legitimation, empires were in most cases legitimized "from above" with reference to religious and dynastic patterns. In light of the competition from the nation-state principle during the nineteenth century, however, empires increasingly found themselves having to formulate "modern" arguments to corroborate their legitimacy besides their traditional ones.[50] The notion of the civilizing

---

[49] Londa Schiebinger, *Plants and Empire: Colonial Bioprospecting in the Atlantic World* (Cambridge, MA: Harvard University Press, 2004); Lars Behrisch, ed., *Vermessen, zählen, berechnen: Die politische Ordnung des Raums im 18. Jahrhundert* (Frankfurt am Main: Campus, 2006); Aleksandra Bekasova, "Izuchenie Rossiiskoi imperii ekspeditsiiami 1760–1780-kh gg: Vzgliad estestvoispytatelei i formirovanie predstavlenii o gosudarstvennykh bogatstvakh," *Istoriko-biologicheskie issledovaniia* 2, no. 4 (2010): 13–34.

[50] Maciej Janowski, "Justifying Political Power in 19th Century Europe: The Habsburg Monarchy and Beyond," in *Imperial rule*, ed. Alexei Miller and Alfred J. Rieber (Budapest: Central European University Press, 2004), 69–82.

mission, which was so important for both empires, was one of these legitimation arguments relating to an (asserted) modernity: the culture of science was ranked among the modern "blessings" introduced to the peripheries by imperial rule.[51]

In addition, disciplines like the historical sciences and archaeology were able to certify historical legitimation arguments with scientific means that were considered "modern." An example of a historiography supporting imperial legitimation narratives is encountered in the work of the Russian historian Vasilii Kliuchevskii, who effectively interpreted colonization as the historic core process of Russian statehood.[52] Both the Habsburg and Russian Empires used narratives placing themselves in the tradition or succession of older empires to legitimize their rule. In Russia, the idea that the tsarist empire held the legacy of Rome and Byzantium was influential. This notion was propagated not only in the ecclesiastic and religious sphere, but also inspired historians to create depictions of Russian history that emphasized parallelisms and analogies with Roman and Byzantine prehistory.[53]

A Habsburg example is once again provided by the archaeologist Patsch, whose interest in the traces of Roman history in Bosnia-Herzegovina likewise coalesced with observations on the present of the Habsburg province. Patsch's example also illustrates, however, that archaeological and historiographic narratives could not always be stabilized in the politically desired legitimizing form since scientists regularly followed their own paths and departed from hegemonial scripts in their analyses. The legitimizing narrative began to crumble in Patsch's interpretation, for the more he dissociated himself from the "distanced Viennese perspective" on Bosnia-Herzegovina, the more he began to miss—in the Habsburg provincial present—the prosperity he considered a Pax Romana based on tolerance to have enabled.

Besides the generation of knowledge with relevance for politics and cooperation in the establishment of legitimizing narratives, science was also constitutive for imperial rule in a third and even more fundamental sense: it contributed to making the initially abstract "imperium" imaginable and perceivable for contemporaries.

Due to the sheer size of their territories and the resulting administrative complexity, empires like tsarist Russia and Austria-Hungary exceeded the individual's comprehension. Did a subject living in Habsburg Galicia feel like he or she was part of a realm extending all the way to Tyrol? In this context, research has shown that empires are dependent on being able to reproduce

---

[51] Boris Barth and Jürgen Osterhammel, eds., *Zivilisierungsmissionen: Imperiale Weltverbesserung seit dem 18. Jahrhundert* (Konstanz: UVK-Verlagsgesellschaft, 2005).
[52] Robert F. Byrnes, *V. O. Kliuchevskii: Historian of Russia* (Bloomington: Indiana University Press, 1995).
[53] Thomas Sanders, *Historiography of Imperial Russia: The Profession and Writing of History in a Multinational State* (London: Routledge, 2015).

themselves anew as conceptional entities time and time again. Assuring a presence on contemporaries' mental maps required the permanent representation of territoriality and statehood in many different contexts and genres[54]—and just like the arts and religion, science offered a spectrum of means to this end. Numerous genres of the representation of scientific knowledge were particularly suitable for visualizing imperial shared identity, allowing the empire to be made tangible as a territory, a conglomerate of people, and a multiplicity of natural spaces.

Maps provided a synoptic view of the respective federated empire as a whole.[55] Tables conveyed impressions of a categorizable heterogeneity of things and persons that together made up the realm. Guido Hausmann discusses the representation of statehood in the works of representatives of political geography in the Russian Empire, demonstrating that the state was increasingly perceived as a dynamic entity over the course of the nineteenth century: while it appeared primarily as a fixed political unit in works from the early 1800s, it was later portrayed as the result of natural, social, and political processes. The contribution by Sabrina Rospert and Mark Hengerer on the Habsburg Monarchy investigates an additional example of a medium transporting conceptions of imperiality: the course catalogues of the Universities of Vienna and Budapest. As Rospert and Hengerer's statistical study shows, these catalogues can be read as representations of the Habsburg realm as an empire due to the various courses referring to the imperial idea.

Conclusion

While discussed separately in this introduction for purposes of simplicity, the imperial imprint of science and the scientific imprint of empire were in fact entwined in myriad ways. Science played a fundamental role for the Habsburg Monarchy and the Russian Empire, and at the same time both empires can be viewed as spaces allowing engagement in and development of scientific practice.

Science made useful knowledge available to imperial politics and administration, helped to legitimize imperial rule, was a key element of the idea of the civilizing mission, and contributed to making the empire concrete and imaginable for its subjects. That science fulfilled these "functions" for empires did not mean that it was primarily "functionalized" for political purposes,

---

[54] Ilya Gerasimov, Jan Kusber, and Alexander Semyonov, eds., *Empire Speaks Out: Languages of Rationalization and Self-Description in the Russian Empire* (Leiden: Brill, 2009).

[55] Encyclopaedias fulfilled a similar function with different creative means. A famous example for the Habsburg Monarchy is the *Kronprinzenwerk*, a 24-volume regional studies encyclopaedia of the empire published from the mid-1880s at the instigation of Crown Prince Rudolph.

however. As has been demonstrated, assuming such functions frequently also served the scientific logic of individual disciplines itself.

But science and empire could also collide due to these very same logical specificities. The activity of researchers could only be politically controlled to a degree—a circumstance that did not prevent corresponding attempts by representatives of the central state, however. One such example is discussed in Andrej Andreev's contribution investigating the activities (some of which were intended to counteract Enlightenment tendencies) of curators appointed by the central government to control universities in the Russian Empire during the first half of the nineteenth century.

As a prestigious cultural practice, science was likewise ideally suited to substantiating nationalist pretensions opposing the imperial order. In this context, the plans for a Czech academy of sciences and the activity of Polish learned societies described in this volume may serve as examples. The "repair and encouragement cartography" of political geographers during the late nineteenth and early twentieth centuries was also an expression of such nationalist aspirations.[56]

When examining the relationship between science and empire in the Habsburg Monarchy and tsarist Russia, one encounters a multitude of relevant disciplines and areas of knowledge of which this volume provides a broad overview. The difference between the humanities and natural science—a differentiation that would only stabilize itself during the final decades of the nineteenth century—plays only a subordinate role in this regard. Although the two fields of learning were slowly beginning to diverge in terms of their methods in the period under scrutiny in this book and fulfilled different functions in the imperial context, a specific greater affinity to "imperium" cannot be ascribed to either of them.

A significant such affinity in the Habsburg and Russian contexts *can* however be attributed to all branches of science developing from the cameralistic and political science doctrines of the eighteenth century, like ethnography and soil science. These fields of knowledge referred to spatially unevenly distributed resources, thereby touching upon an essential structural characteristic of imperial states in the nineteenth century: their expansive relationship to space.

Perhaps one of the most important conclusions of this volume is that seafaring and land-based empires were not dissimilar in terms of the relationships between science and empire observable within them. This applies to the fundamental, imperiality-constituting role of science as well as to the mutual functional entanglement. In this sense, the European continental empires were no less "modern" than their sea-going and colonial counterparts.

---

56 Quoted from a draft of Peter Haslinger's contribution to this volume.

Nevertheless, there are also areas in which the land empires do appear somewhat special. The great attention that scientists paid to the gradual spatial variation of nature and culture was less present in overseas empires. Continental empires also exhibited a more pronounced structural ambivalence between imperial and nation-state patterns of order, which affected the practice of science in manifold ways. Many of the features of science as practiced in empires that have been carved out in the following essays are not dissimilar from those found in nation-state systems. Overall, the situation in continental empires seems to have been less clear-cut than in colonial empires: the boundaries between "own" and "foreign," between nation and empire, must often have appeared blurred to researchers. Indeterminateness and ambivalence are not an altogether bad substrate for scientific creativity, however.

* * *

Martin Schulze Wessel approached me in early 2015 with the idea of jointly organising a conference on the topic of the interdependency between science and imperial orders in eastern Europe. I was in the process of writing my dissertation dealing with related questions at the time, and the opportunity to discuss these issues in a broader setting was thus very welcome. The annual convention of the Collegium Carolinum, hosted in cooperation with the Graduate School for East and Southeast European Studies Munich-Regensburg, took place on 5–8 November 2015 in Bad Wiessee. This volume presents revised versions of the conference contributions. First and foremost, I wish to thank Martin Schulze Wessel not just for the initial impulse, but for his considerable contribution to the overall success of the project. Neither the conference nor the resulting volume would have been conceivable without his outstanding conceptional, organizational, and editorial input. I would also like to thank Ulrike Lunow for her competent organisation of the conference and the participants who presented, moderated, and discussed with great dedication. My gratitude likewise extends to the translators of the contributions, to Jacqueline Friedlander for proofreading, to Amadeus Neumann for preparing the book for printing, and to Helena Zimmermann for her layout and typesetting work. Finally, I wish to thank Johannes Gleixner, Stephanie Weiss, and Kathrin Krogner-Kornalik for handling the editing of the book.

Translated from German by Stephan Stockinger.

Martin Franc

AN ACADEMY OF SCIENCES IN PRAGUE?
Purkyně's Akademia and Josef Hlávka's Emperor Franz Joseph Czech Academy for Sciences, Literature, and Arts

From the perspective of the Habsburg Monarchy at the time, the establishment of a Czech academy of sciences in 1890 may have seemed a somewhat overdue task—after all, an academy of sciences had been founded in Budapest, for example, as early as 1825. The Imperial Academy of Sciences (Kaiserliche Akademie der Wissenschaften) in Vienna, whose members included many scientists who carried on their work in the Czech lands,[1] had been established in 1847. The establishment of a Croatian, or to put it more accurately, a Yugoslav academy of sciences in Zagreb[2] had been initiated and financed by Bishop Josip Strossmayer in 1860.[3] Its existence was officially sanctioned by Emperor Franz Joseph I in 1866. But the main inspiration behind the creation of a national Czech academy of sciences was the elevation of Cracow's learned society to the status of an academy of sciences in 1871.[4] Josef Jireček (1825–1888), a Czech politician and scientist who held the post of Austrian minister of culture and education at the time, and his brother Hermenegild, a high-ranking official at the same ministry, played a significant role in this process. In 1880 it would be the same Josef Jireček who, as the chairman of the Royal Bohemian Society of Sciences (Královská česká společnost nauk,

---

[1] E.g., chemist Karl Joseph Napoleon Balling (1805–1868), historian Konstantin von Höfler (1811–1897), historian František Palacký (1798–1876), Slavicist and archeologist Pavel Josef Šafařík (1795–1861), archeologist and historian Jan Erazim Vocel (1802–1871), and geologist Franz Xaver Maxmilian Zippe (1791–1863).

[2] Milan Moguš et al., *150 godina Hrvatske akademije znanosti i umjetnosti 1861–2011* (Zagreb: Hrvatska akademija znanosti i umjetnosti, 2011).

[3] On the Czech reaction to the establishment of the Croatian academy of sciences, see a letter by an anonymous chaplain sent to the editor of *Národní listy* in January 1861 along with 100 gulden as the initial contribution to a collection for the founding of a Czech academy of sciences; "Zaslán byl redakci naší dnes následující přípis...," *Národní listy*, 31 January 1861, 1. Later, alumni of the Prague seminary would subscribe to the idea of founding a Czech academy of sciences by making donations totaling almost fifty-six gulden.

[4] See, e.g., Stanisław Czarniecki and Janusz Wiltowski, *W stulecie utworzenia Akademii umiejętności* (Warsaw: Państwowe wydawnictwo naukowe, 1972).

KČSN), would demand the elevation of that society to the status of a Czech academy of sciences, explicitly invoking the elevation of the Cracow academy in his supporting arguments.[5] His request, however, was rejected—one justification for the refusal being that the activities of any potential Czech academy already fell under the purview of the Imperial Academy of Sciences in Vienna. The fear that elevating the KČSN to the status of an academy would lead to the definitive Czechization of this formally bilingual institution played an important role in this issue. A new round of negotiations centered on the founding of a Czech academy of sciences had already begun in the second half of the 1880s, prompted by Josef Hlávka, an architect and builder, widely considered the most important patron of Czech science of all time.

In this essay, I would like to deal with what I consider to be the two most important and most interesting plans for the creation of a Czech academy of sciences in the second half of the nineteenth century, both of which have been carefully documented and much studied in the existing literature.[6] The first plan I examine is the proposal put forward by famed Czech natural scientist Jan Evangelista Purkyně[7] during the early 1860s. The second plan I look at (a

---

[5] This document of 14 March 1880 is published in Jiří Beran, *Vznik České akademie věd a umění* (Prague: Ústřední archiv ČSAV, 1989), 87–97. The entire matter was dealt with recently by Tomáš W. Pavlíček in his presentation "Idea akademii narodowej w transferze kulturowym między Wiedniem, Krakowem a Pragą: Próby reformy Czeskiego Królewskiego Towarzystwa Naukowego. Josef Jireček i Josef Hlávka" which he gave at a conference titled 200-lecie Towarzystwa Naukowego Krakowskiego, held on 9–10 December 2015 at the Polish Academy of Arts and Sciences in Cracow. In it he analyzes the later interactions between J. Jireček and J. Hlávka and calls into question the hypothesis that the Czech-German national conflict was the only reason for the failure of this idea.

[6] I cite specific relevant literature where necessary. For both proposals, I have drawn greatly from Jiří Beran's heavily factual studies and published sources. Jan Janko and Tomáš Hermann, among others, have studied J. E. Purkyně extensively. The founding of the Czech Academy of Sciences and Arts and the figure of J. Hlávka have also been the subject of writings by Jiří Pokorný and Magdaléna Pokorná (the history of the KČSN), etc.

[7] Jan Evangelista Purkyně (17 or 18 December 1787 – 28 July 1869), Czech natural scientist and philosopher. After graduating from the Piarist grammar school in Mikulov, he joined the Piarist order in 1804. He taught at the grammar school in Litomyšl. In 1807 he left the order and completed his philosophy studies in Prague. He later worked as a tutor for aristocratic families. He successfully completed his studies of medicine in Prague in 1818. Between 1818 and 1823 he held the position of coroner at the medical faculty of the university in Prague. Between 1823 and 1850 he was professor of physiology at the university in Wrocław (where he founded the institute of physiology in 1839). Between 1850 and 1869 he was professor of physiology at the university in Prague. He made a significant contribution to Czech national life; he was involved in the establishment of a great many associations and organizations. In 1853 he founded the natural science journal *Živa*. He was one of the co-authors of cell theory. He discovered many anatomical structures and processes

plan that was eventually to bear fruit) was proposed by the aforementioned Josef Hlávka[8] at the end of the 1880s. It would of course be an extremely difficult task to compare the two plans directly against one another. Purkyně set out a no less than monumental vision in which the academy would be, in the broad sense of the word, an institution whose standing was envisioned as occupying an equal status as the state and the church. In contrast, the academy of sciences and arts planned and eventually established by Hlávka was envisioned as no more than a standard institution of a type already present in the Habsburg Monarchy, which possessed many features reminiscent of the above-mentioned Academy of Learning (Akademia Umiejętności) in Cracow. Although both proposals focused on the topics of "national science" and a "national academy," the main figures behind them significantly differed in the definitions they used for the term "national." According to Purkyně's definition, the Emperor Franz Joseph Czech Academy for Sciences, Literature, and Arts that was eventually to come into being could not truly be considered a national academy. One reason for this assertion was the official support the institute received from the ruling Habsburg dynasty. In Purkyně's view any truly national academy of sciences could only be funded exclusively by the people. He took much inspiration from

---

(including Purkinje cells in the cerebellum, Purkinje fibers, Purkinje images in the eye, and the Purkinje shift). He worked mainly on the physiology of the senses, producing many publications on histology and embryology. He was also active in promoting modern research methods and technical innovation. See Jan Janko, "Jan Evangelista Purkyně," in *Bohemia docta: K historickým kořenům vědy v českých zemích*, ed. Alena Míšková, Martin Franc, and Antonín Kostlán (Prague: Academia, 2010), 176–178; Vladislav Kruta and Mikuláš Teich, *Jan Evangelista Purkyně* (Prague: Státní zdravotnické nakladatelství, 1962); Václav Žáček, *Jan Evangelista Purkyně* (Prague: Melantrich, 1987).

[8] Josef Hlávka (15 February 1831 – 11 March 1908), after completing secondary school, studied at Prague Polytechnical Institute and completed a degree in architecture from the Academy of Arts in Vienna in 1854. While still a student, he began working for František Šebek's construction company, where he learned the bricklaying trade. In 1855 he was awarded a three-year travel stipend, thanks to which he took the chance to explore central and southern Europe. After his return he took over the company from František Šebek, who had bequeathed it to him. Between 1860 and 1869 he devoted himself to a successful career in construction, during which he built, for example, the Lazarist Church in Vienna, the Vienna Court Opera, the palace of Archduke Wilhelm in Vienna, the Bohemian Maternity Hospital in Prague (based on his own design), and a building complex for the residence of the Bukovinian metropolitan in Chernivtsi. In 1869 he began to suffer from serious health problems due to overwork and became paralyzed in his lower limbs. He only began walking again around 1880. After his recovery, his activities concentrated on philanthropy, mainly through providing support for students and science. He served as a member of the imperial council between 1883 and 1891 and as a parliamentarian in the Cisleithanian House of Lords from 1891 on. See Jiří Pokorný, "Josef Hlávka," in *Bohemia docta*, ed. Míšková, Franc, and Kostlán, 179–184.

the idea of the "national collections," which had financed the construction of the National Theater in Bohemia, for example.

In 1850 Purkyně returned to Prague from the city of Wrocław, where he had spent several years conducting successful research, to take up a post as a university professor of physiology. Already a world-famous natural scientist in his sixty-third year upon his return, he was thereafter to concentrate his attention mainly on establishing an institute of physiology at the university. At the same time, he also became enthusiastically involved in the national movement, ignoring the attention he began to receive from the police as a consequence. One report from 1864 declared that "Professor Purkyně is never so tired as to not, anywhere that he may, take steps towards creating a Czech National Academy, which has become an idée fixe with him."[9] He was never to abandon the idea of creating such an institute, even as a consequence of his organizational involvement in other institutions: he helped found the Sokol gymnastics movement, the Hlahol Choir Association, the People's Fund: the Society for Publishing Inexpensive Books in Czech (Spolek pro vydávání laciných kněh českých – Matice lidu), for which he went as far as taking on the post of chairman, and also contributed to establishing the Artists' Forum (Umělecká beseda) and the Academic Readers' Association (Akademický čtenářský spolek), two associations he wished to use as bases upon which to build the academy he dreamed of.

Purkyně set out his ideas for the creation of an academy in a number of studies published between 1861 and 1863 in *Živa*, a journal he had himself founded,[10] as well as in a separate monograph entitled *Akademia*, published in 1861. Even from today's perspective, his proposals still seem radical; they reflect the extremely optimistic atmosphere of the early 1860s after a decade of severe repression of all original and creative ideas. Purkyně defined three social (or national) "forms"—the state, the church, and science. In a broad sense, he understood the concept of the academy as "science incarnated in human society," that is to say, as the body that represented science as one of the three basic "forms" of the nation.[11] This academy was to have two main tasks—it was to act both as the administrator of the educational system and as an academy in the narrower sense, that is, as a monumental research facility made up of a variety of scientific institutes, whose task, of course, would be to produce the nation's future scientific workers. Purkyně linked scientific progress and research together to be pursued under the umbrella

---

[9] See, e.g., Jaroslav Jedlička, "Jan Evangelista Purkyně v národní kultuře a v politice," *Časopis lékařů českých* 108 (1969): 946–952, here 946.

[10] Jan Evangelista Purkyně, "Akademia," *Živa* 9 (1861): 96–103, 218–227; *Živa* 10 (1862): 37–47, 245–267, 334–344; Jan Evangelista Purkyně, *Akademia* (Prague, 1861). In the text that follow, unless stated otherwise, I draw from these sources in my efforts to present Purkyně's ideas about what his proposed academy of sciences should look like.

[11] Purkyně, "Akademia," 100.

of the academy; he viewed universities and colleges as mere teaching institutes. Thus, he effectively called into question the established Humboldtian intercoupling of education and science as the guiding principle of the modern Central European university.[12]

Purkyně was also skeptical about the potential for transforming the aforementioned Royal Bohemian Society of Sciences, which had existed since the mid-eighteenth century, into a new Czech academy of sciences. He never objected to the idea of doing so on the grounds of the strong emphasis put on bilingualism within the society; his problems with it stemmed rather from the substantial restrictions on its activities and from its reluctance to work more closely with the Museum for the Kingdom of Bohemia (Muzeum Království českého), the institution that later became known as the National Museum (Národní museum). However, Purkyně did criticize the society for its cosmopolitanism and a certain (and not unrelated) tepidness on national issues that characterized what was ultimately a very traditional institution. In the 1850s the KČSN went through a really deep crisis stemming from a lack of resources to finance its activities and from the withdrawal of the society's *spiritus agens*, historian František Palacký,[13] after 1848.[14] During this period, Palacký for the most part did not even attend KČSN meetings. At the same time, a number of problematic political figures, such as Leopold von Sacher-Masoch, Prague's director of police, and Alexander von Bach, the Austrian minister of the interior, became members of the society. It must also be noted, however, that a rift between Palacký and Purkyně opened up in the 1850s; they had conflicting views about where the focus of scientific work in the Czech lands should be directed. Purkyně took a critical view of Palacký's project to compile an encyclopedia on the grounds that he saw the task as being a waste of the meager scientific capacity of the Czechs in the conditions of the time. Crucially, Purkyně failed to vote for Palacký to remain a member of the museum board.[15]

---

[12] On Humboldtian ideas and their reception in the Czech lands, see, e.g., Karel Šima and Petr Pabian, *Ztracený Humboldtův ráj: Ideologie jednoty výzkumu a výuky ve vysokém školství* (Prague: Sociologické nakladatelství/SLON, 2013).

[13] For parallels in the work of Purkyně and Palacký, compare Tomáš Hermann, "Torzo a komplexita v romantické vědě Purkyně a Palackého: Útržky a Krásověda," in *O duši Země a romantické vědě*, ed. Tomáš Hermann and Václav Cílek (Prague: Academia, 2010), 211–254, here 211.

[14] On the history of the KČSN in this period, see Magdaléna Pokorná, "Královská česká společnost nauk," in *Bohemia docta*, ed. Míšková, Franc, and Kostlán, 58–144 (on the KČSN in the 1850s, esp. pp. 93–94).

[15] See Jan Janko, "České země a Purkyňova koncepce vědeckých institucí: Dokončení," *Československý časopis historický* 36 (1988): 697–714.

In a proposal of his from the early 1860s, Purkyně envisioned that the academy of sciences would be built upon the foundation of the Museum for the Kingdom of Bohemia, primarily because of the possibility of linking the academy of sciences with museum collections, a library, an archive, and a number of other similar facilities.[16] Purkyně therefore saw the potential initial shoots of the academy to be found within the various societies, committees, and commissions which had emerged under the patronage of the Museum for the Kingdom of Bohemia and which maintained a connection with it. As I have already mentioned, however, Purkyně also viewed other organizations as being possible islands from which the academy of sciences could develop. Such bodies included not only the Academic Readers' Association, but also the aforementioned Artists' Forum, which Purkyně himself had helped found in 1863. As a result of Purkyně's efforts, a scientific department was established within the organization, in which he frequently lectured. The Artists' Forum defined itself in a sense as a counterweight to the increasingly conservative and to some extent bilingual Museum for the Kingdom of Bohemia. In this context, it is also worth mentioning Purkyně's fundamental contribution to the contemporaneous establishment of the Association of Czech Doctors (Spolek lékařů českých), yet another Czech-language organization that focused on a particular group of professionals.[17]

Purkyně meticulously distinguished between various different types of academies. He thoroughly analyzed the forms of academies and learned societies throughout the world, expressing his greatest affinity for the French model, though he did not consider it absolutely ideal for—or even applicable to—the conditions faced by the Czech nation. He dreamed of a true national academy as a third form of national existence to stand alongside the church and the state. He viewed academies connected to the state or with a university as belonging to a much lower echelon of existence. He considered royal or imperial academies no more than learned societies, or mere factories for the production of detailed studies on scientific matters, without possessing any form of executive power. Accordingly, he demanded the establishment of a truly independent academy that would be "on a par" with the church and the state. Purkyně envisioned his conception of an academy of sciences as being financially independent, and therefore wanted to grant his institute a number of monopolies in science-related industries and trades. Unfortunately, how-

---

[16] As Jan Janko once aptly noted, Purkyně did not demonstrate a sufficient understanding of developments at the museum and viewed it primarily as a scientific institution, ignoring its role in enlightening the public. On top of this, Jiří Beran has also pointed out that J. E. Purkyně paid insufficient attention to the growing conservatism within the Museum for the Kingdom of Bohemia's leadership. See Jiří Beran, "Purkyňova Akademia: Pokrokovost a historická podmíněnost velké koncepce," *Věstník ČSAV* 97 (1988): 118–125.

[17] See, e.g., Jedlička, "Jan Evangelista Purkyně v národní kultuře."

ever, this idea is not comprehensively explicated and analyzed in his proposal as the issue lay outside his main interests. Purkyně anticipated that an international academy for the entire world would be founded in the future, and that it would be capable of grappling with exceptionally challenging matters, such as the study of the cosmos. In his opinion, however, such a worldwide academy of sciences would not be designed to entirely supplant the role of the national academies. In his view, the category of a national academy could be distinguished from the general academy in that the latter dealt exclusively with pure science. He considered the Imperial Academy of Sciences in Vienna to be an example of such a general academy. But national academies of sciences remained an appropriate institution for the purposes of nations contained within such multinational empires as the Central European Habsburg Monarchy and, in particular, for nations whose scientific language was underdeveloped (as was the case of the Czechs at the time). Purkyně considered the main aim of a national academy of sciences was to act as the "living nation." He also placed great emphasis on its tasks in relation to the education of new scientific workers. In his writings he referred to this narrower concept of the academy of sciences as a "standing scientific army." According to Purkyně's vision, any national academy of sciences would have to emerge as a body founded and financed by the Czech nation itself. He anticipated that people would in the future have the opportunity to buy one-thousand-gulden shares in the institution. But this funding was to serve merely as the academy's initial capital. Purkyně wanted to entrust its ongoing development and management to the Bohemian Diet, or rather to a commission established by that assembly made up of leading scientists and other experts. He expected that this commission, in collaboration with the academic board—a self-governing body of scientists—would manage all of the academy's affairs. In his book *Akademia*, he described in detail the structure of the new academy of sciences and the tasks of each of its research institutes as well as the appearance of their central buildings, which he planned to have built on Charles Square in Prague. He anticipated fourteen institutes would be created—institutes of philosophy, mathematics (or, in Purkyně's words, the Institute of Pure Mathematical Sciences), linguistics, literature and philology, history, legal sciences and political science, anthropology and ethnography, chemistry, physics, physiology and anatomy, mineralogy and geognostics, botany, zoology, and paleontology—in addition to special "shared facilities," that is, a library, archives, and collections, as well as, for example, a printing press, a bookstore, a gasworks, and a telegraph service.

For this essay, the chapter dealing with the new academy's official language of discourse is particularly interesting. At the time, Purkyně was one of the very few natural scientists who lectured in Czech at the university in Prague, which at the time had still not been split into Czech and German universities. His students included the sons of leading figures in the national movement, such as Slavicist Pavel Josef Šafařík, Slavicist and poet František

Ladislav Čelakovský, and lawyer and politician Josef František Frič.[18] In 1863 he was one of the main agitators for the idea that Czech-language lectures should be introduced at the faculty of medicine, an activism that the police noted unfavorably.[19] Purkyně was well aware that in the early 1860s it would be impossible to put together a fully-fledged academy made up of only Czech-speaking scientists. In a 1988 study, Jan Janko correctly emphasizes that, despite all its grandeur and largesse, from the perspective of national society, Purkyně's project begins to look like a fallback plan in which the academy of sciences would be expected to fill the role of Czech institutes that either did not exist or functioned only poorly at the time.[20] At that moment in time, he had to make do with a plan for the academy that conceived it as a largely patriotic organization in which Czech-speaking and German-speaking scientists cooperated in harmony and endeavored together to build up a reputation for the institution. This setup, however, would clearly require members of the academy of sciences to be actively bilingual. He therefore did not perceive German-speaking scientists from Bohemia as being part of a German or Austrian scientific culture, and instead primarily emphasized the values of Bohemian patriotism.

Purkyně's book elicited some interesting reactions, both immediately after it was published and in the subsequent decades. For example, in a January 1864 letter to Purkyně, the aforementioned Bishop Strossmayer, the primary advocate of the founding of the Croatian Academy of Sciences and Arts, supported the Czech professor's vision of Prague as the scientific center for all of Austrian Slavdom (going so far as to refer to the city as "the Slavic Athens").[21] Thus, the academy of sciences in Prague was not understood as a purely national Czech institution but rather as a Slavic counterpart to the imperial

---

[18] Ladislav Josef Čelakovský (29 November 1834 – 24 November 1902) was the son of Purkyně's colleague from Wrocław University, František Ladislav Čelakovský. He would later become professor of botany at Prague's Czech university. Antonín Frič (30 July 1832 – 15 November 1913), son of Josef František Frič, who was to become a geologist, paleontologist, and, eventually, director of the Museum for the Kingdom of Bohemia, served as Purkyně's assistant from 1861 on. Another of Purkyně's charges, one who came along with him from Wrocław, was the exceptionally talented Julius Sachs, who would later become the co-founder of modern plant physiology and hold a professorship at the University of Würzburg. A dramatic rift would later develop between the two. It is unclear what role the old professor's attempts to Czechify his student and personal assistant played in their relationship.

[19] Jiří Beran, ed., *Soupis pramenů k životu a dílu Jana Evangelisty Purkyně uložených v československých archivech* (Prague: Ústav československých a světových dějin ČSAV, 1987), 73. The demand for lectures in Czech was made officially by the Association of Czech Doctors.

[20] Janko, "České země a Purkyňova koncepce," 700–711.

[21] Ivan Dorovský, *Balkán a Mediterán: Literárně historické a teoretické studie* (Brno: Masarykova univerzita, 1997), 145.

academy in Vienna, whose Slavic members formed no more than a small minority.

In the early 1850s, Purkyně's ideas, especially his vision of the academy of sciences as an institute that combined the representative function of a learned society with that of conducting independent scientific activity within a network of research institutes, were proclaimed with particular energy. His ideas were seen as proof of just how socially progressive the man had been. As part of a propaganda strategy that sought out suitable Czech predecessors of some of the ideas promoted within the Sovietization effort going on at the time, the figure of Purkyně was taken up as a domestic model for the Czechoslovak Academy of Sciences, which was founded in 1952.[22] And then in the late 1960s, a period in which the concept underlying the Czechoslovak Academy of Sciences was becoming subject to widespread criticism, officials defended their approach of combining the functions of a learned society with those of research institutes by evoking Purkyně's vision, since the alternative of falling back on Soviet models had become a move that could be seen, to put it mildly, as highly questionable.[23] When in April 1968 a group of former members of the KČSN and the Czech Academy of Sciences and Arts issued a "protest resolution" in which they harshly attacked the Czechoslovak Academy of Science in its current form as an element of discontinuity in the Czech (and Czechoslovak) system of science, the following passage evoking Purkyně's vision for the academy appeared in the response of the ČSAV's presidium, primarily as a result of the influence of Ivan Málek:

> As far then as fears that the Czechoslovak Academy of Sciences is not the "continuator and developer" of our great traditions are concerned, I believe that it is not our place to respond and all the less the place of our prophetical critics to respond. After all, the authors of the protest resolution clearly missed that the form that the Czechoslovak Academy of Sciences adopted worthily fulfills Purkyně's ideal as embodied in his text Akademia. Indeed, in connection to the protest resolution, it would seem expedient of its authors to return to J. E. Purkyně's intellectual legacy and to consider the criticism [leveled at him] by our then lead-

---

[22] Respect for progressive national traditions was thought of at the time to be one of the "gifts" of Soviet science. See Ivan Málek, *Učíme se od sovětské vědy: Sborník článků a projevů* (Prague: Nakladatelství ČSAV, 1953), 6–10, here 9–10.

[23] One of the main promoters of Jan Evangelista Purkyně as the intellectual father of the ČSAV was microbiologist Ivan Málek (1909–1994), a member of the presidium of the ČSAV in 1953–1969 and director of the ČSAV's Biological Institute (which was known as the Microbiological Institute from 1962 onward) in 1953–1970. Among his contributions were the foreword and notes he wrote for a new edition of Purkyně's *Akademie* published in 1962; see Jan Evangelista Purkyně, *Akademia* (Prague: Nakladatelství ČSAV, 1962). See also Ivan Málek, "Návrh J. E. Purkyně na Akademii," in *Zdeňku Nejedlému Československá akademie věd: Sborník prací k sedmdesátým pátým narozeninám*, ed. Václav Husa (Prague: Nakladatelství ČSAV, 1953), 371–390 and Ivan Málek, "Purkyňovy úvahy o organizaci vzdělání, výchovy a vědy," *Časopis lékařů českých* 108 (1969): 953–956, here 953.

ing national scientific institute [the reference here is to the KČSN, whose dissolution in favor of the establishment of the Czechoslovak Academy of Sciences was heavily criticized by the authors of the resolution M.F.].[24]

Nonetheless, the real, direct impact of Purkyně's plans was of no real significance in the 1860s. The generally underdeveloped nature of Czech science had certainly played a key role in Purkyně's vision back in his time, along with the gradual fading of the general élan that had once gripped the whole of society, an élan whose eclipse had become particularly manifest after the defeat in the Austro-Prussian War, and the huge disappointment at the introduction of Austro-Hungarian dualism. Moreover, by the second half of the 1860s Jan Evangelista Purkyně was of an advanced age (he was to eventually die in 1869), and no immediate replacement could be found with a similar enthusiasm for organization and at the same time with the benefit of the scientific renown that he possessed. Certainly, his far-reaching vision for an academy, in its holism and depth, had a certain utopianism about it, not to mention a hint of Romanticism. Many researchers have indeed noted that Purkyně was familiar with contemporary utopian literature and found inspiration in it.[25]

Although Purkyně considered the Museum for the Kingdom of Bohemia to be primarily a scientific institution, matters were to develop along a different path.[26] The museum, of course, still focused mainly on various efforts to enlighten the general public. Nonetheless, in 1864 the Committee for the Natural Scientific Study of the Czech Lands (Komité k přírodovědnému zkoumání českých zemí) was established under the aegis of the museum largely on the initiative of the aforementioned Antonín Frič. It was set up to be a joint research institute of the Museum for the Kingdom of Bohemia and the Patriotic-Economic Society (Vlastenecko-hospodářská společnost). It was also the only remnant of an extraordinarily complex plan, which was destined only to be implemented in its main points under the completely different social and political conditions of the second half of the twentieth century.

The transformation of the Cracow learned society into a fully-fledged academy with official support from the highest authorities within the Habsburg Monarchy contributed to a new focus on efforts to create a national academy of sciences (and also of the arts) by means of restructuring and expanding the

---

[24] The protest resolution and the response from the presidium of the ČSAV are published in Miroslav Šmidák, *Institucionální vývoj Československé akademie věd v letech 1960–1969 očima jednoho z přímých aktérů* (Prague: Masarykův ústav a Archiv AV ČR, 2011), 257–266.

[25] Jaromír Loužil, "Purkyně přírodovědec a naturfilosof," in *Jan Evangelista Purkyně: Útržky ze zápisníku zemřelého přírodovědce & O duši Země a romantické vědě*, ed. Tomáš Hermann and Václav Cílek (Prague: Academia, 2010), 107–131.

[26] František Palacký, too, clearly defended this idea as early as the 1840s. See Beran, "Purkyňova Akademia," 124–125.

An Academy of Sciences in Prague

existing Royal Bohemian Society of Studies. Clearly, the forces responsible for these initiatives anticipated a further process of Czechization in the context, although it was to be a gradual and nonviolent process. The institution's firmly established bilingualism did not present any particularly serious problem, especially since its German-speaking members had been largely inactive in the body for quite some time and so did not raise any significant objections to it being transformed into a Czech academy of sciences (and the arts). After all, the original plans had not anticipated the creation of an institution defined along such sharp national and linguistic lines as the one that would eventually emerge. The official request to transform the KČSN into an academy of sciences, as I have already said, was submitted by the society in 1880 and was signed not just by its chairman, Josef Jireček, and other functionaries, but also by Adalbert Carl von Waltenhofen zu Eglofsheimb, one of its German-speaking members. The application did not make any mention of the language issue at the proposed new institute; at its core, besides a slightly overlong enumeration of the various professional and scientific activities undertaken by the society since its foundation, was a demand for regular state subsidies. KČSN officials were forced to wait until 1883 for an answer which, when it eventually came, only satisfied them in part. Minister of Culture and Education Sigmund Conrad von Eybesfeld put the entire matter to the Imperial Academy in Vienna, whose job it was to bring together under its roof all the scientists of the monarchy, including those working in the Czech lands.[27] Its officials refused to comment on the petition, claiming it was more a political issue than a scientific one. Eybesfeld himself was quite skeptical about transforming the KČSN into an academy of sciences, mainly as a result of the influence of the German-speaking members of the KČSN. At the same time, he was also critical of the activities of the Habsburg Monarchy's other national academies in Budapest, Cracow, and Zagreb. According to him, these institutions focused exclusively on national issues and contributed little to overall scientific knowledge. He also denied that there was any parallel between the Cracow society and the KČSN, mainly on the grounds of their differing situation in relation to the competing languages in each case—whereas in Prague what was at issue was German, with its highly developed culture of science, in Galicia the competition related to the Ruthenian language, which lagged markedly behind its Polish counterpart as a language of science. Conrad could only guess whether the KČSN was likely to transform into a similarly strictly national institute or bring Czech- and German-speaking scientists together under the banner of a sort of Bohemian patriotism; nonetheless, he correctly felt things would develop along the lines of the former hunch, that is to say, resulting in a clear Czechization of the entire academy. In any case, he felt that transforming the society into an academy of sciences would not increase the institute's scientific output and would only be-

---

[27] At that point in time, however, it was fulfilling this role only in part.

come, at the time of the transformation, a potential source of disputes and conflicts. In this context, he also mentioned the recent division of the university in Prague into Czech and German sections, a process that had been accompanied by a range of disagreements and squabbles. On the other hand, though, Conrad supported the KČSN's efforts to be awarded regular state subsidies and proposed the first tranche of such funding be paid out in 1884 to the amount of 5,000 gulden (up until then it had been mainly the Bohemian Diet that had funded the KČSN). His proposal was to meet with the emperor's approval.[28]

But in the 1880s a truly rapid and dramatic change occurred in the situation. The establishment of a Czech university in Prague (1882) significantly expanded the potential membership base of any new institute of the type of an academy of sciences, even though not all professorial positions at the new university were filled by leading scientific figures. In the second half of the 1880s, the Czech academic community was gripped by a conflict over the authenticity of what were called *Manuscripts* (Rukopisy),[29] two presumably medieval epic texts written in Old Czech. This controversy had a significant impact on, among other matters, discussions on the creation of a Czech academy of sciences and, later, on the personnel who were to make up its membership. A group known as the "realists," headed by linguist Jan Gebauer and by philosopher and later first president of Czechoslovakia T. G. Masaryk, stimulated substantial discussion on the issue. Even before any important articles devoted to the issue of the manuscripts were published in *Atheneum* magazine, a publication associated with T. G. Masaryk's circle, it printed a discussion in several parts[30] addressing the need to publish high-quality academic literature and scientific journals in Czech, to organize activities aimed at mass public enlightenment, to ensure the high-quality translation of foreign publications, to contribute to the expansion of knowledge on foreign languages in the Czech lands, and to promote Czech science abroad. It also broadly touched upon the issue of an academy of sciences and clearly set out the

---

[28] This document of 18 July 1883 is published in Beran, *Vznik České akademie věd a umění*, 98–107.

[29] This term referred to the *Dvůr Králové* and the *Zelená Hora Manuscripts*, two documents purported to be ancient medieval texts alleged to have been discovered in 1817–1818. Linguist Josef Dobrovský questioned their authenticity immediately upon the first announcement of their discovery. The dispute about their authenticity would last for several decades, during which time skeptics were often condemned as showing a lack of patriotism. Modern research methods have since unambiguously proven that the documents were literary hoaxes. On the topic of academic discussions on the manuscripts, see Dalibor Dobiáš et al., *Rukopisy královédvorský a zelenohorský a česká věda, 1817–1885* (Prague: Academia, 2014).

[30] Reactions to these articles were published in other periodicals, such as in the daily *Pokrok*.

group's demand that such an institution be established.[31] The scientists in Masaryk's circle, however, took a critical stance toward the idea of transforming the KČSN into an academy of sciences. Personal conflicts certainly played a role in this controversy; the leadership of the learned society was made up of staunch adversaries of the realists, who were also their most important opponents in the heated debate over the manuscripts.[32] For understandable reasons, however, a range of the arguments being made by members of the realists began to appear in the public discourse. Although some authors continued to highlight the KČSN's bilingualism, this aspect of the body's character was essentially no longer applicable, especially since Ernst Mach had left the KČSN in April 1884 (the official reason he gave was his heavy workload at other institutions, including at the Imperial Academy in Vienna) and other important German-speaking scientists in the Czech lands, including mathematician Heinrich Jacob Karl Durége, geologist, paleontologist, and polar explorer Gustav Karl Laube, and botanist Moritz Willkomm, had also left in the same year. Even though historian Konstantin von Höfler and philologist Alfred Ludwig, for example, remained members of the venerable learned society, no new German scientists were to join, and Czech-speaking academics clearly began to dominate; scholars of Czech nationality were particularly well represented among associate members, most of whom were younger scientists.[33] Masaryk, who believed that the new academy of sciences (and arts)[34] should systematically ensure the publication not just of purely scientific works but also of popular science literature and textbooks (following the model of the Hungarian academy of sciences, for example), had severe doubts about the KČSN's ability to transform into a new institute and justified his

---

[31] At the same time, he emphasized that without the establishment of a Czech academy of sciences, the further development of Czech institutes of higher learning (including the establishment of a second Czech university) would be unimaginable.

[32] There were of course exceptions even to this rule, such as astronomer and theoretical physicist August Seydler (1849–1891), who was a full member of the KČSN and also an editor of the *Atheneum* journal.

[33] In the fourth part of the discussion, Masaryk reflected upon the reaction of KČSN leadership to Ernst Mach's departure. He also addressed the vestigial character and pointlessness of the principle of bilingualism and focused on the fact that an institution which takes an utraquist approach in language terms cannot meet the demands placed on national academy of sciences. He also questioned the need for German-speaking scientists to have their own academy in Bohemia—as many of them were already being drawn to the Imperial Academy in Vienna. Tomáš Garrigue Masaryk, "Jak zvelebovati naši literaturu naukovou? Článek čtvrtý," *Atheneum* 3 (1885/1886): 76–79. This article is also published in Kateřina Piorecká, *O českou literaturu naukovou: Diskuse o úloze a organizaci českých humanitních věd v letech 1885–1900* (Prague: Academia, 2012), 70–74.

[34] In the last article in the series in particular, Masaryk mentioned the benefits of involving artists in the new academy in order to increase the intellectual potential of the entire institution.

position mainly by pointing out the KČSN's separateness from the nation.[35] As the debate continued, his ideas were backed up by Czech historian and later Czech national minister (Landsmannminister) Antonín Rezek, whose reasoning included mention of the hesitancy of a substantial section of the youngest generation of scientists toward the learned society and also of the traditional institute's insufficient funding levels that even regular state subsidies would struggle to resolve. On the other hand, he also pointed out the problems that the KČSN had had to face throughout its rich history. The issue of the money the body needed to operate played an extraordinarily important role in the debate on the foundation of an academy of sciences. With exquisite precision, Rezek calculated that the annual interest payments, including Hlávka's supporting donation, would exceed 12,000 gulden. Although he did not consider this figure particularly enormous, it did occur to him that an academy of sciences starting from scratch would take many years to accrue the necessary assets to support its costs. Nonetheless, despite all his reservations about the KČSN's leadership, he recommended building the academy upon its foundations, all the while expressing serious fears that, given the situation in the KČSN's management, many people would consider his proposal idealistic and difficult to translate into reality.[36]

In any case, a key figure in the endeavor to create a Czech academy of sciences—the aforementioned Josef Hlávka—acknowledged Rezek's ideas. Hlávka's importance lay in the fact that he possessed something that other Czechs contemplating the foundation of leading national scientific institutes simply did not—an enormous fortune, which he was in a position and indeed actively wished to sacrifice for this purpose.

A man who was to become the greatest patron of Czech science in history, Hlávka had amassed his fortune during the great construction boom in Vienna in the 1860s. Then, after overcoming a long and serious illness, Hlávka turned his focus to philanthropy, mainly as a patron of the sciences, concentrating in particular from the 1880s onward intensively on supporting Czech-language science and education. He viewed the establishment of a Czech academy of sciences and arts as one of his main goals. He initially planned for it to grow out of the KČSN, for which cause he made a relatively large donation (20,000 gulden) on the occasion of the KČSN's anniversary in 1884, the money being intended to finance the publication of original scientific literature in the Czech language. In subsequent years, he increased the size of this donation many times over. But, as it turned out, the relatively inflexible KČSN encountered significant problems in undergoing massive reform and was un-

---

[35] Tomáš Garrigue Masaryk, "Jak zvelebovati naši literaturu naukovou?," *Atheneum* 2 (1884/1885): 270–275; Piorecká, *O českou literaturu naukovou*, 45–52.

[36] Antonín Rezek, "Jak zvelebovati naši literaturu naukovou?," *Atheneum* 3 (1885/1886): 41–46; Piorecká, *O českou literaturu naukovou*, 53–60.

able to put his one-off donations to effective use. Nonetheless, it seemed that despite all obstacles an agreement would be reached between Josef Hlávka and KČSN chairman (and friend of Hlávka's) Josef Jireček, even though the Czech academic community was still entirely in the grip of the aforementioned controversy about the authenticity of the manuscripts. The vast majority of KČSN members in general defended the authenticity of these allegedly ancient works of literature. This fact certainly did not bother the conservative Hlávka; controversies arose mainly because of the society's unwillingness to part ways with bilingualism and fully embrace the Czech language. The general assertion can also be made that the traditionalist KČSN was rather unwilling to allow itself to be manipulated by the authoritarian Hlávka, whose influence over the body was not entirely decisive. This incomplete authority is one of the reasons that toward the end of 1887 he wrote to the then governor (Statthalter) of Bohemia, Alfred Kraus, that he intended to donate 200,000 gulden, a massive sum at the time, for the establishment of a "böhmische Akademie für Wissenschaft, Literatur und Kunst."[37] At this early phase it was not entirely clear whether the name of the academy would include the words "in Prague," in tune with the title of the Imperial Academy in Vienna. It also remained open whether the institution's geographical ambit would be limited to the Kingdom of Bohemia or whether Czech-speaking scientists from Moravia would also be involved.[38] Though Hlávka began to lean in favor of establishing a new institution, this emphatically did not imply that he no longer anticipated KČSN's involvement in its foundation. Now, however, he expected the society to join the new institute and provide its first two "classes." The academy would be rounded out with two new classes, one focused on philology and the other on the arts. He tenaciously advocated this model in spite of all of the objections made by, among others, the Bohemian Diet and several influential collaborators in the project. One of his reasons for doing so was a fear that German-speaking researchers might gain the upper hand within the KČSN again (perhaps with the government's help). Moreover, Hlávka rejected the idea that Prague should host two learned societies at the same time (or even three if an exclusively German-language institution were also to be founded). In the end, the Bohemian Diet—and Minister of Culture and Education Paul Gautsch in particular—vigorously opposed Hlávka's efforts and the ČAVU's eventual statute made no provision at all in relation to the KČSN issue. Thus the two learned societies ended up existing side by side

---

[37] The letter from Josef Hlávka to A. Kraus of 15 December 1887 is published in Beran, *Vznik České akademie věd a umění*, 108–110.
[38] The Austrian government rejected this idea outright. In practice, however, leading Czech-speaking scientists from Moravia were named or elected as ČAVU members even in the academy's earliest days. For example, Vincenc Brandl (1834–1901), the prominent Moravian regional archivist, was among the first group of people appointed members of the ČAVU by the emperor himself.

in Prague, with significant overlaps in personnel.[39] While the KČSN was supposed to largely concentrate on representing Czech science abroad and on publishing purely scientific books, the Emperor Franz Joseph Czech Academy for Sciences, Literature, and Arts (Česká akademie císaře Františka Josefa pro vědy, slovesnost a umění, ČAVU) was expected to devote itself more to topical issues of interest to the general public. This division of responsibilities, however, was not maintained. It would make little sense to enumerate in great detail the complicated political ups and downs that surrounded the creation of the ČAVU and the official recognition it was granted from the emperor as they have already been described more than adequately elsewhere and because I think it better to focus on the plan proposed for the academy. At the same time, it is clear that its plan was not the work of Hlávka alone, though he did have a say in every point of it, a say he expressed with exceptional force and obstinacy. Albín Bráf, a lawyer, economist, and Old Czech politician, played an important role in creating the academy of sciences in its final form.[40] It was he who drafted the new organization's statute. Indeed, he was an exceptional figure in Czech society of the time—one who contributed to the foundation of several important Czech institutions in the late nineteenth and early twentieth centuries. The fact that he was the son-in-law of the leader of the Old Czech Party, František Ladislav Rieger, after whose death Bráf would become head of the Old Czechs, only added to his especially privileged position. Of course, Bráf's involvement meant a strengthening of the links between the new academy and one particular political party, a relationship that would later constitute a significant barrier to those who worked at the ČAVU.

---

[39] Setting aside artists, who were not represented in the KČSN at all, practically all other members of the ČAVU who were appointed directly by the emperor (that is to say, its first flush of members) were also KČSN members, although some merely had the status of corresponding members. J. Hlávka became an honorary member of the KČSN in 1888. The fact that both institutions were essentially ruled by the same clique of scientists was noted with great indignation by many, including Eduard Albert, a professor of surgery at the University of Vienna and J. Hlávka's collaborator in establishing the ČAVU. See Helena Kokešová, *Eduard Albert, 1841–1900: Český intelektuál ve Vídni* (Prague: Vyšehrad, 2014), 175–176.

[40] Albín Bráf (27 February 1851 – 1 July 1912) was a Czech lawyer, economist, and politician. In 1874 he graduated in law from the university in Prague. In 1874 he became professor of economics at the Czechoslavic Business Academy in Prague. In 1877 he completed his habilitation in economics at the Czech Technical University. In 1882 he became an extraordinary professor of economics at the Faculty of Law of the university in Prague and was made a full professor of political economics at the same institution in 1890. He served as a member of the Bohemian Diet between 1883 and 1895, and he was a member of the House of Lords from 1905. In 1909 and again in 1911–1912 he served as the Austrian minister of agriculture. His father-in-law was František Ladislav Rieger, after whose death he became leader of the Old Czech Party.

An Academy of Sciences in Prague

Clearly, Hlávka's and Bráf's plans, as compared with Purkyně's, were far less clearly based on a large-scale vision, and the drawn-out, complicated negotiations that resulted in a series of diverse compromises became the channel through which the ČAVU was established. In my view, it is impossible to accurately determine the extent to which the academy's final shape matched up with Hlávka's original idea. His vision was certainly very vague and there were certainly many issues that the builder and patron did not attempt to deal with from the very start. He was undoubtedly unhappy about the existence of two parallel institutions, whose simultaneous existence—given the limited capacities of the Czech scientific community—made absolutely no sense. The situation lasted until 1952, when both institutions were finally absorbed into the newly founded Czechoslovak Academy of Sciences. Hlávka was, however, in all likelihood satisfied with the structure set out for the ČAVU. The first two classes of the academy were conceptually drawn from the KČSN's structure and reflected the original idea of combining the two institutions. As J. Beran correctly points out, historians as well as lawyers played an important role in the first class, having traditionally held a weaker position within the KČSN.[41] Once again, this situation was associated with efforts to make better use of the ČAVU in promoting national enlightenment and in propaganda—it was lawyers and legal historians who were best placed to make the case for the right to statehood. Moreover, in the second half of the nineteenth century, the legal sciences were developing very rapidly in the Czech lands. All the branches of natural science and all technical and mathematical fields were grouped together into the second class, but on this occasion the medical and biological sciences had been allocated more room than they had in the KČSN. Concentrating so many scientific disciplines into one class demonstrated the weakness of the existing Czech academic community and would in the future become a plentiful source of disputes and conflicts. For example, attempts at founding an independent technical class after 1918 ended in failure and the creation of a new independent institution—the Masaryk Academy of Work. The underrepresentation of the natural, technical, and mathematical sciences contributed to the feeling that the ČAVU was a rather anachronistic institution. A new third special philological class was created, whose focus was to be on national issues.[42] It is worth noting that major long-term funding was pro-

---

41 Beran, *Vznik České akademie věd a umění*, 42.
42 The PAU in Cracow, which was a major inspiration for the founding of the ČAVU, was divided into the same three classes when it was founded, just as its Czech counterpart later would be—a historical and philosophical class, one dedicated to mathematical and natural science, and a class for philology. Geologist A. Alth's efforts in 1871 to simplify the academy's structure, creating just a mathematics and natural science class together with a humanities class (the model that the KČSN would essentially follow) were not successful. Czarniecki and Wiltowski, *W stulecie utworzenia Akademii umiejętności*, 32.

vided to support philological research at the ČAVU, and the institute's largest project was eventually to become the *Dictionary of the Czech Language*.

The greatest difference between the KČSN and the ČAVU was, of course, the fact that the latter was to include a fourth class, made up of artists.[43] The new class was headed by the ČAVU president himself, Josef Hlávka, and it is worth pondering whether it was the concern of the academy's patron to find a suitable position for himself within the new academy that played a decisive role in the incorporation of artists into the institution.[44] This combination of science and art under the roof of a single academy was not, however, considered shocking at the time; the idea had been mooted in a number of texts from the second half of the 1880s. For example, in 1886, the very figure of T. G. Masaryk mentioned the notion in a series of articles published in *Atheneum*, in which he also pointed out that artists had no involvement in the other academies of science then in existence in Austria-Hungary.[45] Paradoxically, Josef Durdík's article "O akademii české" (On a Czech academy), published in *Časopis lékařů českých* (Journal of Czech Physicians) in 1888, had a significant influence on the composition of classes at the ČAVU, even though in many regards the text was the source of irritation to Hlávka.[46] Durdík, a philosopher by training, appealed to the structure of the top scientific institutes in France, in a defense of the idea that the new academy should play a role similar to a combination of the Académie française (which, among other activities, published a dictionary of the French language), the Académie des inscriptions et belles-lettres, and the Académie des beaux-arts. In contrast, according to Durdík's vision, the KČSN would have the same tasks as the Académie des sciences and the Académie des sciences morales et politiques. This meant that the KČSN would consist of a first class (a mathematics and natural science class, according to Durdík's conception) and a second class (philosophy and history), whereas the academy would also have a third "liter-

---

[43] Not all artistic disciplines were represented at the academy. Its membership included no actors and only a few musicians (except musicians who happened to be composers as well).

[44] On the creation of the ČAVU's fourth class, see Jiří Beran, "K činnosti IV. třídy České akademie věd a umění v letech 1890–1918," *Práce z dějin Československé akademie věd* 1 (1986): 13–84.

[45] "Would it not be befitting to accept artists too as members so that our intellectual life in its entirety would find in the academy its organ? Although many academies are not arranged in such a way, for our peculiar situation we need a peculiar institution." Masaryk, "Jak zvelebovati naši literaturu naukovou? Článek čtvrtý," 79. Although the Polish Academy of Arts and Sciences and the Croatian Academy of Sciences and Arts both currently have sections focused on the arts, those sections were founded in the twentieth century.

[46] Josef Durdík, "O akademii české," *Časopis lékařů českých* 27 (1888): 433–435. The article and J. Hlávka's handwritten notes to it are published in Beran, *Vznik České akademie věd a umění*, 171–179.

ary" class ("for the Czech language and literature") and a fourth class dealing with the arts. Ultimately, with a few minor adjustments, the French model was adopted in the end, although not in a form that would divide the power to influence the arts and sciences between the KČSN and the ČAVU.

Unlike Purkyně's academy, in principle the ČAVU did no more than provide support to scientific research at other organizations and possessed practically no institutes of its own. What it did at a maximum was to create committees focused on a variety of fields and subfields, none of which ever conducted any concrete research themselves. The first true research institute to be created within the ČAVU was to emerge only in the early twentieth century—the Office of the Dictionary of the Czech Language. The ČAVU's Institute of Economics, however, remained in operation for a long time, though its relationship with the academy is beyond the scope of this essay to discuss. This arrangement seriously limited the ČAVU's ability to coordinate the conduct of scientific research as a whole across the Czech lands. It is clear that the task of creating a comprehensive network of research institutes was beyond the capabilities of even such a wealthy patron as Josef Hlávka. Besides, it is unlikely that the Czech-speaking scientific community would have been able to maintain such a system. The ČAVU played an important role as a publisher of scientific works as well as producing some books on matters artistic.

Turning to the make-up of the academy's initial membership, it can only be asserted that the hopes of those younger scientists associated with *Atheneum* magazine who had opposed recognizing the authenticity of the manuscripts, were to be deeply disappointed. Despite the fact that Hlávka did not see eye to eye on many issues with the conservative clique that ruled over the KČSN, that group was nevertheless to succeed in dominating the ČAVU.[47] Thus, for example, Masaryk was to become a member of the ČAVU only in 1918 as the president of the Czechoslovak Republic (and even then only in an honorary capacity). As I have already mentioned, the ČAVU became a bastion of the Old Czech Party, from which the most prominent members of the Young Czech faction were to be effectively cut off. Thus, for example, the famous writer, poet, and journalist Jan Neruda was not among its first members. Indeed, such factors beyond the realm of science and art would continue to affect the selection of members for decades, a fact that did not do much to further the institution's general reputation, naturally enough.

Hlávka envisioned the ČAVU as a state institution, the K. u. k. Tschechische Akademie des Kaisers Franz Joseph für Wissenschaft, Literatur und

---

[47] See, e.g., the letters from T. G. Masaryk to Eduard Albert of 11 May 1890 and 26 June 1890, in which he sharply criticized the scientific credentials of some of the figures nominated for the body. See Helena Kokešová and Vlasta Quagliatová, eds., *Korespondence TGM: Korespondence T. G. Masaryk – staročeši* (Prague: Masarykův ústav a Archiv AV ČR, 2009), 172–174, 193–195.

Künste (and thus in Purkyně's view as an inferior category of academy). But the great patron's idea proved impossible to implement, due in part to Gautsch's opposition; the only means by which the academy could be founded was through a decision of the Bohemian Diet, which in the late 1880s was still heavily dominated by the Old Czech Party. At the time, Hlávka was focusing his efforts on the fortieth anniversary of Franz Joseph I's coronation, and in 1888 he managed to gain the emperor's approval for holding negotiations on the establishment of a Czech academy. Negotiations leading to the *punktace* (i.e., a national compromise in Bohemia) between the Viennese government and the Old Czechs in January 1890 were to be a turning point in the effort.[48] The establishment of the ČAVU under the protection of a member of the ruling dynasty (its first official patron was the emperor's younger brother, Archduke Karl Ludwig) was considered a major concession. The conditions of the compromise, however, were to be a turning point in Czech politics, as they led directly to the decisive defeat of the Old Czechs in the 1891 elections. Voters at the time did not consider the establishment of an academy of sciences and arts to be a particularly important concession to the Czech lands or to be of much advantage to them.

If one compares the ČAVU as it came into being against the academy originally proposed by Purkyně, it becomes clear that it was a far less independent and self-confident institution, to some extent due to Hlávka's dominance in its early days (he was to remain president of the body until his death in 1908). In this regard, Hlávka's constant emphasis on loyalty to the Habsburgs—something that would have been inconceivable for Purkyně—played an important role. For Purkyně, however, it was not necessary to think too deeply about the overall social and political context in his plans. He was quite free to formulate a radical, far-reaching vision. In contrast, what Hlávka achieved was the creation of a functioning, real-world institution, which, despite the major weaknesses, problems, and compromises it suffered, was destined to play an exceptionally important role in science in Bohemia (and later in Czechoslovakia) for well over sixty years. Unfortunately, due to the tumultuous social and political changes of the twentieth century, the ČAVU was to be in a position to fulfill its major duties without serious complications only until the outbreak of World War I.

Translated from Czech by Nicholas Paul Orsillo.

---

[48] On the *punktace* and its various impacts, see Elizabeth Wiskemann, *Czechs and Germans: A Study of the Struggle in the Historic Provinces of Bohemia and Moravia* (Oxford: Oxford University Press, 1938); Bruce M. Garver, *The Young Czech Party 1874–1901 and the Emergence of a Multi-Party System* (New Haven, CT: Yale University Press, 1978); and Pieter M. Judson, *Guardians of the Nation: Activists on the Language Frontiers of Imperial Austria* (Cambridge, MA: Harvard University Press, 2006).

Maciej Janowski

GENERAL SCIENTIFIC ASSOCIATIONS IN PARTITIONED POLAND
Socializing the Intelligentsia, Preserving the Nation,
and Cultivating the Sciences

Warsaw

The Enlightenment period in Poland, important and groundbreaking in as many dimensions as it was, did not manage to create a national academy of sciences or a scientific society. The German *Bürgertum* of Gdańsk (Danzig), under Polish rule until 1793, had, like so many German local communities in the Age of the Enlightenment, a learned association of their own; at a certain moment this association, in order to augment its importance, aimed at obtaining the title of a royal scientific society from the king of Poland; however nothing came of it. Warsaw in the Age of Enlightenment was a rapidly growing city which reached about one hundred thousand inhabitants in the early 1790s. Admittedly, this number included the nonpermanent inhabitants who dwelled in Warsaw during the Reformatory Diet of 1788–1792; still, the growth of the city is imposing and made Warsaw an important cultural and political centre. The events of 1788–1794, with the diet session and later the Kościuszko uprising, demonstrated that the city had accumulated the critical mass that makes a specific type of urban politics possible. The king, Stanislas Augustus, managed to gather around himself a group of intellectuals, mainly from the dissolved Jesuit order, whom he treated at the so-called "Thursday dinners;" these meetings, however, were never institutionalized, and it is only in 1800, five years after the final partition, under Prussian rule, that the Warsaw Society of the Friends of Learning was formed.

The Warsaw Society was one of the most important institutions of late Enlightenment Warsaw. Its ideology oscillated between Enlightenment classicism and sentimentalism. It existed until 1831, with the adjective "royal" added after the Napoleonic wars (which the society was very proud of). It met in one of the picturesque canon houses close to the cathedral where, according to the impressions of contemporaries, it breathed the air of peace and seclusion in spite of its relatively central place on the map of the city. In the year 1823, however, the society obtained a new and magnificent building, financed by its long-time president, one of the leading figures of Polish Enlightenment culture, Stanisław Staszic. This building stood right in the centre of cultural Warsaw, close to the newly-established Warsaw University at the Krakowskie Przedmieście (literally "Cracow Suburb"—the name refers to the fact that it was an old road leading from the medieval city gates to the south, in the direction of Cracow). This location had its significance. By unveiling the monument of Copernicus in front of the building, the Royal Society created a strong "scientific" visual landmark in Warsaw and contributed to providing the Krakowskie Przedmieście, with its cafés and institutions of learning, a sort of hub for the Warsaw intelligentsia—with an atmosphere it has retained until now.

While the scientific ambitions of the Warsaw Society were great, including among others, the writing of a collective multivolume *History of Poland* (a task that was finally accomplished, under very different political and intellectual conditions, by the Institute of History at the Polish Academy of Sciences in the Communist period), its central role seems to have been in two other fields. In terms of cultural influence, it was a centre for the late Enlightenment/sentimentalist culture that already nurtured more and more harbingers of Romanticism (although it was eventually Wilno/Vilnius with Adam Mickiewicz, not Warsaw, that became the cradle of Polish Romanticism). After the third partition and the collapse of the state, yet before the Napoleonic victories aroused new hopes (i.e., in the years 1795–1807), the public mood was of necessity sentimental, pondering on the past glory and vanity of hopes. At the same time, this mood already anticipated the coming of Romanticism. A programme of collecting folk songs was presented at the Warsaw Society meeting, and the preservation of the Polish language as the most important remnant of the defunct state was seen as one of the central tasks—"during the destruction of the nation, to preserve from destruction the national language," as Staszic put it.[1]

The second function rather belongs to the sphere of social instead of cultural history: by its very existence the society provided a forum for the nascent Polish intelligentsia—a social group whose origins can be traced back to the mid-eighteenth century but whose first important stage of growth (both in numbers and in importance) occurred only under Napoleonic rule with its development of bureaucratic state institutions that followed French and Prussian models. The society created its conventions, hierarchies, and savoir-vivre: thus it "ennobled" the intelligentsia, bridged at least partially the gaps between it and the aristocracy, and imbued it with a feeling of its own value. It created a social venue with its public meetings, which were strictly formalized in a semi-theatrical form. Although some participants were very critical of their dullness and the very detailed character of the lectures,[2] it seems that

---

[1] Quoted in Aleksander Kraushar, *Towarzystwo Królewskie Przyjaciół Nauk, 1800–1832: Monografia historyczna*, vol. 2, *Czasy Księstwa Warszawskiego, 1807–1815*, part 1 (Kraków: Gebethner, 1901), 190.

[2] See the critical voices of Wojciech Gutkowski and Fruderyk Skarbek, quoted by Aleksander Kraushar, *Towarzystwo Królewskie Przyjaciół Nauk, 1800–1832: Monografia histo-*

their social function did not suffer from it, as they were creating a set of cultural practices[3] rather than providing a high level popularization of science. The Warsaw Royal Society of the Friends of Sciences and the Warsaw University were abolished by the Russian authorities after the defeat of the 1830-1831 uprising, as an element of a wave of repressions.

Stefan Kieniewicz discerns three distinctive periods of institutional development of science in nineteenth-century Warsaw.[4] The first one is the period discussed above, with Warsaw University and the Society of the Friends of Learning. The second one is marked by the reopening of the Polish university (named the Main School, Szkoła Główna) in the post-Crimean War "thaw" in the Russian Empire in the first years of the reign of Alexander II. In this period, thus, it was the new university, not a scientific society, that organized the Warsaw academic milieu. Up to then the Warsaw intellectuals used for some time to gather under the aegis of the periodical *Biblioteka Warszawska* (Warsaw Library), whose Wednesday editorial meetings, from the 1840s, acted as a sort of proxy for scientific debates. In 1859, with the "thaw" of the first years of Alexander II's rule, they tried to reorganize themselves into a full-fledged scientific association[5] with four departments (historical, literary-philosophical, social sciences, and natural sciences), five members each, who should co-opt three additional members. The authorities procrastinated until the outburst of the uprising in January 1863, which made any dreams of associations irrelevant. The university was Russified and existed as a Russian educational institution until 1915.

Throughout the following decades of the nineteenth century, Warsaw provided an interesting example of a big city (more than 100,000 in 1791, around 800,000 before 1914) where the spontaneous meetings and salons of the intelligentsia played the functional role of the non-existent associations in what may be called a "privatized public sphere." The only scientific institution permitted by the authorities after the defeat of the 1863-1865 uprising was the Mianowski Fund (named in honour of a professor at the Main School before 1863), which granted scholarships and financed the publication of scientific works. By the force of circumstances, it adopted some of the roles of a scientific association. At times, the editorial boards of some journals played a

---

ryczna, vol. 2, *Czasy Księstwa Warszawskiego, 1807-1815*, part 2 (Kraków: Gebethner, 1904), 232-233 and 336-337.

[3] Hanna Jurkowska, *Pamięć sentymentalna: Praktyki pamięci w kręgu Towarzystwa Warszawskiego Przyjaciół Nauk i w Puławach Izabeli Czartoryskiej* (Warsaw: Wydawnictwa Uniwersytetu Warszawskiego, 2014).

[4] "Trzy etapy rozwoju nauki w Warszawie w XIX w.," in Stefan Kieniewicz, *Historyk a świadomość narodowa* (Warsaw: Czytelnik, 1982), 143-161.

[5] Stefan Kieniewicz, "Od Towarzystwa Przyjaciół Nauk do Towarzystwa Naukowego Warszawskiego," *Rocznik TNW* XLIX (1986): 17-34, later reprinted in: *Towarzystwo Naukowe Warszawskie* (Warsaw: Towarzystwo Naukowe Warszawskie, 1997), 7-33.

similar role. Magdalena Micińska, in her various studies,[6] has investigated the social life of Warsaw between the 1870s and 1890s as a proxy for the non-existing institutionalized network. She lists the regular contacts of historians, doctors, social scientists, and others at meetings that formally were regular salon meetings but in fact were consciously organized in such a way as to act like a scientific association.

Cracow

In the same period, the situation in Cracow was very different. The university, which was going through various ups and downs throughout the nineteenth century, was the central intellectual institution here. The scientific association connected with the university, meanwhile, was not an important independent factor in the cultural life of the city. It was in the second half of the nineteenth century that things changed. In 1857 the Cracow Scientific Society loosened its links with the university and reorganized itself as an independent body. As the specialists argue, this transformation was of more importance than the creation of the Academy of Learning (Akademia Umiejętności) on the basis of the Scientific Society in 1872.[7] The constitutional period of the Habsburg Monarchy since the 1860s allowed various previously impossible public activities to take place. The Academy of Learning, established in Cracow in 1872, could thus assume the role of a "normal" academy of sciences, acting as an organizer of Polish scientific life in the Habsburg Monarchy. At the same time it had another function, being seen by the Polish educated classes as a proxy representing the whole Polish academic and cultural life from all parts of the partitioned country. In the context of the celebrations for its twenty-fifth anniversary, the activities of the academy were summarized as follows:

> By its structure and its position the Cracow Academy has taken an outstanding place among the scientific centres of Europe, as a rampart of our contemporary existence, and on the borderland to our closest neighbours has proudly unveiled our national standard. Science sometimes is international, that is true, but the ways of its expression, the direction of research and some branches are not cosmopolitan. Therefore, the Cracow Academy too, through its activities and publications, speaks not only to the Polish lands but through its

---

[6] Magdalena Micińska, *At the Crossroads, 1865–1918: A History of the Polish Intelligentsia*, part 3, ed. Jerzy Jedlicki, transl. Tristan Korecki (Frankfurt am Main: Peter Lang, 2014), esp. subchapter 4.3 "The Means of Social Influence," 106–118.

[7] Cf. Piotr Biliński, "Zerwanie więzów łączących Towarzystwo Naukowe Krakowskie z Uniwersytetem Jagiellońskim w 1856 roku," in *Towarzystwo Naukowe Krakowskie w 200-lecie założenia, 1815–2015: Materiały konferencji naukowej 9–10 grudnia 2015*, ed. Jerzy Wyrozumski (Kraków: Akademia Umiejętności, 2016), 27–38; Piotr Hübner, "Od Towarzystwa Naukowego Krakowskiego do Akademii Umiejętności," in *Towarzystwo Naukowe Krakowskie*, ed. Wyrozumski, 39–46.

relations with abroad it proclaims to the fringes of the civilized world that we persist, we act, and so we belong to a species that cannot be denationalized or swallowed.[8]

This phraseology is typical: semantically empty (the clumsiness of the English translation is not an accident but revealing of the lacunae of meaning in the original's pompous style), appealing mostly through established images of a national rhetoric taken from the military sphere (*rampart, standard*). The cited speech invoked the Romantic concept of a national science and, at the same time (and perhaps without awareness of a potential conflict of the two ideas), it saw precisely this national character as a means to acquire international prestige. The text gives a typical impression of popular opinions about the role of the learned societies among the intelligentsia. It is not by chance that its author, Mieczysław Offmański (1866–1945), was a popularizer rather than a researcher. In the context of more professional debates, a language that stressed the importance of science for its own sake was more common.

The academy did its best to unify Polish scientific life across the borders of the partitioning powers. After the 1905 revolution, with the ensuing liberalization of the tsarist regime, it was aided in this task by the Warsaw Scientific Society, which had initially fifteen members, eight of whom were members of the Cracow Academy—an important indicator of a conscious attempt to keep the unity of the Polish culture beyond the partitions. It is equally interesting that seven out of the fifteen developed in their professional life closer or looser bonds with the Russian Imperial Warsaw University. This shows that links between Polish and Russian intellectual life were not totally absent.[9] The details of this connection still await their researchers. The new society obviously attempted to continue the tradition of the old Royal Warsaw Society. One of its members, the outstanding pedagogue and historian of pedagogical thought Bogdan Nawroczyński, noticed the analogy between the statute of the first department of the new society (dealing with the humanities) and the programme set by Stanisław Staszic, stressing the study of the natural and cultural conditions of the Polish lands as the primary goal of Polish science. Indeed, there were clear similarities to Staszic's ideas in his Statistics of Poland and other writings from a century earlier. The call for statistical (in the broadest sense of the term) research which would result in producing an overall picture of the Polish territories was voiced not only by Staszic; it was being repeated in every

---

[8] Mieczysław Offmański, "Towarzystwo Naukowe Krakowskie obecnie Akademia Umiejętności" (n.p., 1897), reprinted in *Od Towarzystwa Naukowego Krakowskiego do Polskiej Akademii Umiejętności: Refleksje jubileuszowe Mieczysława Offmańskiego, Tadeusza Sinki, Stanisława Wróblewskiego, Stanisława Kutrzeby*, ed. Piotr Hübner (Kraków: PAU, 2002), 27–41, here 32.

[9] Leszek Zasztowt, "Towarzystwo Naukowe Warszawskie w I połowie XX wieku," in *Towarzystwo Naukowe Warszawskie—sto lat działalności*, ed. Ewa Wolniewicz-Pawłowska and Włodzimierz Zych (Warsaw: Towarzystwo Naukowe Warszawskie, 2009), 29–38.

generation throughout the nineteenth century, and it had a central place in the journalistic writings of the Warsaw positivists of the 1870s and 1880s. The "positive facts" were to serve as foundations for future social and economic policy. Thus, the new society's first department (humanities) envisaged in its programme:

> Polish science, in order to become a useful factor of the universal-human [scientific, MJ] work [...], must, first of all, work upon the material provided by the biological nature of our homeland and on the questions resulting from the needs and conditions of national life.[10]

With this declaration, its authors consciously positioned themselves as followers of an important intellectual tradition.

There were, however, important novelties. Even the name of the society itself implied that science was being professionalized. Instead of becoming, as was the case before 1831, a society of the "Friends of Sciences," the new institution adopted the name Warsaw Scientific Society (Towarzystwo Naukowe Warszawskie)—clearly to stress the professional character and distance itself from the enlightened *dilettanti*. At the meeting of the department of mathematical and natural sciences, the prevailing opinion was that "we need a higher scientific institution whose task is to pursue pure science," to avoid the danger of limiting themselves to chewing over the ideas of others which, in turn, would cause all scientific activities to be compromised, as "thought, without the incentive of scientific activity, slowly [...] vanishes."[11]

Poznań, Wilno/Vilnius, and Other Local Learned Societies

Apart from the institutions named above, which had ambitions to cover the whole of the Polish territory, there were various provincial scientific associations, the most important of which being the society in Poznań (Posen), a central Polish scientific institution for the Prussian part, with, however, not much influence beyond it. It was established in 1857 and had to fight the enmity of the Prussian authorities who, right at the start, forbade teachers at public schools to become members of the society; some years later, the ban was extended to state officials. Thus, a big part if not the majority of the Polish intelligentsia of Poznań was excluded from the association. It could now only count on the clergy, landowners, and a handful of members of the free professions and private officials. The society was hosted by other Polish institutions and got its own building only in 1881. The main achievement

---

[10] Quoted in Bogdan Nawroczyński, *Towarzystwo Naukowe Warszawskie: Materiały do jego dziejów w latach 1907–1950* (Warsaw: Towarzystwo Naukowe Warszawskie, 1950), 18.

[11] Speech of Teodor Dunin at the general meeting of the Department of Mathematical and Natural Sciences of the Warsaw Scientific Society on 21 November 1908, quoted by Nawroczyński, *Towarzystwo Naukowe Warszawskie*, 19.

was its museum, which exceeded local significance; the 1914 guidebook to "the lands of ancient Poland, Lithuania and Ruthenia" (in effect, Poland in its 1772 frontiers) by Mieczysław Orłowicz stresses the importance of the museum for the whole of historical Poland, not just the Poznań province.[12] The German-language guidebook by the leading German historian of the city, Adolf Warschauer, basically agrees, describing the collections of the museum as *bedeutend* (significant) and *beachtenswert* (remarkable).[13]

In a different situation, a scientific association was created in Wilno/Vilnius. The capital of the historic Grand Duchy of Lithuania was located within the Russian Empire itself (not in the Polish Kingdom). Its population was in large part Polish, its elites mostly so. The city looked back to a strong intellectual tradition in the first half of the nineteenth century, being the seat of an important Polish university. After the closing down of the university (1832), a medical academy still remained until it was closed down, too, in the early 1840s. After a short intellectual revival in the period of the "thaw" of the early years of Alexander II's rule, the intellectual tradition was suppressed (if not completely broken) by the wave of repressions against Polish culture. These repressions were much stronger in the Russian Empire than in the Kingdom. The Scientific Society in Wilno was established in the wake of the 1905 revolution and the ensuing liberalization of the regime (registered in January 1907). The statute read: "Nourishment of sciences, learning and letters in the Polish language, and especially studying the country from natural, ethnographic, historical, economic and statistical points of view."[14] The analogy with the above-quoted programme of the first department of the new Warsaw Society is clear. The important difference is that in Warsaw, as we have seen, the exact and natural sciences played a much bigger role that in Wilno, whose intellectual position had been provincialized to a large degree since the 1820s.

On an even more local level, there are four other associations to be listed. In the Russian part, the constitutional period of the Kingdom of Poland (1815–1831) witnessed the establishment of two local associations in Lublin (1818) and Płock (1820). Both had their statutes modelled after the Warsaw Society of the Friends of Learning, and both were established on the initiative of professors of the local grammar schools; both were interested primarily in local history. An interesting aspect is the position of the bishop of

---

[12] Mieczysław Orłowicz, *Przewodnik po ziemiach dawnej Polski, Litwy i Rusi* (Warsaw: Podróżnik Polski, 1914, reprint Warsaw: Stanisław Krysiński, 1990), 133.
[13] Adolf Warschauer, *Führer durch Posen und Umgebung*, 9th ed. (Posen: J. Jolowicz, 1917), 33–35, accessed 10 September 2016, https://polona.pl/item/3568635/23/.
[14] Quoted in Stanisław Kunikowski, *Towarzystwa Naukowe ogólne w Polsce w XIX i XX wieku* (Włocławek: Lega. Oficyna Wydawnicza Włocławskiego Towarzystwa Naukowego, 1999), 36.

Płock, Adam Prażmowski (also an amateur archeologist and member of the Warsaw Society), as a protector of the society.

Both the Płock and Lublin societies shared the fate of their Warsaw counterpart and could not continue their activities after the defeat of the 1831 uprising. Their reestablishment, as in the case of Warsaw, came after the 1905 revolution; the Płock society was reactivated under the same name in 1907 and exists until now (which makes it, if we take the claim to continuity seriously, the oldest existing scientific association in Poland).

Two more Polish scientific associations have to be mentioned. In Toruń (Thorn) in Pomerania in 1875 an association was formed as a reaction to the German celebrations in Toruń of the 400th anniversary of Copernicus' birth in 1873. It had three departments (historic-archeological, theological, and medical-biological), but its activities were confined mainly to the study of local history.[15]

Przemyśl in Galicia was the last city before World War I to establish a Polish learned society—the Society of the Friends of the Sciences (1909). Like other societies, it undertook collecting museum items, and Mieczysław Orłowicz deemed its "small museum" worth mentioning in his guidebook from 1914 (as he did with the Toruń society, but not with that in Płock).[16] The Przemyśl museum, described in detail in another guidebook by him from a slightly later date, could boast of no less than five rooms overflowing with historical and archaeological artefacts, seventeenth-century Flemish paintings, folk costumes, and other ethnographical items, patriotic relics (a sword from the 1863 uprising) as well as a collection of minerals.[17] The museum's library, kept in the same room as the collection of minerals, counted five thousand volumes in 1914.[18] The activities of the Przemyśl society developed mainly in the later period, which exceeds the scope of our story.

The cases of Przemyśl and Toruń may make us ponder on the common characteristics of places that saw the organization of scientific societies. One would be tempted to say that the scientific associations were established either in those cities that were main centres of Polish intellectual life (Warsaw, Cracow) or in those that were on ethnic frontiers. Both Toruń and Przemyśl were located almost exactly on the Polish-German and Polish-Ukrainian ethnic frontier, respectively. Poznań and Wilno/Vilnius would count for both —they were important cultural centres and at the same time ethnic frontier

---

[15] Kunikowski, *Towarzystwa Naukowe*, 45–48.
[16] Orłowicz, *Przewodnik*, 86 (Przemyśl), 159 (Toruń).
[17] Mieczysław Orłowicz, *Ilustrowany przewodnik po Przemyślu i okolicy* (Przemyśl: nakładem Zjednoczenia Towarzystw Polskich w Przemyślu, 1917), 59–62.
[18] This we can learn from yet another one of Orłowicz's guidebooks: Mieczysłąw Orłowicz, *Ilustrowany przewodnik po Galicyi* (Lwów: Książnica Polska, 1919, reprint Krosno: Ruthenus, 2003), 260. The original edition was ready for print in July 1914, but the publication was delayed by the war; it depicts, therefore, the state of Galicia on the eve of the Great War.

places. (Wilno was actually a Polish ethnic island beyond the proper ethnic Polish territory.)

Scientific Associations in Partitioned Poland: Social, Political, and Cultural Functions

To see how learned societies socialized the intelligentsia, we have to look at the transformation of this stratum throughout the analysed period—a transformation that followed the general pattern of the transformation of the educated strata in Europe but at the same time displayed a certain specificity. Researchers studying the transformations of the German *Bildungsbürgertum* (educated middle class) have stressed the somewhat paradoxical phenomenon that the *Bildungsbürgertum* was the strongest as a political force and a social ideal before it gained in strength as a social group; Jerzy Jedlicki has established a similar proposal for the history of the Polish intelligentsia, stating that—counter-intuitively—the 1850s were the decade of its greatest blossoming. This is explained by the fact that later the intelligentsia became much more numerous and various subgroups developed different aims, interests, and priorities in the process.[19] Therefore, one should expect to find the socializing function of the scientific associations to be strongest either in the earlier part of the nineteenth century or in smaller cities; that is, in the cases when the educated stratum was relatively less numerous. It seems that such a pattern can indeed be discerned: if there is a line of development, it leads in the direction of professionalization (although the feeling of responsibility for the national culture as a whole was never abandoned by the scientific societies).

A further noteworthy factor in the activities of the presented associations are the private donors, their social position, and their role in sustaining—and influencing—the activities of the associations. We see the aristocrats, at times also the new bourgeoisie, constantly donating funds and buildings, for example, Józef Potocki, who financed the house for the new Warsaw Society in 1911. On the other hand, we see an identifiable tendency of self-organization among the intelligentsia which tried—with a certain degree of class pride—to be independent of the old elites. The relations between the intellectuals and the donors (who often aspired to the rank of protectors) went through a complicated trajectory throughout the analysed period. Generally speaking, the role of the aristocracy as a benefactor was smaller in 1914 than one hundred years earlier, although still significant.

The limits of the self-help of the intelligentsia were set clearly enough by the limits of the appeal of the scientific institutions to the broader educated

---

[19] Cf. Jerzy Jedlicki, *The Vicious Circle, 1832–1864: A History of the Polish Intelligentsia*, part 2, ed. Jerzy Jedlicki, transl. Tristan Korecki (Frankfurt am Main: Peter Lang, 2014), passim.

circles.[20] The whole nineteenth century abounds in reflections on the passivity of the intelligentsia, their lack of interest in national science and culture. Critical voices deploring the minor social role of the Warsaw Society of the Friends of Learning have already been mentioned. In Poznań, Wilhelm Bogusławski complained in 1894 that there were some tens of thousands of members of the intelligentsia in the Grand Duchy of Poznań (the Poles did not acknowledge the new German name of *Provinz Posen* and preferred to refer to the Grand Duchy of Poznań, as the region had been called after the Congress of Vienna), while the meetings of the Poznań Society of the Friends of Learning hardly ever attracted more than fifty people. The museum, in spite of its importance, had no more than only one thousand visitors yearly.[21] Such complaints cannot be discarded, but they should not be taken at face value. The socializing function was performed only for a small fraction of the culturally most active intelligentsia members, to be sure, but this testifies to the unrealistic expectations of the nineteenth-century enthusiasts of associations rather than to the weakness of these associations. The blossoming of various associations all over nineteenth-century Europe led to a sort of a "civil society utopia," especially in countries where (as was the case in Poland) the elites felt suppressed by foreign governments. This utopia envisaged a network of—to use an anachronistic term—NGOs which would render the state, apart from the army, foreign politics, administration, and justice, practically irrelevant. Fulfilling this dream was plainly impossible, but the importance of the civil society institutions should nonetheless not be underrated.

All these institutions fulfilled various tasks, depending on the historical situation. They were hubs for the social life of the intelligentsia, they were national institutions and last, but not least they organized, as far as their funds permitted, scientific research. They belong to the sphere of "non-govern-

---

[20] The problem of mutual relations between the nobility/aristocracy and the intelligentsia as the elites of Polish society belongs to the most interesting problems of Polish cultural and social history of the nineteenth and early twentieth centuries. Eminent sociologist Józef Chałasiński was the most important representative of what may be called a "continuity thesis" between nobility and intelligentsia. See his *Vergangenheit und Zukunft der polnischen Intelligenz* (Marburg: Herder Institut, 1965). By contrast, Warsaw social historian Ryszarda Czepulis-Rastenis presented a different view, in which the intelligentsia developed as a new social force, opposing the old noble elites. Cf. her books *"Klassa umysłowa:" Inteligencja Królestwa Polskiego, 1832–1862* (Warsaw: Książka i Wiedza, 1973) and *Ludzie nauki i talentu: Studia o świadomości społecznej inteligencji polskiej w zaborze rosyjskim* (Warsaw: PIW, 1988). The controversy has persisted until now. E.g., in his *Prześniona rewolucja* (Warsaw: Krytyka Polityczna, 2014), the philosopher Andrzej Leder more or less endorses Chałasiński's views while Jerzy Jedlicki and Magdalena Micińska, in their books quoted above, are closer to Ryszarda Czepulis-Rastenis.

[21] Wilhelm Bogusławski, *Wizerunek czynności i zasług Towarzystwa Przyjaciół Nauk Poznańskiego*, offprint from *Charitas: Księga zbiorowa wydana na rzecz Rzymsko-Katolickiego Tow. Dobroczynności przy kościele św. Katarzyny w Petersburgu* (St. Petersburg, 1894), 401–438, here 435.

mental" activities which are traditionally known in Polish historiography under the general label of "organic work." The concept is a crucial one for Polish nineteenth-century history. Without entering into details, it is only worth reminding the reader that "organic" was not necessarily meant in the conservative sense (as in some versions of organicism, where the "organic" structure of society is invoked in order to legitimize the rule of a "head" over its "members"). There was a liberal sense of "organic work," too, which understood the organic metaphor as an expression of the duty of the intelligentsia to take care of the social advancement of the "people" as a whole—if only some particular strata benefitted from progress, this line of thinking went, society would not develop in an organic way.[22]

To sum it up, any attempt at a deeper understanding of the social function of the associations demands that they be put into the broad context of organic work—of all the non-governmental activities of nineteenth-century society. It seems that the formal qualification of a "scientific" society was not crucial to the social function of any given institution. Of course, being an association gives a certain air of formality in interpersonal relations, but even here the frontier line does not have to be very clear (as we could see above by the examples from Russian Poland). The borderlines between general scientific associations, societies of practitioners of a given profession (which I have decided to exclude from the picture), and other societies of the cultural or artistic type are even more impossible to delineate. The co-operatives and other economic associations are, it can be argued, a different story (and strictly political organizations, whether legal or illegal, are outside of the scope of the organic work-type activities altogether). Even there, at a closer look, the divisions would be far from clear.

Such a broad background panorama of non-governmental institutions will not be sketched out here; it should be mentioned only to make clear that the present topic is but a small part of a broad picture, therefore any generalized conclusions may be erroneous. For example, we do not see L'viv (Lemberg, Lwów) on our map of Polish scientific associations, but it would be absurd to infer that L'viv was not important as a cultural/scientific centre. To the contrary, in the late nineteenth and early twentieth centuries it was one of the most important centres of various Polish associations and scientific as well as cultural institutions (and, by the way, the home of the central Ukrainian scientific association, the Shevchenko Society). L'viv hosted also various disciplinary scientific associations, such as the Historical Society (Towarzystwo Historyczne), established in 1886 (the Polish Historical Society, since 1925). One can plausibly argue that it was precisely because L'viv was a haven of a

---

[22] On the concept of organic work, cf. "Problem pracy organicznej, 1840–1890 [1958]," in Kieniewicz, *Historyk a świadomość narodowa*, 34–58; *Droga do niepodległości czy program defensywny? Praca organiczna: Programy i motywy*, ed. Tomasz Kizwalter and Jerzy Skowronek (Warsaw: Pax, 1988).

blossoming sociocultural life that a general Polish scientific association was not considered necessary; whereas in the Ukrainian case, a relatively weaker level of institutional development in the sphere of culture and science called for a general association that would cover all branches of intellectual life and could serve as a focal point for the not very numerous Ukrainian intelligentsia of the city. By the same token, if there was no branch of the above-mentioned Historical Society in Cracow until just before World War I (1913), this was not due to any underdevelopment of historical science in the city. To the contrary, historians were so dominant both at the Jagiellonian University and the Academy of Learning that a special association of historians was not deemed necessary. If there were no Polish scientific associations in Warsaw or in Wilno between 1831 and the early twentieth century, this was not due to a lack of intellectual potential but to the political situation. Thus, a scrutiny of a few associations of "friends of sciences" cannot provide us with any general picture of Polish society in the nineteenth century. What it can do, however, is to bring before our eyes a sample of various types of activities and functions which a stratum of the intelligentsia could perform in a largely unfavourable political climate.

Jan Surman

HABSBURG HIGHER EDUCATION IN THE EYES OF FOREIGNERS: LUDWIK TĘGOBORSKI AND WILLIAM WILDE*

Nineteenth-century Habsburg universities have been a subject of analysis from many angles, ranging from a comparison of their structural and legal conditions through works concentrating on the professoriate and the students to issues of daily academic life. Historians scrutinized, in particular, issues connected with universities as factors supporting cultural dominance and gender hegemony, stabilizing empires, and their peculiar measures to sustain the volatile stability.[1] These analyses have, however, only in the rarest cases dealt with international comparisons and transimperial entanglements. This is quite surprising, because for mere reasons of geographical and political overlapping such entanglements, "croisements," or "carrefours" structured the academic space of Europe and beyond. One can think, for instance, of the German Confederation, which was a state encompassing manifold states but whose analysis has hitherto been divided into a "German" narrative and an "Austrian" one, rarely taking the political, social, and confessional variety of the "German" part into account.[2]

---

* The book chapter was prepared within the framework of the HSE University Basic Research Program and funded by the Russian Academic Excellence Project "5-100."
1 For a recent overview, see Friedrich Stadler, ed., *650 Jahre Universität Wien: Aufbruch ins neue Jahrhundert*, 4 vols. (Vienna: V&R unipress, 2015). See also Gary B. Cohen, *Education and Middle-Class Society in Imperial Austria, 1848–1918* (West Lafayette, IN: Purdue University Press, 1996); Jan Havránek, "Nineteenth Century Universities in Central Europe: Their Dominant Position in the Science and Humanities," in *Bildungswesen und Sozialstruktur in Mitteleuropa im 19. und 20. Jahrhundert/Education and Social Structure in Central Europe in the 19th and 20th Centuries*, ed. Victor Karady and Wolfgang Mitter (Vienna: Böhlau, 1990), 9–26; Sonia Horn and Ingrid Arias, eds., *Medizinerinnen* (Vienna: Verlagshaus der Ärzte, 2003); Christof Aichner and Brigitte Mazohl, eds., *Die Thun-Hohenstein'schen Universitätsreformen 1849–1860: Konzeption, Umsetzung, Nachwirkungen* (Vienna: Böhlau, 2017); Jan Surman, *Universities in Imperial Austria, 1848–1918: A Social History of a Multilingual Space* (West Lafayette, IN: Purdue University Press, 2019).
2 See Andreas Hofmann, "Suprastaatlichkeit, Interstaatlichkeit und Transstaatlichkeit: Ein Drei-Ebenen-Modell zur Beschreibung zwischen staatlicher Beziehungen im Deutschen Bund," in *Transkulturalität, Transnationalität, Transstaatlichkeit, Translokalität: Theoretische und empirische Begriffsbestimmungen*, ed. Melanie Hühn et al. (Münster: Lit, 2010).

Trans-imperial comparison, to pick one of these as yet underexplored strains of research, has its own history. Analysis of the universities, academies, etc. of neighbouring empires was quite common in the nineteenth century and were not only a tool of information but also a vehicle of reform. Parisian philosopher and politician Victor Cousin, for instance, analysed Prussia and the Netherlands before starting to carry out reforms in France, and as a minister of education in 1840, he also intensively worked to apply some of his findings practically.[3] The ways in which education and scholarship systems abroad were studied are manifold and include both private and political projects, ranging from scholars going from one country to another to experience its educational system to government-sponsored research trips in order to study scholarly institutions comparatively.[4] Expectedly, such works were ordered mostly in the periods preceding reforms or already during reform processes, strengthening the view that officials and scholars had taken a close look at what was happening in other states and empires. Sometimes they even tried to influence academic policy in other countries, like the famous case of Alexander I's attempt to curtail students in the German Empire.[5] Damiano Matasci has recently impressively demonstrated that the analysis of such reports can shed light on how reforms previously described only through "national" lenses were enmeshed in networks of information exchange and were in fact more transnational than commonly assumed.[6] And it is just a matter of time until further works take this direction.

---

[3] Walter Brewer, *Victor Cousin as a Comparative Educator* (New York: Teachers College Press, 1971); Pella Kaloyannaki and Andreas M. Kazamias, "The Modernist Beginnings of Comparative Education: The Proto-Scientific and the Reformist-Meliorist Administrative Motif," in *International Handbook of Comparative Education*, ed. Pella Kaloyannaki and Andreas M. Kazamias (Dordrecht: Springer, 2009), 11–35, esp. 25–30.

[4] For example, Henri Hauser, *L'enseignement des sciences sociales: État actuel de cet enseignement dans les divers pays du monde* (Paris: A. Chevalier-Marescq, 1903). Habsburg voyages of this sort included Joseph/Józef Dietl's journey through the hospitals of Europe in 1846 or Václav Vladivoj Tomek travelling to European institutes for historical source analysis in 1848.

[5] Alexander I sponsored the member of the Russian consulate, Alexandru Sturdza, who proposed that German universities should be tightly controlled and changed into completely state-controlled schools. See Sturdza's *Mémoire sur l'état actuel de l'Allemagne* (Paris, 1818). Alexander I had the text distributed among the participants of the Congress of Aix-la-Chapelle (Aachen) in 1818 and it met with some sympathy, from, among others, Habsburg chancellor Klemens Metternich, but had no direct influence on university policies. See Carl Brinkmann, "Die Entstehung von Sturdzas 'État actuel de l'Allemagne,'" *Historische Zeitschrift* 120 (1919): 80–102. See also Alexander Martin, "Die Suche nach dem juste Milieu: Der Gedanke der Heiligen Allianz bei den Geschwistern Sturdza in Russland und Deutschland im napoleonischen Zeitalter," *Forschungen zur osteuropäischen Geschichte* 54 (1998): 81–126.

[6] Damiano Matasci, *L'école républicaine et l'étranger: Une histoire internationale des réformes scolaires en France, 1870–1914* (Lyon: ENS Éditions, 2015).

With Matasci's theoretical assumptions in mind, this essay will follow the assessment of Habsburg education, especially higher education, by Russian diplomat Ludwik Tęgoborski (Liudvig Tengoborskii, also Louis de Tegoborski)[7] who anonymously published a booklet, *De l'instruction publique en Autriche* (On public education in Austria), in Paris in 1841.[8] Tęgoborski's book was one of several which inquired about culture and education in the Habsburg lands in the early 1840s, a time when serious reforms were underway. The issue of the establishment of an academy of sciences in Vienna and the reform of gymnasia and universities were on the agenda of Habsburg intellectuals and politicians from the late 1830s, which resulted in an exceptional number of publications and internal documents on educational reform.[9] While historians have scrutinized the Habsburg documents on this topic, foreign writings have found little or no place in the narrative on educational reform. These books, however, do not only provide insight into the various assessments of Habsburg intellectual culture but may also be seen as voices in the reform movements, whether this was intended by their authors or not. Importantly, while Habsburg education after the revolution of 1848 was scrutinized from many angles, the *Vormärz*, in the Habsburg Empire, defined mostly through the reign of Franz I (reigned 1804–1835) and Ferdinand I (1835–1848) and the government of Chancellor Klemens von Metternich (1821–1848), is not. It is mostly seen through the lens of the 1848 revolution and claims raised in propagandistic writings justifying the reforms of 1848–1849. Tęgoborski, writing before the revolution, offers a fresh view on Habsburg education, quite contrary to rather gloomy descriptions coined by his Habsburg contemporaries.

This fresh view is indeed necessary given how research has regarded higher education in the Habsburg Empire before 1848.[10] In comparison to the post-1848 period, the *Vormärz* is much understudied and overviews focus on the relations between politics and the university, accentuating censorship, the lack of research at universities, and the various steps taken to liberalize the

---

[7] Since Tengoborskii used "Tęgoborski" as an alias beginning with his second publication, I will use this version of his name.

[8] Ludwik Tęgoborski, *De l'instruction publique en Autriche, par un diplomate étranger, qui a longtemps résidé dans ce pays* (Paris, 1841).

[9] See, for instance, Richard Meister, *Entwicklung und Reformen des österreichischen Studienwesens*, Part II, *Dokumente* (Vienna: ÖAW, 1963). See also Franz Exner, *Die Stellung der Studierenden auf der Universität: Eine Rede gehalten an der k. k. Universität zu Prag* (Prague, 1837).

[10] The situation is in fact similar in several countries. For Germany, see in this regard Andreas C. Hofmann, "Deutsche Universitätspolitik im Vormärz (1815 bis 1848) zwischen Zentralismus, 'Transstaatlichkeit' und 'Eigenstaatlichkeitsideologien'" (PhD diss., University of Munich, 2013).

situation.¹¹ Nevertheless, it was made clear that many of these oppressive measures backfired completely, with universities becoming a place where liberal professors and students were fostered.¹² Only marginal attention was paid to single faculties and the ways in which they were constantly reformed, providing possibilities for further innovation,¹³ and recently focus has somewhat shifted towards the question of continuities across the nineteenth century.¹⁴ As for the state of science in the monarchy, most research has inquired about how provincial and noble institutions provided venues for research, often sheltering its members from political pressure.¹⁵ The exception is medicine, where research has highlighted how the reforms of the Josephine era brought about the flourishing of research at universities, which was even exported abroad.¹⁶ Another strain of research has argued that late Enlightenment traditions developed further among Habsburg scholars, but also outside universities.¹⁷ Still, the general framework was defined by the keyword *Ver-*

---

[11] Peter Stachel, "Das österreichische Bildungssystem zwischen 1749 und 1918," in *Geschichte der österreichischen Humanwissenschaften*, vol. 1, *Historischer Kontext, wissenschaftssoziologische Befunde und methodologische Voraussetzungen*, ed. Karl Acham (Vienna: Passagen, 1999), 115–146; Hedwig Kadletz-Schöffel, "Metternich und die Wissenschaften" (PhD diss., University of Vienna, 1989); Eva S. Widmann, "Vormärzliches Studium im Spiegel autobiographischer Quellen," in *Österreichische Bildungs- und Schulgeschichte von der Aufklärung bis zum Liberalismus*, ed. Gerda Mraz (Eisenstadt: Institut für Österreichische Kulturgeschichte, 1974), 118–137.

[12] Clemens Novak and Martin Haidinger, *Der Anteil der organisierten Studentenschaft an der Märzrevolution 1848 in Wien und die Bedeutung dieses Ereignisses für die Korporationen von Heute* (Vienna: Akademischer Corporations-Club, 1999); Wilhelm Brauneder, *Leseverein und Rechtskultur: Der Juridisch-Politische Leseverein zu Wien, 1840–1990* (Vienna: Manz, 1992).

[13] Erika Rüdegger, "Die philosophischen Studien an der Wiener Universität 1800 bis 1848" (PhD diss., University of Vienna, 1964); Paula Sutter Fichtner, "History, Religion, and Politics in the Austrian 'Vormärz,'" *History and Theory* 10, no. 1 (1971): 33–48.

[14] Christine Ottner, Gerhard Holzer, and Petra Svatek, eds., *Wissenschaftliche Forschung in Österreich 1800–1900: Spezialisierung, Organisation, Praxis* (Vienna: V&R unipress, 2015).

[15] Claudia Schweizer, "Migrating Objects: The Bohemian National Museum and Its Scientific Collaborations in the Early Nineteenth Century," *Journal of the History of Collections* 18, no. 2 (2006): 187–199; Rita Krueger, *Czech, German, and Noble: Status and National Identity in Habsburg Bohemia* (Oxford: Oxford University Press, 2009), 161–191.

[16] Erna Lesky, *The Vienna Medical School of the 19th Century*, trans. L. Williams and I. S. Levij (Baltimore, MD: Johns Hopkins University Press, 1976). For an example of international acknowledgment, see Marcel Chahrour "'A Civilizing Mission?': Austrian Medicine and the Reform of Medical Structures in the Ottoman Empire, 1838–1850," *Studies in History and Philosophy of Science. Part C: Studies in History and Philosophy of Biological and Biomedical Sciences* 38, no. 4 (2007): 687–705.

[17] Esp. Christine Ottner, "Historical Research and Cultural History in Nineteenth-Century Austria: The Archivist Joseph Chmel, 1798–1858," *Austrian History Yearbook* 4 (2014): 115–133.

spätung (delay),[18] both regarding institutional innovation and scholarly development. And this keyword was first shaped by the negative assessment of the *Vormärz* by the post-1848 propaganda[19] and by a lack of comparative research which could pinpoint more clearly where exactly the monarchy lagged behind. Without attempting to compensate for this lack, this essay points to possible counternarratives based on assessments by contemporaries.

In what follows, I scrutinize Habsburg education through the lens of Tęgoborski's book. After situating him biographically, I will concentrate on his assessment of higher education and scholarship, written in French, and thus, in Tęgoborski's text, *l'enseignement supérieur*, including gymnasia, universities, polytechnics, and academies of sciences. Although Tęgoborski's book also included remarks on primary education, and this was also the part of the Habsburg educational system he appreciated most, his insights into higher education are among the most thorough we know from the time. Then I will compare Tęgoborski's notes with another foreign analysis of Habsburg scholarliness, this one published by Irish oculist William Wilde in 1843. Although they had different foci—Tęgoborski provided an overview of different forms of education, while Wilde's interests were mostly in the training of physicians—in some parts they markedly overlap, pointing to advantages and disadvantages of Habsburg education as compared with both authors' experiences.

Ludwik Tęgoborski and his Time

The fact that the first mention of the name Tęgoborski in this essay was followed by two alternative versions of his name already indicates that he was an imperial subject of several cultural backgrounds and allegiances. His life is indeed a very good example of what David Lambert called "imperial careering" and what Tim Buchen and Malte Rolf denoted as an "imperial biography,"[20] a career leading him from Warsaw, via Paris and Vienna, to the imperial court in St. Petersburg.

---

[18] Walter Höflechner, "Österreich: Eine verspätete Wissenschaftsnation?," in *Geschichte der österreichischen Humanwissenschaften*, vol. 1, *Historischer Kontext, wissenschaftssoziologische Befunde und methodologische Voraussetzungen*, ed. Karl Acham (Vienna: Passagen, 1999), 92–114.

[19] The book on which many narratives of *Vormärz* Habsburg universities are based is *Die Neugestaltung der österreichischen Universitäten über Allerhöchsten Befehl dargestellt von dem k. k. Ministerium für Kultus und Unterricht* (Vienna, 1853), a centrepiece of propaganda campaigns for the Habsburg educational reforms of 1848–1849.

[20] David Lambert and Alan Lester, eds., *Colonial Lives across the British Empire: Imperial Careering in the Long Nineteenth Century* (Cambridge: Cambridge University Press, 2006); Tim Buchen and Malte Rolf, eds., *Eliten im Vielvölkerreich: Imperiale Biographien in Russland und Österreich-Ungarn, 1850–1918* (Berlin: De Gruyter, 2015).

Ludwik Tęgoborski (1793–1857) was born in Warsaw, where he also spent his early adult years as a civil servant, dealing particularly with economic questions in the western province of the Russian Empire, the Kingdom of Poland. Here he was working for Prince Franciszek Ksawery Drucki-Lubecki, an anti-*laisser-faire* interventionist minister of the treasury in the then partially autonomous province. From 1828 on Tęgoborski served as the Russian consul general in Danzig (Prussia, now Poland), where, among other things, he was responsible for organizing provisions for the imperial army during its efforts to suppress the Polish November uprising of 1830–1831.[21] From the shores of the Baltic Sea he was sent, in 1832, to the capital of the Habsburg Monarchy in Central Europe, Vienna, to represent the Russian tsar during the negotiations on the constitution of the Free City of Cracow.[22] Except for a short interlude in Paris, he remained in the Habsburg capital for almost two decades. As a privy councillor (Geheimer Staatsrat), he belonged to the elites of the civil services and was instrumental in a number of treaties between the two empires, regulating, for instance, river trade. In 1846 he was called back to St. Petersburg and continued his political and publicist career at the imperial court.[23]

Born in the former Polish-Lithuanian capital, Warsaw, and a son of Walerian Tęgoborski, chamberlain to Polish-Lithuanian King Stanislas August Poniatowski, Tęgoborski escapes easy categorization into national categories like Russian or Polish. His cooperation with Drucki-Lubecki, a member of the conservative Polish faction in the Kingdom of Poland who considered economic and social development paramount for national independence, allows for describing him as a statist, Polish Russian subject. In 1833, *Gazeta Polska* wrote of him as "our degenerate/treacherous compatriot [wyrodny ziomek],"[24] and his participation in the Russian suppression of the November uprising and then in drawing the constitution of the Free City of Cracow certainly did not help much to win over Polish nationalists. However, for some he was still a member of the Polish culture—for his adversaries in a derogatory way (e.g., *halb-Pole*), for others, also Polish-language writers, simply somebody of Polish origin.[25]

---

[21] Chris D. Monday, "L. V. Tengoborskii i russkaia ekonomicheskaia mysl'," *Ekonomicheskaia istoriia Rossii XX veka: Problemy, poiski, resheniia* 2 (2000): 53–68, here 53 f.

[22] The Habsburg, Prussian, and Russian empires were the so-called protectors of the free City of Cracow after the November uprising, largely controlling it.

[23] See Monday, "L. V. Tengoborskii:" M. M. Savchenko, "Tengoborskii (Tęgoborski) Liudvig Valerianovich," in *Ekonomicheskaia istoriia Rossii s drevneishikh vremen do 1917 g. Entsiklopediia*. vol. 2, eds. L. M. Epifanova, B. Iu. Ivanov, and A. A. Komzolova (Moscow: ROSSPEN, 2009), 900–904.

[24] Anonymous, "Pogłoski dzienne," *Gazeta Polska*, 13 July 1831, 3.

[25] Leon Rogalski, *Leona Rogalskiego Historya literatury polskiéj*, vol. 2 (Warsaw, 1871), 293–295; Lesław Łukaszewicz, *Rys dziejów piśmiennictwa polskiego, uzupełnione i doprowadzone do r. 1859* (Poznań, 1859), 450–451. Łukaszewicz sees Tęgoborski as the most important Polish economist of the early nineteenth century.

In his years in Vienna, the most important period of his life in the context of this essay, he was not only the Russian Empire's representative at the Habsburg court but became better known as a national economist, writing highly assessed yet also controversial books like *Des finances et du crédit public de l'Autriche* (On finances and public credit in Austria; 1843) or *Uebersicht des österreichischen Handels in dem eilfjährigen* [sic] *Zeitraume von 1831–1841* (A Survey of Austrian trade over a period of eleven years, 1831–1841; 1844). After going back to Russia, he continued his economic-statistical studies. His multivolume work *Études sur les forces productives de la Russie* (Studies on Russia's productive forces; 4 vols.; 1852–1855) was translated into Russian[26] and English.[27] These translations made Tęgoborski's work famous, and they remained an important statistical source until nowadays.[28] In the Habsburg Empire, his *Des finances* produced a stir among economists, caused not only by his critical attitude toward the structure of the Habsburg budget but also since he was seen as a foreigner trying to give advice to the leaders of a country which was not his own. A book by the Viennese journalist and politician Adolf Wiesner (1807–1867) in which he openly attacked Tęgoborski[29] even led to a series of polemic articles in the authoritative daily *Augsburger Allgemeine Zeitung*.[30] Later on Tęgoborski also did not shy away from criticism, for instance speaking of servitude as an economically counterproductive form of slavery in his work on the Russian economy. In modern Russian historiography, he is regarded as a liberal, free-trade economist.[31]

On Public Instruction...

Although Tęgoborski decided to publish *De l'instruction publique* anonymously, it seems that some of the readers knew who the author was. At least

---

[26] Liudvig Tengoborskii, *O proizvoditel'nykh silakh v Rossii*, vols. 1–3, trans. I. V. Vernadskii (Moscow, 1854–1858).
[27] M. L. de Tęgoborski, *Commentaries on the Productive Forces of Russia*, vols. 1–3 (London, 1855–1856).
[28] For a nineteenth-century assessment, see, for instance, *Putnam's Monthly* vol. 6 (October 1855): 442 f.
[29] *Adolf Wiesner, Russisch-politische Arithmetik: Streiflichter auf das Werk des russischen Geheimrathes M. L. von Tegoborski: "Ueber die Finanzen, den Staatskredit, die Staatsschuld, die Hülfsquellen und das Steuersystem Oesterreichs, mit Rücksicht auf Preußen und Frankreich"* (Leipzig, 1844). Wiesner attacked Tęgoborski with manifold arguments, beginning with his servile attitude toward the Russian emperor and ending with errors in calculations.
[30] Adolph Wiesner, *Zwanzig Spalten über ein Pamphlet: Streiflichter auf eine sogenannte Kritik betreffend die Schrift "Russisch-politische Arithmetik" in Nl. 217, 223, 224, 225, 226, 227 der Augsb. Allgem. Ztg.* (Leipzig, 1844).
[31] Monday, "L. V. Tengoborskii"; Chris D. Monday, "Pol'skii rasprostranitel' russkoi idei: Liudvig Tengoborskii", *KLIO* 2 (2000): 226–242.

in French works of the time his name was openly mentioned, and he himself feared no repercussions. When *Des finances* was published in 1843, now under Tęgoborski's name, *De l'instruction publique* figured on the title page as a previous work by this author. In Vienna he was also known as the author of publications on educational matters, even before his book was printed.[32] And the sources he used, both for his book on education and for subsequent studies on finances, show that even under Metternich's restrictive regime information could travel freely, although exclusively among the high aristocracy.

Among the large number of written and oral sources Tęgoborski used, although he rarely quoted them, one deserves special mention, since it was also of an inter-imperial comparative kind. In his books Tęgoborski drew several comparisons with the work of Saint-Marc Girardin, a French historian and journalist who, in the early 1830s, studied German educational systems.[33] From that source he might also have taken a generally positive attitude toward Habsburg education, which was markedly different from numerous writings of the period. Even Girardin was conscious that his assessments contrasted with those of other analysts and critics of Habsburg education, stating at the beginning of the chapter on Austria that, in comparison with the "bad reputation"[34] Habsburg imperial educational policy enjoyed abroad, his piece might indeed sound like an apologia.[35] Girardin highlighted Habsburg achievements, especially in primary and secondary education and stated that the benevolent emperor had had considerably success with the economic and social development of the country, which justified the policy of the inhibition of free thinking.[36] While he was not really fond of the university system, he studied the Vienna Polytechnic in detail. He regarded it as an exemplary institution which, in comparison to Habsburg universities (but also to the French *École polytechnique*), even allowed freedom of teaching (Lehrfreiheit), thanks to the ideas and protection from politicians of its omnipotent principal, Johann Joseph Precht.[37] This assessment of the Vienna Polytechnic will resound in Tęgoborski's description, to which I would like to turn now.

---

[32] See Hammer-Purgstall's memoirs, transcribed in Joseph von Hammer-Purgstall, *Joseph von Hammer-Purgstall: Erinnerungen und Briefe*, vol. 3, ed. Walter Höflechner and Alexandra Wagner (Graz: Adeva, 2010), accessed 7 March 2019, http://gams.uni-graz.at/hp/pdf/Typoscript-47.pdf.

[33] Saint-Marc Girardin, *De l'instruction intermédiaire et de son état dans le midi de l'Allemagne*, 2 vols. (Paris, 1835–1839).

[34] Ibid., vol. 2, 169–415, here 169.

[35] Ibid.

[36] Tęgoborski, *De l'instruction publique*, esp. 172–175.

[37] Ibid., 257.

Façades and Statistics: Gymnasia

In the Habsburg Empire, gymnasia constituted a necessary point of entry to university education, thus an analysis of their description helps to position Tęgoborski's outlook on higher education. In comparison to popular education (l'éducation populaire), of which the Russian diplomat spoke with high praise, his description of education at gymnasia included a number of critiques. The main reason was its social exclusiveness, which he contrasted to more socially inclusive education in Prussia—an archetypal model of successful education in most of the text. According to him, in the Habsburg Empire gymnasia were highly attuned to social class, privileging the bourgeoisie and offering more specialized education aiming at the training of civil servants, while in Prussia gymnasia were more generally accessible, thus enabling a broader populace to achieve higher education.[38]

According to Tęgoborski, even in the function of preparing students for a future career in the Habsburg civil service, gymnasia failed because, apart from knowing Greek and Latin, their students only knew bits and pieces of the subjects necessary for bureaucratic duties. They were barely educated in geography or statistics and even their mathematical skills were not sufficient for practical work.[39] After the whole cycle of seven years of education, they were not even able to complete simple arithmetical operations, because mathematical schooling had been subordinated to the needs of teaching logics.[40]

Thus, according to Tęgoborski, Habsburg gymnasia could not be regarded as an establishment where graduates could complete their education and be ready for jobs, but rather they served primarily as a preparation for universities. Even in this function he saw their programmes as not being adequate. Students' knowledge of Latin, for instance, was not sufficient to follow university courses given in this language, which, according to Tęgoborski, forced more and more university professors to teach in German or, in the Italian provinces (i.e., Lombardy and Veneto), in Italian.[41] This points to a more general problem to which Tęgoborski paid a great deal of attention, the low quality of students coming from gymnasia. While most gymnasia graduates felt compelled to go to universities to have the capacity for civil service careers, there were no selection procedures for sifting out those who were not gifted enough, thus lowering the level of the universities. And many of the students who went to university also belonged to higher social classes, this educational path thus not befitting their social status in the first place.[42]

Tęgoborski underscored the gloomy picture of the low level of gymnasia education with glimpses into the practice of teaching. He analysed not only

---

[38] Ibid., 123.
[39] Ibid., 125.
[40] Ibid., 132.
[41] Ibid., 188.
[42] Ibid., 128.

the programme of teaching and the contents of classes but also recorded his talks with several gymnasium professors.⁴³ And it was the practice of teaching to which he expressed his most serious objections, apart from his not at all favourable comments on the programme. Since professors mostly read their lectures—in the literal meaning of the word "read"—and were not really entering into stimulating conversations, even the brightest students could not develop their talents beyond the average.⁴⁴ One of the reasons for this *désolante médiocrité* of professors and their unwillingness to engage in intellectual discussions with students, according to Tęgoborski, was the lack of special education for teachers—candidates for teaching positions only had to complete preparatory philosophical studies (Philosophicum) and attend additional courses in pedagogy.⁴⁵

This structural calamity was caused, according to Tęgoborski, by religious obeisance, which the Russian diplomat saw more as a stabilizing factor than an obstacle for teaching, but more importantly was caused by the lack of any intellectual mission gymnasia might have. While he did not comment more generally on this topic, it is quite easy to see that the fact that the road to a career in civil service led through the gymnasium and then the university was hurting both those more gifted and those less intelligent due to this (limited) mass character and lack of alternative ways to provide higher education. To conclude the topic of gymnasia: in Tęgoborski's narrative a fish rots from the tail, not from the head down. It was the gymnasia which produced mediocre students who then lowered the educational level at university. And this affected scholarship in general as well.

Universities, the Polytechnic, and the Theresianum: Tęgoborski and Academic Education

Having finished gymnasia, students moved to universities. In his book, Tęgoborski analysed only the University of Vienna, the most powerful university in the monarchy. Although Habsburg universities were officially structurally equal, which justified Tęgoborski's choice of Vienna as an exemplum, there were some differences, and Tęgoborski indeed mentioned some of them. The privileged position of Vienna was guaranteed by several privileges, from the right to nominate professors of their choice to financial treats. Structural measures also sustained this supremacy—teaching in the capital city meant having access to better collections to work with and also being near to the court (which, however, also meant tighter supervision). Viennese professors also enjoyed other honours: the principal/president of the Viennese medical faculty, for instance, was at the same time *proto-medicus* for the whole mon-

---

⁴³ Ibid., 132.
⁴⁴ Ibid., 134.
⁴⁵ Ibid., 145 f.

archy, supervising all of its sanitary institutions and being the court and government expert on medical questions.[46] Tęgoborski did not delve deeply into issues of the political significance of universities but remarked, for instance, that universities were not only responsible for the close supervision of gymnasia, but their representatives (delegates of four faculties) also constituted the majority on provincial committees supervising all types of education in a province, the so called *Studien-Concesse*.[47]

Tęgoborski's thorough analysis, which cannot be reproduced here in detail, rarely presented general opinions and instead focused on details. And when Tęgoborski decided to include some opinions, he limited himself to very neutral wording. For example, when summing up the chapter on theological studies at the universities, he wrote that it was in perfect harmony with the educational premises.[48] He also refrained from detailed analyses of certain issues—for instance, of the right of persons of the Jewish faith to attend universities[49] or the issues of students of different Habsburg linguistic backgrounds.

From all the topics Tęgoborski analysed, to me three seem to be of particular importance. First, the treatment of students, second, the university structure, and third, professorial issues. Taken together, they not only demonstrate several peculiarities of Habsburg higher education but also provide insight into how Tęgoborski saw the role of higher education in general and where he identified Habsburg deficiencies.

In the era of Metternich's phobia about the free exchange of knowledge, something he in fact shared with quite a number of the German governments who signed and observed the Carlsbad Decrees of 1819,[50] students were regarded as a possible source of endangerment of the political order. Thus, both in legislation and in Tęgoborski's description they had a very prominent place. Tęgoborski remarked that the Habsburg government intended to keep Habsburg students—like the university system itself—separate from those abroad, inhibiting even exchanges with other German states. Habsburg subjects were allowed to study abroad only with special permission, but the grades achieved did not entitle them to civil service jobs, thus being completely worthless. Similarly, foreigners could study at Habsburg universities only with special permission. The only exception to this rule were Habsburg

---

[46] Ibid., 253 f.
[47] Ibid., 274 f.
[48] Ibid., 221.
[49] The only mention of this is in ibid., 230, where Tęgoborski writes that Jews were not permitted to sit examinations in canon law and church law.
[50] The Carlsbad Decrees were German legislation aimed at countering revolutionary tendencies, especially among student organizations (Burschenschaften). They expanded censorship, strengthened political control over the universities, and led to the removal of some liberal professors.

Lutheran and Calvinist students, who could, under special conditions, study at a few selected universities in other parts of the German Confederation.[51]

As was the case in other German states, Habsburg students had no right of association, an issue linked directly to the Carlsbad Decrees. But they were also not subordinated to the university as was the case in the corporation model of universities. If students considered themselves mistreated or believed they had received inappropriate grades, they could appeal to the *Studiendirektor* or even to the provincial governments.[52] Tęgoborski also mentioned that such cases did in fact happen, and that some faculties were reprimanded and told to hold their exams in an objective and impartial way.[53]

There was also one other measure concerning students Tęgoborski felt obliged to discuss in more detail to underscore the benevolence and social agenda of the emperor. While Habsburg students paid for university attendance—18 florins per year at the philosophical faculty, 30 at the other faculties[54]—this money was neither to be used for financing the university nor to improve the wages of professors but had to be used by universities for funding students who could otherwise not afford to study. Tęgoborski mentioned 354 scholarship holders in 1832, but with a downward tendency, falling to 233 in 1838. At the same time, however, the number of students exempted from paying the fees rose. In sum, around half of Viennese students were either scholarship holders or exempted from fees which, according to Tęgoborski, showed that the Habsburg government exercised a *protection paternelle* toward the *classes pauvres*.[55]

With the different handling of philosophical faculties in the question of attendance fees, a Habsburg peculiarity is already marked. In all Habsburg universities of the time, the philosophical faculty served to offer preparatory instruction for further studies at three other faculties (juridical, medical, and theological) and not as a faculty with equal rights, although Tęgoborski also mentioned some modifications which had been made in the past decades. Comparing the plan of 1805 (when philosophical preparatory studies lasted three years) with 1824 (two years), Tęgoborski also mentioned that a certain specialization had been achieved this way and that facultative courses at the philosophical faculty were becoming more oriented toward the practical needs of those students who, for example, wanted to work as teachers. One chair at the philosophical faculty attracted Tęgoborski's special attention—religious studies (Religionslehre). Tęgoborski was rather critical of including

---

[51] Ibid., 220. On this point, see also Surman, *Universities in Imperial Austria*, 33.
[52] Ibid., 259 f.
[53] Ibid., 158 f.
[54] On 23 June 1841, one could buy a twenty-year-old *Ausbruchwein* (a sweet, quite pricey wine) from Tuscany for one florin, and for ten florins a half-year subscription to the daily newspaper *Wiener Zeitung*; Wiener Zeitung, 23 June 1841, 1 and 18.
[55] Ibid., 266.

this chair, since in his eyes the treatment of religious issues on a par with other scholarly disciplines had a harmful effect on moral education. While the laws carefully regulated what could and could not be taught by the professor of religious studies, in practice it would be hard to avoid controversial questions which, according to the regulation, were to be delegated to the lectures in philosophy.[56]

Students could naturally be dismissed from the university after having failed the annual exams twice, and especially the first exam, after the first year of philosophical studies, was to be treated exceptionally.[57] This was indeed quite problematic, since students who failed twice could not change to other universities or colleges, such as the school for surgeons and country doctors (Medicinisch-chirurgische Lehranstalt).[58] However, if they "lacked the talents" but were morally excellent, they could attend facultative courses at the university.[59] Moral behaviour and passing religion exams with good marks were essential, counting as much as other subjects.[60]

For the supervision of moral and political issues at the university, there was an additional authority at the Habsburg universities apart from rectors and deans: *Studiendirektoren*. There was one *Studiendirektor* for each faculty and one for grammar schools in each province. Their main concern was the legal and moral supervision of the faculty and the students, but they also took part in the meetings of the academic senate, proposed new chairs, and were responsible for all disciplinary questions. Moreover, they corresponded with the provincial authorities, sending them the programmes of courses for approval and lists of students and professors with details even including comments on students' progress.[61] They also played a decisive role in the *Concours*, the Habsburg approach to choosing new professors. The *Concours* was a formal exam of candidates[62] for professorial posts, with an oral presentation and an essay (Elaborat) on a chosen topic. The faculties would then discuss all the candidates and, also taking moral behaviour into account, compose a list, ordered from the most to the least suitable—if there were many candidates, the list could also include only the three candidates the faculty considered best.[63]

---

[56] Ibid., 183.
[57] After having failed the first annual exam, students could repeat the year. On the particular attention professors were supposed to place on the moral and intellectual qualities of students at the first annual exam of the Philosophicum, see ibid., 202.
[58] Ibid., 246.
[59] Ibid., 202.
[60] Ibid., e.g., 141.
[61] Ibid., 277, see also 179.
[62] Foreigners were banned from attending the *Concours*. Ibid., 270.
[63] Ibid., 267.

However, according to Tęgoborski, the *Concours* hardly fulfilled its function.[64] While he considered it an appropriate procedure for choosing teachers at gymnasia, where scholarly quality did not really count, it affected competition at universities. Since it also allowed mediocre scholars to compete with outstanding professors, those already distinguished by their scholarly works would feel compromised if they had to take part in such a procedure. While the government had the possibility of nominating outstanding scholars without a *Concours*, this did not happen frequently—and even if it did, the *partisans de l'ancien systeme* (Tęgoborski did not clarify whether these were politicians or professors) loudly opposed it.[65]

Tęgoborski's remark on the essential difference between professors at gymnasia and at universities was, however, relativized in several instances. In fact, one can say that it contradicted Tęgoborski's statement that the very idea of Habsburg universities was to educate civil servants and not scholars. Accordingly, for instance, the lectures themselves were not very different from those at gymnasia. Not only were the professors obliged to teach strictly according to textbooks (Vorlesungsbücher)—only in the rarest cases they were allowed to use their own manuscripts, if approved by the *Studiendirector* and a governmental board—but if they did not comply with this, the *Studiendirektoren* were obliged to intervene and report this misbehaviour to superior authorities.[66] This supervision went even further: while professors were also allowed to give private lectures, these were also to be checked by the *Studiendirektoren* and should follow individually approved programmes.[67]

Especially at the theological faculty there was also an additional instance of supervision: the Catholic Church. Catholic authorities not only had a say in professorial nominations and the control of the textbook manuscripts but were also obliged to check the dissertation theses submitted to the theological faculty.[68] They had, however, no legal means of persecution in this situation, but any supposed aberration was supposed to be communicated to the civil authorities.

In sum, Tęgoborski was not quite pleased with the conditions at the universities. According to him, they did not forge scholarly qualities, although they fulfilled the important social function of moral disciplining. Tęgoborski held, however, two other Viennese establishments of higher education in high esteem: the Polytechnic and the school for the nobility, the Theresianum.

The description of the Theresianum (Theresianisch-Leopoldinische Akademie), the Piarist Order-directed college for the nobility, prepared noble offspring for higher public and governmental service. The fact that Piarists

---

[64] Ibid., 314.
[65] Ibid., 314 f.
[66] Ibid., 263.
[67] Ibid., 264.
[68] Ibid., 219.

controlled the Theresianum gave Tęgoborski the opportunity to comment once more on church-state connections, although now in a more critical way than concerning the gymnasia or universities. Acknowledging the importance the Piarists had had for primary and secondary education in the empire, Tęgoborski criticized their influence on higher education in general—the Theresianum had the status of an *Akademie* (academy). Here education should "have a closer connection to the needs of a modern society"[69] and students were supposed to be familiarized with "the needs of the century and with changes to the social economy happening almost every day."[70] Being under the governance of a Catholic order clearly did not allow, according to Tęgoborski, the Theresianum to react adequately to these necessities.

In general, Tęgoborski saw the Theresianum with a less critical eye than the universities and foretold that with a few changes of staff and direction this institution could become "one of the most distinguished institutions in Europe."[71] He particularly stressed that the twelve-year education was all encompassing and that the facilities were perfectly adequate to the intended function—that is, the close supervision of the moral development of the young aristocrats and noblemen. He particularly appreciated strong limitations on contacts with commoners and between students of different school years, which one could see as a result of the elitist education Tęgoborski had received in Warsaw. This supervision was indeed very close, and Tęgoborski quoted Saint-Marc Girardin's work, which described how the students of the Theresianum were not even allowed to read *belle-lettres*, dramas, comedies, and all works which might "irritate the spirit" and "expose the soul to false sentimentality."[72]

But neither the university nor the Theresianum but the Viennese Polytechnic was, according to Tęgoborski, the finest Viennese school of higher education. As was the case in Girardin's publication, in Tęgoborski's eyes it "ranked first among the institutions concerned with practical education [l'instruction usuelle et pratique]."[73] Tęgoborski was equally interested in the programme of studies and the ways in which the Polytechnic communicated with the general public and how it invigorated the commerce and industry of the Habsburg lands. In terms of communication with the public, he mentioned yearly séances discussing new developments at the Polytechnic and its practical functions, and its annual journal, the *Jahrbücher des kaiserlich-königlichen polytechnischen Institutes in Wien*. In terms of commerce and industry, the Polytechnic also functioned as a Society for the Invigoration of the National Industry (Verein zur Beförderung der Nationalindustrie, sometimes

[69] Ibid., 321.
[70] Ibid., 322.
[71] Ibid., 330.
[72] Ibid., 325.
[73] Ibid., 339.

referred to as Gesellschaft zur Aufmunterung der Künste und Gewerbe), although Tęgoborski did not make explicit what the function of this society was.[74] Still, according to him, the creation of a new society which would include scholars and practitioners and which would be close to the government was a promising enterprise.[75] This society was not supposed to compete with the Society for the Invigoration of National Industry, but they should stand side by side, also joining forces with the already existing Lower Austrian Commercial Club (Niederösterreichischer Gewerbeverein).

However, somewhat similarly to other areas of higher education, Tęgoborski remarked that the professors at the Polytechnic did not receive a salary adequate to their achievements and their quality, and that this prevented the institution from appointing the best scholars. Especially in comparison with private industry, the payment was low, the social esteem insufficient, and future advancement limited. He agreed that there were many exceptional professors but, backed by the opinions of the chairholders he had talked with, not all professors were optimal appointments.[76] This was well in line with what Tęgoborski wrote about scholars in the Habsburg Empire in general: they were talented, but the lack of the support, or even of adequate recognition by the state, held them back. Especially in comparison with other states, notably Prussia, the Habsburg Monarchy had much to improve in this regard. A further point emphasized by Tęgoborski were conflicting ideas about education held by the bureaucrats governing the universities on the one hand and by the scholars on the other, which resulted in frequent conflicts between the two groups. As Tęgoborski argued, "science and scholarship are rarely taken care of satisfactorily by the bureaucrats."[77] However, on the bright side, complaints voiced by "enlightened men with good intentions" to the government had been heard and some positive changes concerning these matters were to be expected.[78]

One of the matters that, according to Tęgoborski, had to be changed in the near future was the issue of an academy of sciences in Vienna. The lack of such an institution was one of the most serious points of his criticism, well in accordance with his disproval of Habsburg politicians' low regard for scholarly progress in general. The lack of an academy meant that there was "no centre of rivalry and mutual encouragement, no Areopagus which could judge on the merits of production of the genius and would give opinions concerning all things which affect the progress of sciences."[79]

---

[74] Ibid., 348, 350 f.
[75] Ibid., 353.
[76] Ibid., 349, 350 f.
[77] Ibid., 317.
[78] Ibid.
[79] Ibid., 316.

However, as Tęgoborski mentioned with approval, there were already preparations for the establishment of an academy, and enlightened men and the government had achieved agreement about its structure and functions. Even if one swallow does not make a summer, a change of atmosphere concerning the scholarship was in the air. For instance, there was already an academy of medical sciences, the Royal and Imperial Society of Physicians (k. k. Gesellschaft der Ärzte), created in 1839.[80] Tęgoborski saw it as a first step toward allowing independent scholarly organizations.[81] Since the k. k. Gesellschaft was not merely allowed but the government subsidized it with both money and some buildings, this signalled for Tęgoborski that the Habsburg government did not principally inhibit the "progress of enlightenment" but was just cautious and therefore slow in this matter.[82] There was still, however, opposition from the lower bureaucracy, which played an important role in the political system of the monarchy, but Tęgoborski envisaged an agreement to come.[83]

To sum up briefly, Tęgoborski's narrative was less positive the higher the education. The Habsburg Empire had adequate gymnasia, the respectable Theresianum and Polytechnic, second-rate universities, and was mediocre when it came to academies. This listing—deliberately—follows a "modern" understanding of education as forging scholarship. Some, the most widely quoted writings composed after 1848, followed precisely this narrative. However, if read as Tęgoborski intended—and I followed Tęgoborski in the order in which the different types of educational and scholarly institutions appears in the discussion above—educational facilities were subordinated to their social functions and did quite well in this regard. And the Theresianum and the Polytechnic stood higher in Tęgoborski's hierarchy than the university. Nevertheless, Tęgoborski also saw that where scholarship was concerned, the Habsburg Empire had no institutions forging it: those which might do so were either non-existent (an academy) or attuned to different needs (the universities).

It has to remain an open question to what extent Tęgoborski's narrative was shaped by his willingness to make his book a statement within the current debates about education in imperial Russia.[84] Unlike the abovementioned Cousin, Tęgoborski was not directly involved in educational policies, but still, as a diplomat he might have hoped to improve his career. While em-

---

[80] Its statutes had been approved on 17 October 1839. See Salomon Hajek, *Geschichte der k. k. Gesellschaft der Ärzte in Wien von 1837 bis 1888* (Vienna, 1889), 2.
[81] Ibid., 317.
[82] Ibid., 318.
[83] Ibid.
[84] Joanna Schiller, *Universitas rossica: Koncepcja rosyjskiego uniwersytetu 1863–1917* (Warsaw: Wydawnictwo Instytutu Historii Nauki PAN, 2007). On the situation in the Russian Empire, see also Rebecca Friedman, *Masculinity, Autocracy, and the Russian University, 1804–1863* (New York: Palgrave Macmillan, 2005).

phasizing issues like the lack of autonomy within higher education and great productivity where there was freedom, issues debated within Russian academia and by politicians before hint at this direction but would need more research to be substantiated. Of course, the narrative was shaped by his experiences and an interest in corroborating them and in gaining some insights into the history of Habsburg higher education. Now I would like to turn to a book by another foreigner who visited the Habsburg Monarchy in the 1840s, William Wilde.

Tęgoborski and Wilde: "Regards Croisés"

William Wilde, better known as the father of a more famous son, Oscar,[85] and as a figure in the Irish cultural renaissance,[86] visited the Habsburg lands in the early 1840s while on a tour of medical institutions on the continent. As a renowned ophthalmologist, his interest in Habsburg educational institutions lay mostly in the education of physicians, but in his book, published 1843 as *Austria: Its Literary, Scientific, and Medical Institutions*, he intended to describe the variety of educational and scholarly establishments of the empire. Tęgoborski's book was one of his sources, although they apparently had never met. In his publication Wilde also referred to a broader range of (mostly) German-language articles and books, supplementing the information taken from printed sources with administrative documents and personal talks with Viennese scholars.[87] Thus Wilde was well aware of the debates of the time. The fact that he prominently and very emotionally stressed the lack of an academy of sciences in Vienna and claimed the necessity of its creation, both with a scholarly interest in mind and by stating political arguments, not only proves that his book reflected the debates of the time well, but that it can even be read as a voice in the Viennese debates themselves. At the very same time that he was in Vienna (and also when he published his book), Joseph Hammer-Purgstall, one of Wilde's liaisons in the Habsburg capital, was spending hours trying to persuade Metternich and other politicians, as well as Habsburg aristocrats, that an academy should be created.[88]

---

[85] The German translation of his book on Austrian educational institutions is titled *Oscar Wildes Vater über Metternichs Österreich* (Oscar Wilde's Father on Metternich's Austria); Irene Montoye, ed., *Oscar Wildes Vater über Metternichs Österreich: Ein irischer Augenarzt über Biedermeier und Vormärz in Wien* (Frankfurt am Main: Peter Lang, 1989).

[86] See Thomas George Wilson, *Victorian Doctor Being the Life of Sir William Wilde* (New York: L. B. Fischer, 1946) and the articles in the special section of the *Irish Journal of Medical Science* 2/185 (2016): 275–307, concerned with the life and works of Wilde.

[87] William Wilde, *Austria: Its Literary, Scientific, and Medical Institutions* (Dublin, 1843), xvii–xx.

[88] Höflechner and Wagner, *Joseph von Hammer-Purgstall*, vol. 3, 1959–2168.

In his book Wilde focused on topics Tęgoborski did not extensively cover, especially the details of medical studies. Many parts central to Tęgoborski were briefly examined in Wilde's narrative and served only to provide background information for following the structure of Habsburg education. For instance, Wilde's statistical overview of primary schooling and gymnasium education—notably mentioned as the second best on the continent, topped only by Prussia—was less than ten pages long and included mostly rudimentary statistics.[89]

Wilde's description of higher education was in fact limited to the description of the Vienna Medical Faculty. He saw it as one of the leading institutions in Europe and praised the celebrities teaching there. Student education was, however, not entirely adequate; for instance, the students were offered only a very few opportunities of making dissections, thus limiting their anatomical education.[90] However, slightly surprisingly given the amount of criticism directed at the time to compulsory textbooks (Vorlesungsbücher), Wilde positively embraced the idea of standardized, printed compulsory books which would not overburden the students with unnecessary information and manifold names, as he had experienced education in Great Britain.[91]

Wilde also added some information on medical-surgical studies in Vienna, at the Josephinum. As good as overlooked by Tęgoborski, legally the Josephinum had the same status as the university faculties, that is, it could award doctoral degrees but was not officially a part of the Viennese university. Wilde paid even less attention to the Polytechnic. And many of his other remarks were cursory. For instance, when describing the general working conditions of the professors, he mentioned only their regular salary and the fact that they "are virtually appointed by the government," though "a *nominal concours*" is the official procedure.[92]

Apart from different educational foci, Wilde and Tęgoborski differed in their assessment of the imperial situation. While Tęgoborski was careful to omit references to the composite, plurilingual character of the monarchy, Wilde not only did not hesitate to write about the Slavic, Hungarian, or Italian populations but also very attentively and brilliantly presented an analysis of the effects this composition had on scholarship.

This different handling of social and cultural diversities can be demonstrated by Wilde's treatment of the issue of the academy of sciences which he, like Tęgoborski, saw as indispensable for the future progress of scholarship in the empire. Yet, while Tęgoborski in this context referred only to scholarly needs, Wilde translated this question into cultural categories. Contrasting the efforts of Czechs and Hungarians which were visible in the scholarly arena to

---

[89] Wilde, *Austria*, 4–14, his praise for Habsburg schools on 8.
[90] Ibid., 45–46, see also 42 f.
[91] Ibid., 43.
[92] Ibid., 54.

the lack of a parallel German institution, and praising Habsburg's provincial academies, especially the one in Pest, Wilde emphasized the centripetal nature of scholarship in the monarchy, claiming the need for a central academy in Vienna to keep Habsburg scholars together.[93] Commenting on the fear of the government that such an academy would "encourage a revolutionary spirit in the heart of its dominions,"[94] he saw no real threat to the imperial capital, since the burghers of Vienna were, according to him, of a stable and content mood. However, it was Slavs and Hungarians who were running successful academies that threatened the German population of Austria and the ruling house—for Wilde likewise praised the German national character:

> Is it not an unaccountable and unwarrantable neglect of the German race, whose scientific worth and capability is so much underrated in comparison with the Hungarians, Bohemians, and Italians to whom academies are permitted [...]?[95]

The different appraisals of the social stratification of Habsburg education did not end there. The Theresianum, so dearly cherished by Tęgoborski for shielding the nobles from the general society and its distractions, also met with Wilde's criticism. For Wilde, the idea of a purely aristocratic institution was slightly disconcerting. However, it also revealed for him the ways in which the empire remained in tranquillity and in "civil and political ignorance."[96] This was not necessarily a negative statement, since states which had not established such a stability experienced social unrest, according to Wilde.

In his general assessment Wilde was indeed sceptical as to whether Austria would be a scholarly power. The current emperor wanted "bureaucrats" (Beamte) and not "scholars,"[97] the government did not adequately praise scholarly excellence, and if scholars were acknowledged internationally through medals or membership in academies and learned societies, they were not allowed to accept these honours. Habsburg bureaucrats regarded everything through the lens of politics and saw even mineralogy as a political threat.[98] Also, textbooks were over-interpreted by authorities seeking threats to moral and religious beliefs everywhere. For instance, *Physiologia medicinalis* by Pest professor Mihály Ignác Lenhossék (1773–1840) was reprimanded in 1835, after having been in use for a few years. The reason given was the possibility that some of its passages might be interpreted as running contrary to moral and religious beliefs. While it was not forbidden at universities already using it (although professors there were notified that they should interpret doubtful passages in a proper way), it was not to be introduced where it had not yet

---

[93] Ibid., xxi–xxii, 83 f.
[94] Ibid., 84.
[95] Ibid., xxii.
[96] Ibid., 22 f.
[97] Ibid., 99.
[98] Ibid., 98. Wilde referred here to Friedrich Mohs's popular open lectures in this subject, which were forbidden within half a year.

served as a textbook. Since the authorities approved no other textbook for physiology thereafter, this left students without stable standards to follow and, says Wilde, these would be quite necessary since physiology in Austria was not quite up to the modern standards.[99]

The criticism went further. While the collections of the city were excellent, there were neither enough people working on them nor journals which would make them known. In fact, Wilde remarks that after flourishing in the eighteenth century, "the literature of Austria, in quality as well as quantity, appears to have degenerated during the last fifty years."[100] Censorship was the main reason for this decline, but also the fact that Habsburg literature was divided into many languages, which inhibited mutual understanding among scholars.

Conclusion: Cross-vision of the Habsburg "Vormärz"

Thus, Tęgoborski and Wilde agreed in many respects but each presented a different nexus of factors which led to the state of the education and scholarship they described. Tęgoborski was more accepting of the way the Habsburg state was handling its linguistic diversity, or he even ignored it in his analysis, while Wilde was more prone to emphasize cultural conflicts. Also, Wilde stressed more explicitly the social disunion of Habsburg society, although both did acknowledge that the Habsburg state did remarkably—or surprisingly, to allude to Girardin—well in providing education to the poorer classes. (In Tęgoborski's book this was emplotted into an almost utilitarian idea of the social climbing of talented yet poor students). Likewise, the different social status and attitudes towards the empire presented by both writers caused this difference. Tęgoborski was not only an imperial subject but also an imperial bureaucrat, concerned with the stability of monarchic rule. In this, his views alleviated cultural differences and, as the hardly pleasant reactions of Polish nationalists underscore, he viewed culturalized-nationalized rhetoric with the utmost scepticism. Wilde on the other hand expressed the ideas of a nationalist Irish subject of the British Empire. In *Austria as It Is*, cultural difference was the daily routine of the empire and its management of utmost importance—and in this he saw Slavs and Hungarians gaining power while the Germans, whom he clearly considered to be culturally privileged, were losing ground due to the ill-directed cautiousness of the government. Both narratives saw, however, education and scholarship as being paramount for the future well-being of the state, demonstrating that scholarship had established itself well as a vehicle of progress in the first half of the nineteenth century.

---

[99] Ibid., 44. Wilde quotes here from an imperial edict of 1835.
[100] Ibid., 99.

Importantly, while scholarship was crucially gaining importance in the early nineteenth century, both writers saw the Habsburg Empire lagging behind in its idea of progress—or better, in lack of thereof. Conservatism, sometimes quite cherished, especially in Tęgoborski's book, made the government fear liberalism in all its manifestations, including the idea of scholarly progress, that is, international contacts, the acknowledgment of scholarly merits, and research organizations. Instead, Habsburg higher education was directed toward providing stability and it was very successful in this field, even if 1848 proved this impression wrong.[101]

Importantly, and this is what this essay intended to demonstrate, for foreign writers, stability and scholarship were of equal importance or at least two sides of one coin. Given the bad press Habsburg scholarly policy faced from liberals at home and especially the ways it was described after 1848, international contextualization may be very fruitful for rewriting and historicizing *Vormärz* policies. For instance, the unity of gymnasia and universities stressed by both authors sheds a different light on the ideals of education than the accounts focussing only on universities or on universities and learned societies, thus privileging scholarship over education. Changes like this could help to contextualize and even to relativize the bad press and help to question the widespread assumption that education in the Habsburg lands was seriously lagging behind in international comparison. While in Wilde's and Tęgoborski's books it did lag behind other countries in some respects, it did well in others; and while it lagged behind Prussia, a post-1848 topos that gave Habsburg, and later also Austrian historians, intellectuals, and politicians nightmares, it did well if compared to other German states, and the accounts analysed above help us to contextualize the Habsburg educational system in a much more refined way than the accounts and literature based on the propagandistic accounts produced during the reforms of 1848–1849.

---

[101] Cf. Stachel, "Das österreichische Bildungssystem;" Kadletz-Schöffel, "Metternich und die Wissenschaften;" Widmann, "Vormärzliches Studium."

Mark Hengerer/Sabrina Rospert

DID THE UNIVERSITIES OF THE LATE HABSBURG MONARCHY
IMPART "IMPERIAL KNOWLEDGE"?
A Survey of the Course Catalogues of the Universities of Budapest
and Vienna (Summer Semesters 1866, 1886, 1906)*

Introduction

The relationship between empire and science in the nineteenth century has hitherto been approached primarily from the perspective of colonial empires with overseas territories, with variegated research taking place, especially under the influence of postcolonialism.[1] Terms like "imperial knowledge and colonial violence" (Heé),[2] "imperial knowledge" (Thompson),[3] or "colonial

---

* We cordially thank Thomas Winkelbauer for giving us access to his as yet unpublished manuscript "Geschichte des Faches Geschichte an der Universität Wien: Von den Anfängen bis 1875." The data on which this analysis is based were deposited at the University Library of Ludwig-Maximilians-Universität in Munich, DOI: https://doi.org/10.5282/ubm/data.131.

[1] On the relationship between empire and knowledge in the nineteenth century, cf. David Amigoni, *Colonies, Cults and Evolution: Literature, Science and Culture in Nineteenth-Century Writing* (Cambridge: Cambridge University Press, 2007); Catherine Delmas, *Science and Empire in the Nineteenth Century: A Journey of Imperial Conquest and Scientific Progress* (Newcastle upon Tyne: Cambridge Scholars Publ., 2010); there is a noticeable emphasis on Great Britain and its overseas colonies in North America and India, especially in Anglo-American research. Cf. Raymond Phineas Stearns, *Science in the British Colonies of America* (Urbana: University of Illinois Press, 1970); Zaheer Baber, *The Science of Empire: Scientific Knowledge, Civilization, and Colonial Rule in India* (Albany: State University of New York Press, 1996); Christopher Bayly, *Empire and Information: Intelligence Gathering and Social Communication in India, 1780–1870* (Cambridge: Cambridge University Press, 1996); Dhruv Raina and S. Irfan Habib, *Domesticating Modern Science: A Social History of Science and Culture in Colonial India* (New Delhi: Tulika Books, 2004); Brett M. Bennett, *Science and Empire: Knowledge and Networks of Science across the British Empire, 1800–1970* (Basingstoke: Palgrave Macmillan, 2011). For the early modern period, see Arndt Brendecke, *Imperium und Empirie: Funktionen des Wissens in der spanischen Kolonialherrschaft* (Cologne: Böhlau, 2009).
[2] Nadin Heé, *Imperiales Wissen und koloniale Gewalt: Japans Herrschaft in Taiwan 1895–1945* (Frankfurt am Main: Campus, 2012).
[3] Ewa M. Thompson, *Imperial Knowledge: Russian Literature and Colonialism* (Westport, CT: Greenwood, 2000).

knowledge" (M. Bayly)[4] can be found in the titles of corresponding studies. The term "imperial knowledge" itself is not actually defined even by Heé, however—only "scientific colonialism" is defined, namely as a "specific form of colonial policy that consists in facilitating the exertion of rule through the production of knowledge on the annexed territory and its population." The conceptual section on "imperial knowledge" forgoes a definition, although it does point to the undoubtedly important aspects of the production, circulation, and categorization of knowledge.[5]

The term "scientific colonialism" can hardly be applied to the late Habsburg Monarchy, however, for considering its form and development, one cannot speak of annexation or of colonies in the strict sense with regard to most of its territories.[6] With a view to the model of education,[7] it is likewise debatable to what degree the design of academia and the concept of rule were interlaced at all.

Against this background, it is understandable that the problem of the connection between knowledge and empire has primarily been approached by asking about the role of the educational system for both the stability and instability of the empire of the Habsburgs. Current research on this problem points out certain ambivalences: Judson, for example, emphasizes that the politicization of cultural conflicts surrounding the educational reforms during the second half of the nineteenth century threatened the unity of the Habsburg Monarchy.[8] This applies in particular to the language policy featuring targeted assistance for minorities to support imperial unity, which collided with the instrumentalization

---

[4] Martin J. Bayly, *Taming the Imperial Imagination: Colonial Knowledge, International Relations, and the Anglo-Afghan Encounter, 1808–1878* (Cambridge: Cambridge University Press, 2016). See also Ulrike Hillemann, *Asian Empire and British Knowledge: China and the Networks of British Imperial Expansion* (Houndmills, NY: Palgrave, 2009).

[5] Heé, *Imperiales Wissen*, 7 and 14–17.

[6] Although research has frequently discussed the Habsburg Monarchy as a colonial power, this assessment hardly seems tenable; cf. e.g., Johannes Feichtinger, Ursula Prutsch, and Moritz Csáky, eds., *Habsburg postcolonial: Machtstrukturen und kollektives Gedächtnis* (Innsbruck: Studienverlag, 2003). In the peripheral territories of Austria-Hungary in particular, e.g., Dalmatia, Bohemia, or Galicia, the Habsburgs found it difficult to exert administrative power and enforce their rule against the burgeoning nationalistic currents; cf. Pieter M. Judson, "L'Autriche-Hongrie était-elle un empire?," *Annales: Histoire, Sciences Sociales* 63, no. 3 (2008): 564 and 589; Pieter M. Judson, *Guardians of the Nation: Activists on the Language Frontiers of Imperial Austria* (Cambridge, MA: Harvard University Press, 2006), 16–18; István Deák, "Comments," *Austrian History Yearbook* 3 (1967), 1.

[7] On the model of education and its encyclopaedic redefinition in nineteenth-century Austria, see Alois Brusatti, Herbert Matis, and Karl Bachinger, *Betrachtungen zur Wirtschafts- und Sozialgeschichte* (Berlin: Duncker & Humblot, 1979), 69.

[8] Pieter M. Judson, *The Habsburg Empire: A New History* (Cambridge, MA: Harvard University Press, 2016), 292–299. On the system of education, see the seminal work of Helmut Engelbrecht, *Erziehung und Unterricht im Bild: Zur Geschichte des österreichischen Bildungswesens* (Vienna: ÖBV Pädagogischer Verlag, 1995).

of languages as carriers of new, nationalistically indoctrinated generations.[9] Similarly ambivalent was the founding of various universities: the establishment of the German-language University of Chernivtsi in 1875, for instance, frustrated those who were hoping for an Italian-language university in Trieste.[10] The separation of the German-language University of Prague into a German and a Czech university in 1882 resolved an existing conflict, but simultaneously buried all hope for the successful coexistence of the two languages.[11] What was more, the establishment of additional universities reduced the integrative influence of the University of Vienna: from the 1880s onwards, the number of "Austrians" studying at the University of Vienna grew in comparison to that of students from other parts of the monarchy. The previously higher degree of cohesion had strengthened nationalisms, especially during the 1870s, however.[12] The situation in regard to academic subjects and publications was similarly equivocal: while the Vienna Institute for Slavic Studies founded in 1849 played a key scientific role in the Habsburg Monarchy (under the eschewal of "national idealism") and was considered a symbol of the recognition of the "diversity of nations and languages of the monarchy," its graduates often operated in the context of a specifically Slavic nationalism.[13] And the *Kronprinzenwerk*—an attempt at an encyclopaedic presentation of the flora, fauna, geology, and ethnography of every individual crown land with the purpose of displaying the Habsburg Monarchy

---

[9] In the Hungarian part of the empire in particular, the Hungarian language was viewed as a link between the different nations; within the framework of "Magyarization," the Hungarians attempted to take away the non-Hungarian nations' distinct languages and culture and instead integrate them into the Hungarian nation, thereby contributing to its growth; cf. Judson, *Habsburg Empire*, 302–309.

[10] During the 1870s, demands for an own university or at least for a faculty in the corresponding language were voiced not only in the Italian territories but also by Czechs; guided by its belief in the principal superiority of the German language and culture, the government instead decided to establish a German-language university in one of the remotest areas of the Habsburg Monarchy. Chernivtsi, the capital of Bukovina, boasted only 25,000 inhabitants, who were mostly illiterate and spoke a mixture of Ruthenian, Romanian, Yiddish, German, and Polish; cf. Judson, *Habsburg Empire*, 321–327.

[11] Hans Lemberg, "Universitäten in nationaler Konkurrenz: Zur Geschichte der Prager Universitäten im 19. und 20. Jahrhundert," in *Universitäten in nationaler Konkurrenz: Zur Geschichte der Prager Universitäten im 19. und 20. Jahrhundert*, ed. Hans Lemberg (Munich: Oldenbourg, 2003), 28.

[12] Helmut Engelbrecht, *Geschichte des österreichischen Bildungswesens: Erziehung und Unterricht auf dem Boden Österreichs*, Band 4, *Von 1848 bis zum Ende der Monarchie* (Vienna: Österreichischer Bundesverlag, 1986), 242 and 246.

[13] Cf. Rudolf Jagoditsch, "Slavistik an der Universität Wien," in *Studien zur Geschichte der Universität Wien*, Band 3 (Vienna: Böhlau, 1965), 39–43. Generations of young "Slavs" educated at the Institute for Slavic Studies, which was considered the key centre for the study of Slavic languages and culture, went on to awaken the proclivity for their "national" language and history in their respective areas of origin; in this way, the Viennese Slavicists contributed to the national and cultural self-image of the thus-constructed "Slavic nations" within the Danube Monarchy.

and its diversity in an agreeable light—was eventually published only (or at least?) in German and Hungarian.[14]

This emphasis on the ambivalent function of the educational system for the stability of the Habsburg Monarchy naturally suggests narrowing down the area of enquiry and investigating a smaller section of the state system of knowledge production, circulation and categorization. Inspiration for such research may be taken from Thomas Winkelbauer, whose study on the history of the subject of history at the University of Vienna highlighted the formative influence of the division of world history into the four so-called "world monarchies" and the considerable importance of empires for the research of Viennese historians during the nineteenth and twentieth centuries.[15]

Our own approach in this context is not one of institutional history, however, but instead a semantic one: it focuses only on "the teaching" offered by the university as such. The concrete object of investigation are the courses listed in the course catalogues of the universities in the two most important cities of the Austro-Hungarian Empire, Vienna and Budapest. While academic teaching can be assigned to the general realm of production, circulation, and categorization of knowledge, almost everything else to do with it is rather unclear: its reception by the students, the actual contents and attendance at the lectures, possible cancellations or belated additions of courses, accompanying reading, the "diversity of voices" in courses and among their recipients, etc. Irrespective of these aspects, however, the question of whether or to what extent university courses (or more precisely, their titles) can be linked to "imperial knowledge" may be posed. Since this question can only be answered with a view to the semantics of course titles, we assigned the attribute "imperial" to a course if

1. the word "Imperium" or "Reich" was used in the title,[16] and/or

2. the subject of the course was a specific empire or its law, history, or culture and/or

3. the subject of the course was a language or languages assignable to an empire, to countries with "imperial" history, or to the contemporary present, and/or

---

[14] The *Kronprinzenwerk*, a twenty-four-volume encyclopaedia financed with state subsidies, can be viewed as an attempt to emphasize the geographic and cultural diversity of the monarchy (cf. Judson, *Habsburg Empire*, 327–328).

[15] Cf. Thomas Winkelbauer, *Das Fach Geschichte an der Universität Wien: Von den Anfängen um 1500 bis etwa 1975* (Göttingen: V&R unipress, 2018). On historical science at the University of Graz, see Alois Kernbauer, "Grazer Geschichtsforschung von europäischem Rang," in *Kunst- und Geisteswissenschaften aus Graz: Werk und Wirken überregional bedeutsamer Künstler und Gelehrter. Vom 15. Jahrhundert bis zur Jahrhundertwende*, ed. Karl Acham (Vienna: Böhlau, 2009), 559–576.

[16] Country names alone were not classified as "imperial" in our analysis.

4. the course imparted "cohesion knowledge," which is defined as knowledge related to the cohesion between parts of empires.[17]

It is quite obvious and a result of the exploratory character of our investigation that we conceded a certain amount of analytic potential by combining semantic (1, 2) with analytic (3, 4) terminology at the level of our indicators. What is more, the binary code of "imperial" vs. "not imperial" frequently seemed too coarse a framework for the classification of university courses. We therefore also decided to evaluate the "degree of imperiality" and thus differentiate between "latently imperial" and "fully imperial" courses: courses bearing the name of a concrete empire or the term "Reich" or similar words in their title were considered "fully imperial," while those whose titles merely contained a certain reference to a specific empire were considered "latently imperial." In the summer semester 1906, for example, a course on Austrian History was offered at the University of Vienna; since "Austrian" refers (only) to a part of the Habsburg Empire, we classified this title as "latently imperial." The course on Austrian Imperial History,[18] on the other hand, was classified as "fully imperial." We also treated the attribute "German" in political contexts accordingly against the background of the making of the German Empire. Wherever no explicit differentiation is made between "latent" and "full" in the following, the text refers to the sum total of both types.

In order to assess developments across different political circumstances, we chose to analyse course catalogues for the years 1866, 1886, and 1906, which represent the periods before and after the Austro-Hungarian Compromise, the phase of high imperialism, and the crisis-laden run-up to the First World War; the selection of the catalogues of the Universities of Vienna and Budapest allows a comparison between institutions that the working hypothesis assumes were under the influence of differing political preferences—namely for preservation of the empire in Vienna and for imperial nationalization ("Magyarization") in Budapest respectively.[19]

---

[17] In this case, delimitation problems to do with country names were unavoidable. By way of example, let us consider the philosophical faculty of the University of Vienna: the course on Serbo-Croatian offered in all three semesters surveyed was classified as cohesion knowledge; within the Dual Monarchy, Serbo-Croatian was among the minority languages, and its development and spread as standard language for various small nations had been intensively promoted by the Habsburgs since the second half of the nineteenth century. Cf. Robert Greenberg, *Language and Identity in the Balkans: Serbo-Croatian and Its Disintegration* (Oxford: Oxford University Press, 2004), 18–24.

[18] This course was offered by the Faculty of Philosophy in 1906.

[19] The course catalogues on which the quantitative content analyses are based are *Öffentliche Vorlesungen an der k. k. Universität zu Wien im Sommer-Semester 1866*; *Öffentliche Vorlesungen an der k. k. Universität zu Wien im Sommer-Semester 1886*; and *Öffentliche Vorlesungen an der k. k. Universität zu Wien im Sommer-Semester 1906* for the University of Vienna. For the University of Budapest, the following three course catalogues were analysed *Magyar Királyi Tudomány-Egyetem Tanrende: Az MDCCCLXV-VI. Tanév. Nyáry szakára*;

Both universities had four faculties at the time: in addition to the departments of theology, medicine, and law, each included a humanities or philosophical faculty that was also home to disciplines such as mathematics, geology, and geography. Following a suggestion voiced during the discussion of this essay, we also included an analysis of the professional status of lecturers and the duration of the courses offered at the University of Vienna.[20] Percentages without decimal places are generally rounded.

Basic Information

Let us first examine the sheer numbers of university courses offered during the three summer semesters surveyed, as distributed across the respective four faculties.

Table 1: Number of courses offered in Vienna and Budapest (by faculty and semester)

Faculty	Vienna				Budapest			
	1866	1886	1906	Total	1866	1886	1906	Total
Theology	19	24	33	76	14	21	16	51
Law	40	66	75	181	21	56	70	147
Medicine	96	194	368	658	31	92	128	251
Philosophy	111	210	303	624	61	147	204	412
Total	266	494	779	1539	127	316	418	861

We immediately see a considerable increase—most noticeably between 1866 and 1886—in the number of courses offered in Vienna as well as in Budapest. With the exception of the Faculty of Theology in Budapest, which reduced its teaching after 1886, but still offered more courses than in 1866, there is a steady increase across all faculties. In both Vienna and Budapest, the medical and philosophical faculties offered significantly more courses than the theological and juridical faculties.

Magyar Királyi Tudomány-Egyetem Tanrende: Az MDCCCLXXXV-VI. Tanév. Nyáry szakára; and *Magyar Királyi Tudomány-Egyetem Tanrende: Az MDCCCCV-MDCCCCVI. Tanév. Második Felére.* During the three surveyed summer semesters, 1,539 courses were held at the University of Vienna and 861 at the University of Budapest. Not included in our analysis are the so-called *Fertigkeiten* (Proficiencies), which offered students the possibility of acquiring additional skills like fencing, singing, or stenography in addition to their regular curricula.

[20] Such an analysis was unfortunately not possible for the University of Budapest: while the course catalogues included the name of the person giving each course, they did not include the respective occupational status.

Determination of the percentages of courses classified as "imperial" (including "latently imperial") results in the following table:

Table 2: Courses identified as "imperial" (including "latently imperial" courses)

Faculty	Vienna				Budapest			
	1866	1886	1906	Total	1866	1886	1906	Total
Theology	1/19 5%	2/24 8%	2/33 6%	5/76 7%	2/14 14%	3/21 14%	4/16 25%	9/51 18%
Law	25/40 63%	41/66 62%	47/75 63%	113/181 62%	7/21 33%	19/56 34%	26/70 37%	52/147 35%
Medicine	0/96 0%	0/194 0%	1/368 0.3%	1/658 0.2%	0/31 0%	1/92 1%	0/128 0%	1/251 0.4%
Philosophy	28/111 25%	61/210 29%	83/303 27%	172/624 28%	25/61 41%	60/147 41%	76/204 37%	161/412 39%
Total	54/266 20%	104/494 21%	133/779 17%	291/1539 19%	34/127 27%	83/316 26%	106/418 25%	223/861 26%

The first thing to be noted here is that the percentage of courses classified as "imperial" remains quite stable across the years surveyed. In Vienna, the share is 20 percent in 1866, 21 percent in 1886, and 17 percent in 1906; in Budapest, it is 27 percent in 1866, 26 percent in 1886, and 25 percent in 1906. With slightly greater variance, this likewise applies to the ratio between "fully imperial" and "latently imperial" courses, which is roughly 1:2 throughout.

Table 3: Proportion of "fully imperial" and "latently imperial" courses taught in Vienna and Budapest

	Vienna				Budapest			
	1866	1886	1906	Average	1866	1886	1906	Average
Fully imperial	41%	30%	41%	37%	59%	30%	31%	35%
Latently imperial	59%	70%	59%	63%	41%	70%	69%	65%

The three years selected thus indicate no significant development in terms of the frequency of "imperial" courses. Also noteworthy is the fact that the average share of courses classified as "imperial" at the University of Budapest (26 percent) is around seven percent higher than that for Vienna (19 percent). This is partly due to the far smaller offering of courses in medicine in Budapest; the medical courses are almost entirely "non-imperial" at both universities across all three years.

The distribution of the "imperial" university courses across the faculties also shows that the medical and theological departments offered few or no relevant courses and the vast majority of "imperial" courses were instead offered at the law and philosophy departments. Taking all three years surveyed together, we arrive at values of 0.2 percent and seven percent for the Faculty of Medicine and the Faculty of Theology in Vienna respectively, while the value for the Faculty of Philosophy lies at 28 percent and the one for the Faculty of Law at 62 percent. The distribution in Budapest is slightly different: while the value for the medical department is similar at 0.4 percent, the 18 percent for the theological department is significantly higher than the corresponding percentage in Vienna (with the reason for this discrepancy being primarily that the Budapest theological faculty taught more languages of old empires); the values for the Budapest juridical and philosophical departments are closer together at 35 percent and 39 percent respectively. There are no noteworthy changes in the ratios at the individual faculties between the years surveyed; only the Faculty of Theology in Budapest saw an increase from 14 percent in 1866 and 1886 to 25 percent in 1906, but the absolute numbers of courses offered by this department are particularly low and the change therefore is statistically hardly significant.

Before looking more closely at the university courses classified as "imperial," let us investigate two possible factors for the degree of "imperiality" of academic teaching: the status of the lecturers and the temporal duration of the individual courses.

Status of Lecturers

During the discussion following the lecture in Bad Wiessee on which this essay is based, we formulated the hypothesis that membership of the teaching staff in different status groups might be connected to different preferences for "imperial" topics. This hypothesis could be confirmed for Vienna; Budapest could not be investigated in this regard due to the prohibitive amount of effort involved in collecting the necessary data. Analysis of the course catalogues showed that the members of the teaching staff at the University of Vienna could be divided into eight groups: four different groups of *Professoren* (professors; the groups are *o.ö. Prof.*, *a.ö. Prof.*, *a.c.r.p.o. Prof.* and *a.c.r.p.e. Prof.*), *Doktoren* (doctors), *Privatdozenten* (private lecturers with habilitation), *Lehrer* (teachers), and finally, all other teaching staff, whose values are not taken into consideration in this study.[21]

---

[21] The abbreviations *o.ö. Prof.* and *a.ö. Prof.* stand for *ordentlich-öffentlicher Professor* and *außerordentlich-öffentlicher Professor* respectively. Both titles (Amtsbezeichnungen) referred to professors who were permanently employed at the University of Vienna and thus enjoyed civil servant status. The difference between the two groups is that an *ordentlich-öffentlicher Professor* held his own chair at the university whereas an *außerordentlich-öffentlicher Professor* did not enjoy this privilege. An *a.c.r.p.o. Prof.* is essentially the same as an *o.ö. Prof.*, while the title of *a.c.r.p.e. Prof.* is equivalent to that of *a.ö. Prof.*

Collating the courses classified as "imperial" with these groups, we discovered a significantly—and surprisingly—inhomogeneous distribution. The highest percentage of "imperial" courses within a status group was offered by Lehrer (47 percent or 28 of 60 courses); the reason for this is the overrepresentation of this group in the field of philology, which dealt particularly intensively with manifestations of empires, especially with their languages.[22] They are followed by the o.ö. Professoren, who offered a total of 519 courses during the three summer semesters 1866, 1886, and 1906, 24 percent of which were classified as "imperial." In third place are the Privatdozenten with a share of "imperial" courses of 15 percent (82 of 543 courses), closely followed by the a.ö. Professoren (12 percent "imperial" courses or 36 of 295). Significantly fewer "imperial" courses were held by a.c.r.p.o. Professoren at 6 percent (2 of 31).

Closer examination of the developments within the individual status groups reveals interesting results. At the Faculty of Law, the percentage of "imperial" courses in the status group of the o.ö. Professoren increased over the years from 59 percent (1866) to 60 percent (1886) and finally to 73 percent (1906). The same applies to the Faculty of Philosophy, where the share of "imperial" courses taught by the o.ö. Professoren increased from 13 percent (1866) through 24 percent (1886) to 28 percent (1906).[23] Neither in the juridical nor in the philosophical department, however, did the percentages of "imperial" courses in the status group of the o.ö. Professoren increase significantly, nor did they exceed the average of the respective faculty. Among the a.ö. Professoren, on the other hand, the percentages declined between 1866 and 1906.[24] This can be interpreted as a sign that "imperial" topics were slowly shifting from the realm of the more specialized courses, which tended to be held by staff of lower statuses, into the focus of the—naturally not undivided—attention of the professors in ordinary.

Among the Privatdozenten, who offered a total of 543 university courses in the years analysed, we can see with regard to the Faculty of Law that the share of "imperial" courses in this status group across all three years (69 percent) was slightly above the faculty average of 62 percent. At the Faculty of Philosophy, the

---

[22] The value for the Doktoren was even higher at 75 percent, but lecturers with this status only offered a total of four courses. The other extreme in this regard were the a.c.r.p.e. Professoren, who offered a total of three courses, none of which were classified as "imperial."

[23] For the Faculty of Theology, the percentages are not very meaningful due to the low absolute numbers of courses classified as "imperial." It can nevertheless be stated that the o.ö. Professoren at the theological faculty offered practically no "imperial" courses (6 percent or two of 36 courses across all years surveyed). The only course classified as "imperial" at the Faculty of Medicine was offered in 1906 by an a.ö. Professor.

[24] For the Faculty of Law, the values correspond to 50 percent (1866), 56 percent (1886), and 23 percent (1906), with a total of 35 courses; for the Faculty of Philosophy, the values are 83 percent (1866), 20 percent (1886), and 25 percent (1906), with a total of 70 courses offered by a.ö. Professoren.

values were closer together: private lecturers offered 27 percent of their courses as "imperial" courses, whereas the faculty average was 28 percent. In a diachronic perspective, this comparatively strong variance can be explained with the relatively small absolute numbers (only 51 courses at the juridical and 168 at the philosophical faculty by Privatdozenten in total across all three years). The plausible hypothesis formulated during the discussion on the lecture in Bad Wiessee prior to analysis of the corresponding data, namely that the more frequent selection of "imperial" course topics by private lecturers might be owed to their career aspirations, could not be corroborated.

Duration of Courses

The duration of university courses was a further variable to be investigated. This is not because of a highly unequal distribution of the course offering between professors and private lecturers. However, o.ö. Professoren taught an average of 3.42 hours per week at the University of Vienna across all three years analysed, while a.ö. Professoren taught 3.34 hours, and Privatdozenten taught 2.9 hours. This average was the result of wide variation even among the course offerings of o.ö. Professoren.

The o.ö. Professoren at the philosophical faculty offered primarily two-hour courses across the three years in question (104); in addition, they also offered 49 three-hour, 31 four-hour, 33 five-hour, three six-hour, and two eight-hour courses. Upon collating the offered "imperial" and "non-imperial" courses with their respective durations, we found that "imperial" teaching made up 27 percent of the total time. This value is only one percentage point below the faculty average of the number of "imperial" courses held by the professors in ordinary.

The situation is similar for the Faculty of Law, where the average share of hours spent by o.ö. Professoren teaching "imperial" topics was 65 percent— exactly the same as the average percentage of "imperial" courses. It should be noted, however, that the differences in the duration of courses at the juridical faculty were even greater than at the Faculty of Philosophy. For Privatdozenten, the collation of courses with their duration likewise shows no distortive effects at the juridical (69 percent "imperial" time units vs. 69 percent "imperial" courses) or the philosophical faculty (29 percent "imperial" time units vs. 27 percent "imperial" courses). This means that distinguishing between the number of courses offered and their respective duration makes no difference in regard to the exposure of students to "imperial" topics.

Interdisciplinarity

Courses like Marriage Law of the Austrian Empire (Eherecht des Imperii austriaci) offered by Franz Laurin at the theological faculty in 1866 or German

Imperial and Legal History (Deutsche Reichs- und Rechtsgeschichte) offered by Georg Phillips and Johann Adolph Tomaschek at the juridical faculty in the same year raise the question of whether "imperial" topics may have correlated with interdisciplinary approaches. On the other hand, interdisciplinary courses were apparently also offered without such "imperial" connotations—for example, the course Forensic Medicine and Criminal Psychology under Consideration of Existing Legislation (Gerichtliche Medizin und Criminalpsychologie mit Rücksicht auf bestehende Gesetzgebung), held by Hieronymus Beer in 1866 at the Faculty of Law, or the course History and Geography of the Widespread Diseases (Geschichte und Geographie der Volkskrankheiten) offered by Theodor Puschmann at the Faculty of Medicine in 1886. We considered interdisciplinarity worth investigating in this context because the assumption seemed plausible that empires may have been viewed as a topic so important or rewarding that their scientific elucidation justified a more complex form of subject constitution. The concept of interdisciplinarity our analysis was based on was formulated ad hoc: if the title of a course referred to a discipline outside of its own faculty or field, we classified it as "interdisciplinary."[25]

The share of courses thus categorized as interdisciplinary among the entire teaching offering was four percent for the University of Vienna and seven percent for the University of Budapest, with no major temporal developments discernible (Vienna: 1866: three percent, 1886: four percent, 1906: four percent; Budapest: 1866: eleven percent, 1886: seven percent, 1906: seven percent).[26] The distribution of these courses across the faculties exhibits a significant non-uniformity, however: at the theological departments, the mean values for the three years surveyed were 15 percent for Vienna and 20 percent for Budapest, at the juridical departments eight percent (Vienna) and 14 percent (Budapest), at the medical departments four percent (Vienna) and six percent (Budapest), and at the philosophical departments only 0.5 percent (Vienna) and three percent (Budapest). Although the values for the individual faculties differ slightly between Vienna and Budapest, their order is the same in both cases.

In Vienna, a noteworthy interlacing of interdisciplinarity and "imperiality" can be discerned only[27] at the Faculty of Law in courses on the historic dimension of imperial constitutions, for example, in 1866 in the courses on German

---

[25] Exempted from this approach are courses on church history at the Faculty of Theology; they were not classified as interdisciplinary because church history was considered constitutive for the teaching of the subject as well as for the Christian Church as an institution. As classification was disputable in several cases, the data set deposited at the repository contains the respective entries for critical assessment and, if necessary, supplementary statistical analysis with changed data.

[26] When examining the faculties individually, the values for the Faculties of Theology and Law of the University of Vienna fluctuate quite strongly, while those for the corresponding faculties in Budapest change only very little.

[27] The higher values for theology can be explained with the discipline's interest in art history, cultural history, and philosophy.

Imperial and Legal History (Deutsche Reichs- und Rechtsgeschichte) taught by Georg Phillips and Johann Adolph Tomaschek or on Recent German Constitutional History (Neuere deutsche Verfassungsgeschichte) by Hugo Ritter von Kremer-Auenrode, or in 1886 in the course Imperial History (Reichsgeschichte) taught by Heinrich Siegel and Johann Adolph Tomaschek. The situation was different in Budapest, however, where a course on European History of Law (Európai jogtörténet) not classified as "imperial" was taught by József Illés in 1906. In the same year, at the Budapest Faculty of Philosophy János Krcsmárik offered a General Introduction to religious law in Turkish Civil Legislature (A török polgári törvénykönyvnek vallásjogi általános bevezető)—a course classified as "imperial" due to the fact that it dealt with the language of an empire. Overall, while some interdisciplinarity is visible in the course catalogues of the Universities of Vienna and Budapest, its correlation with "imperiality" is weak unless established through engagement with the languages of empires. In Vienna, it is noticeably focused on the study of the German Empire.

"Imperialities"

It is now time to examine the university courses classified as "imperial" in more detail.

Faculties of Law in Vienna and Budapest

At the Viennese Faculty of Law, the high percentage of "imperial" courses resulted primarily from the large number of courses dealing with the laws of various empires. In symbolic first place in all the course catalogues examined was Roman law, including the Pandects, though it was quantitatively surpassed by Austrian law. The law of the German Empire (e.g., 1866: Georg Phillips and Johann Adolph Tomaschek, *Deutsche Reichs- und Rechtsgeschichte*) was not taught very frequently, but still more often than English law (1886, Victor Waldner). We classified courses on the law of parts of the Habsburg Monarchy, for example on Hungarian constitutional law, Hungarian law, and the law of Bohemia and Moravia (1866: Anton Veghy, Johann Adolph Tomaschek) as cohesion knowledge. The Faculty of Law also offered teaching on Statistics of the Austrian Imperial State (1866: Leopold Neumann, *Statistik des österreichischen Kaiserstaates*) and the history of law of the Frankish Empire (1906: Emil Goldmann).

At the Faculty of Law in Budapest, Roman law represented a smaller share of the total offerings than in Vienna (1866 and 1886: Pál Hoffmann, *Római magánjog*; 1886: Ágost Pulszky, *Jog- és állambölcsészet*; Lajos Takács, *Római örökjog*; Gusztáv Schwarz, *Pandekták – római dologbeli jog*; 1906: Tamás Vécsey, *Római magánjog, tekintettel a pandectákra*; Marton Szentmiklósi, *Római jog*; Zoltán Pázmány, *Római kötelmi jog általában*). On the other hand, the relative importance of Austrian law ("cohesion knowledge") was greater—albeit at a lower base level—than that of Hungarian law in Vienna (e.g., 1866: János Baintner, *Ausztriai magánjog*; Sándor Konek, *Ausztriai birodalom*; 1886 Gyula Antal, *Ausztriai általános polgári magánjog*; 1886 and 1906: Gyula Sághy, *Ausztriai általános magánjog*). In 1906, more teaching was offered on the legal relationship between Austria and Hungary (Károly Kmety, *A magyar-osztrák dualizmus*; Ödon Polner, *Magyar közjog (különösen Magyarország és Ausztria kapcsolati és a társországok közjoga)*. Finally, there were courses on English law as well: in 1886 on the English constitution (Gyula Kautz, *Az angol alkotmány jelen állása szerint*) and in 1906 on English social policy (Manó Somogyi, *Angol szociálpolitika*).

Faculty of Philosophy in Vienna

The high percentage of courses classified as "imperial" at the philosophical department of the University of Vienna is owed to its curriculum's intensive engagement with the history, culture, and languages of empires. In 1866, for example, we find Roman history (Joseph Aschbach), the history of the Austrian Empire (Albert Jäger), German Historical Sources (Ottokar Lorenz, *Deutsche Geschichtsquellen*), Territories of the Austrian Imperial State and its Cultural Circumstances (Joseph Lorenz, *Gebiete des österreichischen Kaiserstaates und seine Culturbedingungen*), or Cultural Circumstances of the Russian Empire and its Relations to Western Europe and East Asia (Vincenz Klun, *Culturverhältnisse des russischen Reiches und dessen Beziehungen zu West-Europa und Ostasien*). Having barely exhibited any "imperial" contents in 1886, the science of history offered more "imperiality" in 1906, for example, Austrian Imperial History (for Jurists) (Gustav Turba, *Österreichische Reichsgeschichte* [für Juristen]), History of the Austrian Central Administration (Heinrich Kretschmayr, *Geschichte der österreichischen Zentralverwaltung*), Austrian History (Alfons Dopsch, *Österreichische Geschichte*),[28] and Overview of Russian Constitutional and Administrative History (Hans Uebersberger, *Überblick über die russische Verfassungs- und Verwaltungsgeschichte*) as well as courses on the Roman history of the imperial period and Greek coins of the imperial period (Wilhelm Kubitschek).

Imperial history was important in the teaching of history during the period under scrutiny as well as beyond it all the way up to the Second World War.

---

[28] Dopsch, who was appointed professor in ordinary for general and Austrian history in 1900, was to represent Austrian history especially in the areas of constitutional, administrative, and economic history. His professorship, which was tailored specifically to this purpose, was dedicated to Austrian Imperial History; cf. Winkelbauer, *Das Fach Geschichte an der Universität Wien*, 123 f. with reference to Pavel Kolář, *Geschichtswissenschaft in Zentraleuropa: Die Universitäten Prag, Wien und Berlin um 1900*, Halbband 2 (Leipzig: Akademische Verlagsanstalt, 2008), 301.

Against the background of the tradition of the professors of imperial history or "Reichs-Historie" (meaning the history of the Holy Roman Empire and its institutions), which had been active since the second half of the eighteenth century, the concept of empire was relevant in the work of many influential academics. Theodor Sickel for instance, published a collection of sources from archives and libraries of the "imperium Austriacum" (1858–1882) along with sources on the history of German/Roman kings and emperors; Emil von Ottenhals contributed to the *Regesta Imperii* as well as to a volume in the MGH series *Diplomata regum et imperatorum Germaniae* (vol. 8). Among the most important works by Alfons Huber was the book *Österreichische Reichsgeschichte: Geschichte der Staatsbildung und des öffentlichen Rechtes* (Austrian Imperial History: History of State Formation and Public Law, 1st edition 1895). Austrian imperial history also became a compulsory subject at the Faculty of Law and Political Science in 1893. Oswald Redlich likewise contributed to the *Regesta Imperii* (1898) and published the relevant work *Österreichs Großmachtbildung in der Zeit Kaiser Leopolds I.* (Austria's Development into a Great Power during the Time of Emperor Leopold I, 1st edition 1921). The interest in empires remained high, as evidenced by Wilhelm Kubitschek's *Imperium Romanum tributim discriptum* (1889) and the idea of a *translatio imperii* of the Holy Roman Empire to the Austrian imperial state advanced by Heinrich von Srbik.[29]

Cohesion knowledge dominated in the field of philology in 1866 with teaching in the languages of the monarchy: Slavic languages, "Bohemian," Italian, Hungarian, and Hungarian stenography (Franz Xaver Milosich, Alois Šembera, Adolph Mussafia, Cattaneo Giammaria, Johann Reméle, Johann Markovits) were the most important subjects. French and English as languages of contemporary colonial empires were likewise taught (Georg Lega, Joseph Gischig, Johann Högl) as were languages with an imperial tradition like Persian and Arabic (Jakob Goldenthal, Friedrich Müller, Adolph Wahrmund) or those with an alluring significance in the history of languages like Indogermanic (Anton Boller). Courses on Latin authors were generally advertised with the name of the author they dealt with—we adopted the differentiation between Latin (e.g., Horace) and Roman (law)[30] insofar as we did not classify courses in Latin philology as "imperial." Despite its considerable offerings in classical philology, however, the philosophical faculty achieved a large percentage of "imperial" courses.

A significantly greater number of philological courses were offered in 1886. Cohesion knowledge remained important, but after the Austro-Hungarian

---

[29] Cf. Winkelbauer, *Das Fach Geschichte an der Universität Wien*, passim.
[30] In 1886, however, Tacitus was offered as a "Roman author" (Max Büdinger) and classified accordingly. In 1906 Eugen Bormann offered a course on Literature of the Roman Imperial Period in the Greek Language (Literatur der römischen Kaiserzeit in griechischer Sprache) and Edmund Hauler offered another on the History of Roman Literature (not "Latin"; *Geschichte der römischen Literatur*).

Compromise and the cession of Italian territories, it now related only to the Slavic languages (Johann Leciejewski, Ferdinand Menčík). French and English, and England in general gained massively in significance (Jacob Schipper, Adolph Mussafia, Ferdinand Lotheissen, Wolfram Zingerle, Johann Alton, G. G. Bagster), and the field of Sanskrit, New Persian, and the Indian languages likewise grew (Friedrich Müller, Georg Bühler, Eugen Hultzsch, Jakob Polak). The connection between this realm of older South and West Asian empires and former and still existing empires at the fringes of Europe was established with courses in Greek, Armenian, Lithuanian, Arabic, and Turkish (Joseph Sklenař, Rudolf Mehringer, Josef Karabacek, Adolph Wahrmund). Russian was now taught as well (Johann Glowacki), and the offerings were also extended to include the languages of other old empires: Welsh, Babylonian, and Assyrian-Babylonian. The Ethiopian Empire, the Egyptian Empire (with hieratic script and hieroglyphs), and the Byzantine Empire could likewise be studied (Johann Hanusz, Heinrich Müller, Leo Reinisch, Jacob Krall).

In 1906, the philological teaching was extended even further—not exclusively (Norwegian: Rudolf Much), but certainly substantially in the area of languages of old and new empires. In terms of old empires, Assyrian (Friedrich Hrozný); Old Babylonian and the Armana tablets (David Heinrich Müller); Ethiopian (Maximilian Bittner); Armenian (Maximilian Bittner); Hebrew (August Haffner); Persian and New Persian (Adolf Wahrmund, Maximilian Bittner); Arabic (Rudolf Geyer, Josef Karabacek, Adolf Wahrmund, David Heinrich Müller, August Haffner); and Syrian (David Heinrich Müller) were now offered. Sanskrit continued to be taught as well (Leopold Schroeder). Among the languages of the contemporary colonial empires, English (Jacob Schipper, Rudolf Brotanek, Francis H. Pughe, G. G. Bagster) had surpassed French (Wilhelm Meyer-Lübke, Armand Rey, Marc Gratacap) in importance. The languages of the former Spanish Empire (Wilhelm Meyer-Lübke, Rudolf Behr) and the oldest still existing empire, China, were new in the curriculum (Franz Kühnert).

Languages in the area of cohesion knowledge were also increasingly being taught (again) in 1906: Old Church Slavic (Wenzel Vondrák); South Slavic (Josef Konstantin Jireček); Serbo-Croatian (Milan Ritter von Rešetar); "Bohemian" (Ferdinand Menčik); Romanian (Sextil Puşcariu); Italian (Philipp August Becker, Edgardo Maddalena); Hungarian (Julius Stockinger) and, in the field of interest of the Habsburg Monarchy, modern Greek (Eugen Somarides).

Faculty of Philosophy in Budapest

At the Philosophical Faculty of the University of Budapest, the subject of history contributed only a little in terms of "imperial" courses during the summer semester of 1866, for instance with Archaeology of Barbarian and Roman Statues (Flóris Rómer, *A barbár és római műemlékek régészte*) or various courses on

Hungarian history (Árpád Alajos Kerékgyártó, *Magyarország története*; Ferenc Toldy, *A magyarok története*; Árpád Horvát, *Magyar oklevéltan*). Twenty years later, the few courses on Hungarian history (Árpád Alajos Kerékgyártó, e.g., *Magyarország újkori története*; Árpád Horvát, *A magyar oklevéltan folytatólag*) were supplemented with several on the history of the Roman Empire (Károly Torma, *Római régiségek*; József Hampel, *Római történelem Nagy Constantinus óta*), on German and Spanish history (Aladár Ballagi, *A németalföldi szabadságharcz története*; *Don Carlos története*), or on the history of the resistance against the Turks (Lajos Szádeczky, *A visszafoglalási harczok története*).

In 1906, however, the teaching offered in the field of history was much more heavily oriented towards "monumental" historiography (Nietzsche) with a "Hungarian" and "imperial" character. Source Studies on Germanic Tribes under Roman rule (Gyula Lánczy, *Forrástanulmányok germán népek és a római birodalom*), Hungary in Roman Times, and History of the Romans (Bálint Kuzsinszky, *Magyarország a rómaiak idejében*; *A rómaiak története*), or Roman State Antiquities and The Life and Work of Julius Caesar (József Cserép, *Római államrégiségek*; *C. Julius Caesar élete és művei*) were the titles of courses. Other courses went so far as to unabashedly use the Roman terms for territorial and ethnic entities, for example, Pannonian and Dacian art (József Hampel, *Pannoniai és daciai emlékek*). The history of Hungary was told as one of heroes and of the founding of the empire: Hungarian-Croatian Contacts until 1105 (Antal Hodinka, *Magyar-horvát érintkezések 1105-ig*); The History of Hungary in the Age of the Hunyadis and the Jagiellons (Henrik Marczali, *Magyarország története a Hunyadiak és Jagellók korában*); The Development of Parliament after the Defeat at Mohács (Henrik Marczali, *Az országgyűlések fejlődése a mohácsi vész után*); The Era of Bethlen Gábor (Sándor Mika, *Bethlen Gábor kora*); and Bocskai István and the Treaty of Vienna (András Komáromy, *Bocskai István és a bécsi béke*). Courses dealing with non-Hungarian history were likewise related to empires, for example to Egypt (Ede Mahler, *Az egyiptomiak története*), to German History in the Age of the Hohenstaufen (Antal Áldásy, *Németország története a Hohenstaufok korában*) or to the Ottoman Empire (János Krcsmárik, *A mohamedán házasságkötés kánonja*).

A large number of courses with "imperial" characteristics were offered by the philological branch of the Faculty of Philosophy in 1866. We classified courses on German (Szende Riedl, *A német irodalom története*) and Serbo-Croatian (József Ferenc, *A szerb-horvát irodalomtörténet világi része*) literary history, on the West Slavic nations (József Ferenc, *A nyugati szlávfajok nyelvtani néprajza*), and on the Romanian (Sándor Román) and Italian languages (Antal Messi) as "cohesion knowledge." Turkish (Ármin Vámbéry), French (Alajos Mutschenbacher, Károly Collaud), and English (Lajos Lewis, James Egan) were the languages of contemporary empires on offer.

By 1886, the number of language courses had increased significantly. The languages spoken in parts of Hungary and its neighbouring countries remained important: besides Hungarian, they included German (Gusztáv Heinrich), Romanian and Romanian-Macedonian (Sándor Román), Italian (Antal Messi), and Czech (Oszkár Ásbóth). The languages of ancient Middle Eastern empires and of the Ottoman Empire were also taught: Arabic, Syrian-Chaldean, and Hebrew (Péter Hatala, *Szirus és khald olvasmányok*; *Arab olvasmányok*) as well as Persian, Turkish, Turkish-Tatar, and Kyrgyz (Ármin Vámbéry). The interest in the history of Hungarian was likely the reason for teaching in Lappish, specifically Sami (József Budenz, *Svéd- és finnmarki-lapp nyelv*) and Finnish (József Szinnyei) as well as a course on the comparison of Ugric languages and Yakutian, a Siberian Turkic language (József Budenz, *A magyar-ugor összehasonlító nyelvészetből*; *A jakut nyelv ismertetése*). This interest in the history of languages was further satisfied by courses in Sanskrit and "Indogermanic" (Aurél Mayr). French (Sándor Rákosi) and English (Lajos Lewis, James Egan) were taught as the languages of the two major contemporary colonial empires.

Philology continued to contribute significantly to the "imperiality" of the overall course offerings in 1906. It is noteworthy that courses on Latin topics were now more frequently referred to with the attribute "Roman" (Emil Thévrewk, *A görög és római lantos költészet*; Vilmos Pecz, *Romános középkori görög egyházi költő hymnusai*; Géza Némethy, *A római költészet ezüstkora*); there was also a course on Greco-Roman Music (Géza Molnár, *A görög-római világ zenéje*). Oherwise, the spectrum of teaching on the languages of old and new empires developed in 1886 remained largely the same: Arabic and Syrian (Ignaz Goldziher); Ottoman-Turkish (Ignác Kúnos); Persian and Sanskrit (Sándor Kégl); Armenian (Lukács Patrubány); Hungarian with reference to the history of language and literature (Zsigmond Simonyi, *Régi magyar nyelvészek*; Zsolt Beöthy, *A régi magyar költészet főepikusai*; Zsigmond Bodnár, *A XVI. és XVII. század magyar irodalma*; Cyrill Horváth, *A középkori magyar irodalom története*; Lajos Dézsi, *A régi magyar epikai költészet története*); and Finno-Ugric studies (József Szinnyei, *Finn-ugor összehasonlító nyelvészeti gyakorlatok*). The naming of a course Correct Hungarian (Gyula Zolnai, *A helyes magyarság*) indicates a high level of normative thinking in the area of Finno-Ugric philology, as did the focus on Sándor Petőfi, János Arany, and Mihály Vörösmarty, three Hungarian national poets (Frigyes Riedl, *Petőfi Sándor élete és művei*; *Gyakorlatok Arany lyrai és Vörösmarty epikus költészetéről*). Also in the catalogue were courses on languages of parts of the Hungarian population ("cohesion knowledge"): Slavic languages (Oszkár Ásbóth, *A szláv igék képzése és ragozása*) like Croatian and Serbian, specifically Serbo-Croatian (Ede Margalits), along with Romanian (János Ciocan, József Popoviciu, György Alexics—Alexics also lectured on the influence of Hungarian on Romanian, *Magyar hatás a román nyelvre*); German (Gedeon Petz, Gusztáv Heinrich); and Italian (Péter Zambra). Courses in French (Frigyes Medveczky, Lucien Bezard) and English (Arthur Battishill Yolland) were offered as well.

Faculties of Theology and Medicine in Vienna and Budapest

The medical and theological faculties contributed little to the "imperiality" of the teaching offered by the Universities of Vienna and Budapest. An academic engagement with tropical medicine might have been expected in the Habsburg Monarchy, but there was only a single course on Asian Cholera in Budapest in 1886 (Frigyes Korányi, *Az ázsiai koleráról*) and a course on Animal Plagues and Invasive Diseases with Special Consideration of Bacteriology and Parasitology in Vienna in 1906 (Johann Nepomuk Csokor, *Tierseuchen und Invasionskrankheiten mit besonderer Berücksichtigung der Bakteriologie und Parasitenkunde*).

The situation in theology was slightly different, as the departments afforded some space to the laws and languages of old empires. At the Faculty of Theology in Vienna, for example, Franz Laurin offered a course on Marriage Law of the Austrian Empire (Eherecht des Imperii austriaci) in 1866 as well as another on canon law with explicit reference to Roman law in 1886. In 1906, there was one course each on papal documents (Hirsch) and the Frankish Empire (Rudolf Ritter von Scherer). At the Budapest theological department, languages and history were apparently considered more important: in 1866, Arabic, Syrian, and Chaldean grammar (János Ruzsicska, *Syriai és chaldeai nyelvtan*) along with Hebrew (Ev. János Berger) and (as well as in 1906) Arabic, Syrian, and Chaldean (György Kayurszky, *Szyr és khald nyelvtan*) were part of the curriculum.

Conclusion

Research on the connections between empire and knowledge has hitherto been focused primarily on colonial circumstances. For the late Habsburg Monarchy, attempts to address this lacuna have generally consisted in examining the (ambivalent) effects of education policy. Against this background, we used highly simplified concepts of knowledge and empire to reduce the problem to the question of whether, to what degree, and in which faculties academic teaching offered in the course catalogues of the Universities of Vienna and Budapest for the summer semesters of 1866, 1886, and 1906 can be described as "imperial." We made this classification if the word "empire" or equivalent terms were used in course titles, if courses related to old or new empires and/or their languages, and/or if courses offered "cohesion knowledge" like the law of a part of the Habsburg (or a different) empire. Due to the high importance of semantics, we also assessed an "imperiality" that was only implied in context ("latently imperial"); the ratio between "manifest" (ca. 36 percent) and "latent" (ca. 64 percent) was around 1:2.

The data thus collected shows that around 20 percent of the courses in Vienna and around 26 percent of those in Budapest should be considered manifestly or latently "imperial," but also shows that—although the surveyed time period included the period of high imperialism—there was no noticeable development in the frequency of "imperial" topics. The majority of relevant academic courses were offered at the Faculties of Law and Philosophy. For Vienna, our survey showed that the degree of "imperiality" varied with the occupational status of the teaching staff. The courses offered by *Lehrer*, who frequently taught languages (of empires), exhibited the highest ratio of "imperiality." O.ö. Professoren offered "imperial" courses somewhat more often (with a diachronically increasing tendency) than a.ö. Professoren (with a diachronically decreasing tendency) (24 percent vs. 12 percent overall); the value for Privatdozenten (15 percent) was slightly greater than that for a.ö. Professoren. The significant variance in the duration of courses made no difference in terms of their "imperiality," however. The phenomenon of interdisciplinary teaching played a rather marginal role at both universities, barely correlating with the feature of "imperiality" outside of the teaching of languages—most noticeably in the field of history of law of the German Empire.

Courses at the Faculties of Law in Vienna and Budapest classified as "imperial" dealt primarily with the law of the Roman Empire, that of the Austrian Empire and, to a lesser degree, that of the German Empire. For Vienna, teaching on the law of various parts of the empire (Hungary, Bohemia, Moravia) was considered "cohesion knowledge"; in Budapest, Roman and Austrian law was likewise important, as were matters pertaining to the relations between the two parts of the empire. The large numbers of "imperial" courses at both philosophical departments resulted from the great significance placed on the history of empires: ancient Rome, the Austrian Empire, the German Empire (often with its preceding empires), and the Russian Empire were topics of interest. In Budapest, the Faculty of Philosophy also taught Hungarian history as imperial history, with a historical perspective on its neighbouring countries and the Finno-Ugric family of languages.

Although courses on Latin were only classified as "imperial" if their titles included the term "Roman" (which was rarely the case, albeit with a slightly increasing tendency over time), the teaching of languages of old and new empires (even without Latin) contributed significantly to the presence of "imperiality" in the course catalogues; what is more, the years 1886 and 1906 saw a distinct increase over 1866 in the respective offerings. Both faculties offered courses in the important languages of the two monarchies: various Slavic languages along with Hungarian, German, and Italian formed the core, which was supplemented with French and English as the languages of the major contemporary colonial empires as well as with Turkish and the languages of older empires, particularly those of the Near East. Hence in Vienna, students could study the languages of Egypt, Assyria, Persia, Ethiopia, the Caliphate, the eastern Roman Empire, Russia, Spain, and China. In Budapest, the spectrum in terms of the old empires was not quite as broad, but in exchange, Turkish and Romanian were taught more intensively and Armenian, Arabic, and Turkish were also offered.

The overall result is the following: the teaching offered by the Universities of Vienna and Budapest in the period surveyed transported references to empires to a significant extent primarily in the Faculties of Law and Philosophy respectively; the law, history, and languages of empires were key to the share of "imperial" courses in this context. Especially noticeable is the prominence of the Roman Empire—well ahead of the contemporary colonial empires, the Austrian Empire and the old empires of the Near East: almost a quarter (23 percent) of all university courses classified as "imperial" engaged with the empire whose "best poet" (John Dryden) had perhaps described "imperial" rule the most succinctly, *pacis imponere morem,/parcere subjectis et debellare superbos* (Verg. Aen. VI, 852 f.).[31]

Translated from German by Stephan Stockinger.

---

[31] Roland Gregory Austin, *P. Vergili Maronis Aeneidos liber sextus, with a commentary* (Oxford: Clarendon Press, 1977 [ND 1992]), 263, translates (in the commentary, abridged) the passage as: "these skills you shall have, to set the stamp of civilized usage upon peace, to be merciful to the submissive, and to crush in war those who are arrogant" and interprets the last lines as a "final definition of the fusion of pax with imperium."

Andrej Andreev

BETWEEN SCIENCE AND EMPIRE
Superintendents in Russian Universities during the First Half
of the Nineteenth Century

Historiographical Observations

Researchers who study the history of the university usually recognize that ideas about this institution are subject to change. In fact, across the centuries-long existence of the European university, these ideas could hardly retain one and the same meaning.[1] Certain terms were handed down from generation to generation but between the twelfth and the twenty-first centuries, the meaning of these terms sometimes changed. What the very concept of "university" meant changed fundamentally—moving from the designation of a corporation, a guild of scholars (universitas magistrorum et scholarum), to a new meaning connected with the unity of teaching and research (universitas litterarum).[2]

Naturally, the substance of university posts changed over time. In the Middle Ages, "rector" and "dean" did not mean exactly what they meant in the nineteenth century, and the men who held those offices in the Middle Ages did not perform anything like the same functions that rectors and deans were expected to carry out in the nineteenth century, or indeed the functions they have in the present-day university. But it can be said with equal confidence that the meaning of such concepts and terms changed when ideas about universities were transferred from one country to another. Moreover, such changes let us understand not only the transfer itself but also the adaptation which takes place in each new country against the backdrop of different conditions, in a different environment with its own well-established ideas and developed traditions of education.

---

[1] On the significance of the "history of concepts" for the history of science and, in particular, for the history of higher education, see Michael Eggers and Matthias Rothe, eds., *Wissenschaftsgeschichte als Begriffsgeschichte: Terminologische Umbrüche im Entstehungsprozess der modernen Wissenschaften* (Bielefeld: Transcript, 2009); Reinhart Koselleck, *Begriffsgeschichten: Studien zur Semantik und Pragmatik der politischen und sozialen Sprache* (Frankfurt am Main: Suhrkamp, 2006); Hans Erich Bödeker, *Begriffsgeschichte, Diskursgeschichte, Metapherngeschichte* (Göttingen: Wallstein, 2002).

[2] Hartmut Boockmann, *Wissen und Widerstand: Geschichte der deutschen Universität* (Berlin: Siedler, 1999), 13 f.

All of this is extremely important for an understanding of the process through which the university was established in Russia. Along with the idea of the university, its specific attributes were transferred to Russia, in particular the administrative positions that formed the backbone of the system of administering the corporate body of the university. The words "curator" (kurator), "rector" (rektor), and "dean" (dekan) entered the Russian language and became part of everyday administrative practices. At the same time—as happens by the way in every translation from one language to another—an understanding and a misunderstanding arose. On the one hand, these terms carried the original meanings that came from the countries where the transfer originated (in our case, from universities in the Protestant lands of Germany in the second half of the eighteenth century to the beginning of the nineteenth century). On the other hand, these terms were "misunderstood" in the sense that their initial meanings were replaced by other meanings which issued purely from the life of the Russian state.

This essay will examine how the post of the superintendent (popechitel') was adapted in Russian universities. The superintendent was the official of the Ministry of National Enlightenment (Ministerstvo Narodnogo Prosveshcheniia) who was directly responsible for the connection between "science and empire" during the nineteenth century. Historically, the equivalent of this position was the curator (kurator), who first appeared in German *Reformuniversitäten* in the eighteenth century. In Russian universities, however, this position received perhaps more significance than in its German analogues, from whence it was borrowed.

In fact, the Russian scholarship on the history of universities shows evidence of this. In it, a great deal of attention has traditionally been given to the role of curators and superintendents. It suffices to say that in *A History of Imperial Moscow University* (Istoriia Imperatorskogo Moskovskogo universiteta, 1855), the first generalizing historical work about Moscow University, which was written for its hundredth anniversary, the years in which various curators and superintendents were active were chosen as the starting points for periodization. At the same time, the custom of dividing superintendents into "progressives" who made a positive contribution to the university's development and "reactionaries" whose activity was harmful to the university already arose at the end of the nineteenth century, in the framework of the so-called "liberal historiography."[3]

During the Soviet period all superintendents received the epithet "reactionary" in the literature, and it was argued that the essence of their activity consisted in suppressing the collegial governing bodies and restricting the al-

---

[3] For more detail about how the role and the significance of superintendents has been assessed in the historiography, see S. I. Posokhov, "Popechiteli Khar'kovskogo uchebnogo okruga pervoi poloviny XIX veka: Popytka reabilitatsii," *Universitates-Universiteti: Nauka i prosveshchenie* 39, no. 4 (2009): 40–50, here 40–42.

ready incomplete "university autonomy." The "infringement" of this autonomy was seen as an a priori, inalienable property of the autocratic state, and in this conceptualization the superintendent was the embodiment, the living presence, of this autocratic power. The following quotation from a work written in the 1950s is revealing: "Universities were virtually under the complete control of the superintendent, who was the guardian and keeper of the interests of the autocratic, serf-owing system, in molding and educating the younger generation."[4] Another study from that time claims that the superintendent's main task was to establish a "harsh regime" in order to "transform the university into a hotbed of reaction."[5]

We note that in some contemporary work about Russian universities it is still possible to encounter an interpretation of this kind, although in its softest possible formulation, where there is talk of the superintendents systematically "trampling the professors' rights," "oppression of the university council," "autocratic decisions," "entrenching bureaucratism," and the like. But at the same time, other tendencies are also characteristic of the Russian historiography of the late twentieth and twenty-first centuries. Among these, the personified approach should be singled out, that is, paying attention to the character and world view of a particular university administrator, whose activity is understood by studying his own views on education. A focus on the superintendent's personal efforts to be a guardian of the sciences and the respect which he showed to professors and students are also features of this approach.[6]

---

[4] M. N. Tikhomirov, ed., *Istoriia Moskovskogo universiteta*, vol. 1 (Moscow: Izdatel'stvo Moskovskogo universiteta, 1955), 78–79.

[5] *Khar'kovskii gosudarstvennyi universitet im. A. M. Gor'kogo za 150 let* (Kharkov: Izdatel' stvo Khar'kovskogo universiteta, 1955), 55.

[6] In the contemporary historiography, the figure of Count Sergei Grigoryevich Stroganov is the classic example of this kind of reexamination and the creation of the image of the "ideal" superintendent. Analysis of the work done by the superintendents of Russian universities in the first half of the nineteenth century, especially S. G. Stroganov, has an important place in F. A. Petrov's comprehensive work, *Formirovanie sistemy universitetskogo obrazovaniia v Rossii*, 4 vols. (Moscow: Izdatel'stvo Moskovskogo universiteta, 2002–2003). The first biographical reference book about prerevolutionary superintendents in the historiography came out in Kharkov. See B. P. Zaitsev and S. I. Posokhov, *Popechiteli Khar'kovskogo uchebnogo okruga* (Kharkov: Universitet vnutrennikh del, 2000). In addition, several general books and articles should be mentioned. These include O. I. Zavgorodniaia, *Formirovanie pravogo statusa instituta popechitelei rossiiskikh universitetov v pervoi polovine XIX v.: Istoriko-pravovoe issledovanie* (Moscow: Sovremennaia ekonomika i pravo, 2006); K. A. Il'ina, "Informatsionnye resursy popechitelia uchebnogo okruga: Pervaia chetvert' XIX veka," in *Sankt-Peterburgskii universitet v XVIII–XX vv.: Evropeiskie traditsii i rossiiskii kontekst*, ed. T. V. Burkova et al. (St. Petersburg: Izdatel'skii dom Sankt-Peterburgskogo gosudarstvennogo universiteta, 2009), 37–45; T. V. Kostina, "Akademik S. Ia. Rumovskii i Kazanskii universitet: Istoriograficheskii kontekst," in *Akademiia nauk v istorii kul'tury Rossii XVIII–XX vekov*, ed. Zh. I. Alferov (St. Petersburg: Nauka, 2010), 81–101; I. V. Cherkaz'ianova, "Akademicheskie uchenie v roli gosudarstvennykh chinovnikov: Na primere popechitelei uchebnykh okrugov," in *Akademiia*

Curator or Superintendent?

In this essay, superintendents are analyzed as an official institution that grew out of thinking about how to adapt this institution during the transfer of ideas about universities from Western Europe to Russia. This approach makes it possible to step away from two a priori notions: first that superintendents were an "external" factor through which the state intervened in university affairs and second, that such intervention—which has been characterized as "oppression," "violation of laws," and with other epithets denoting the absence of freedom—reflected the specific nature of the Russian autocratic state. In fact, the opposite is true. A study of the transfer of certain European administrative practices and institutions and the adaptation of these in Russian universities shows that nothing in such interventions and in the position of the superintendent was connected to a specifically Russian style of autocracy.

This post appeared naturally, during the "modernization" of European universities. Universities began to receive additional funding from state (princely) treasuries in that period. Therefore, it was necessary to ensure that this money was spent in a timely and maximally effective manner to strengthen the university's scholarly foundation and educational process in order to make these correspond to the "needs of the state," as they were understood during the Enlightenment. The state handed this task over to the curator, who carried it out and, at the same time, gave the state feedback. That is, he was able to petition the sovereign about the university's urgent needs.

Moreover, the post of curator was not seen in any way as an office that was *external* ("alien," "hostile," and the like) to the university: it immediately became part of university administration, yet at the same time it connected the academic corporation with the highest level of power. This becomes especially clear from the fact that by virtue of the authority he had over the corporation, the curator occupied the place held by the medieval university chancellor, without whom key corporate procedures and ceremonies could not be carried out.[7] At the same time, while the chancellor's fundamental functions

---

nauk, ed. Alferov, 102–134. Cherkaz'ianova's essay includes an appendix listing the superintendents of the educational districts of the Russian Empire for 1803–1917. See also Posokhov, "Popechiteli Khar'kovskogo uchebnogo okruga pervoi poloviny XIX veka;" D. A. Tsygankov, "Universitetskii popechitel' v Rossii: Mezhdu tsennostiami evropeiskoi nauki i gosudarstvennoi sluzhby, vtoraia polovina 18–pervaia tret' 19 veka," in *Polietnichnost' Rossii v kontekste istoricheskogo diskursa i obrazovatel'nykh praktik XIX–XXI vv.*, ed. V. V. Afanas'ev et al. (Cheboksary: Izdatel'stvo Chuvashskogo gosudarstvennogo universiteta, 2010), 515–527. A first (candidate's) dissertation systematically analyzing the institution of the superintendent in Russian universities was successfully defended in 2007. See Ol'ga Ivanovna Zavgorodniaia, "Popechitel'stvo v istorii universitetov Rossii pervoi poloviny XIX veka" (Diss. na soisk. uch. step. kand. ist. nauk, Rossiiskii Universitet Druzhby Narodov, 2007).

7   Laetitia Boehm, "Cancellarius Universitatis: Die Universität zwischen Korporation und Staatsanstalt," in *Geschichtsdenken, Bildungsgeschichte, Wissenschaftsorganisation*, ed.

had a symbolic character, the curator's basic functions were directly administrative. As officials of the Russian Empire, curators first appeared in 1755 when Moscow University was founded. Superintendents replaced curators, and the position of "superintendent" existed in Russian universities from 1803 until 1917.

The name of the position was changed from "curator" to "superintendent" at the beginning of the nineteenth century, at the same time that the Ministry of National Enlightenment was established in the Russian Empire. This was not a simple renaming despite the seeming equivalence of the meaning of the words. (The Latin word *curator* is the equivalent of the Russian *popechitel'*.) Although by the very nature and origin of his position the eighteenth-century curator was a Russian state official, at the same time he acted as an independent figure in relation to other organs of state power. In his actual position he could be seen as a patron—an enlightened guardian of the sciences—and as an authoritative representative of the nobility who was responsible for the university's harmonious connections with the local cultured stratum. For example, the curator encouraged the nobility to contribute financially to the university, that is, to help finance university projects which were independent of the state.[8]

In nineteenth-century Russia, the superintendent was part of the ministry's hierarchical structure. He reported directly to the minister of national enlightenment, whom he was required to inform about his activities, and in this sense he had many more of the traditional attributes of a state official. The obligation to visit St. Petersburg regularly to attend meetings of the Main Administration of Schools (Glavnoe Pravlenie uchilishch) made the superintendent a link connecting the university, no longer with local society, but rather with the central, imperial power. The introduction of the position of superintendent in general should be considered a factor in the formation of a centralized and unified system of governance of Russian universities.[9]

An opposition of this kind, between the superintendent as the representative of the central organs of power in the university and curators as figures connected with the local university environment, showed itself clearly in the case of Dorpat University during the early years of the Ministry of National Enlightenment. An analysis of the relevant documents shows that even the new title for the position appeared in the ministry's norms in order to distinguish the holders of this post from those who had held the previous position of curator.

Laetitia Boehm, Rainer A. Müller, Winfried Müller, and Gert Melville (Berlin: Duncker & Humblot, 1996), 695–713.

[8] For more detail, see A. Iu. Andreev and S. I. Posokhov, eds., *Universitet v Rossiiskoi imperii XVIII–pervoi poloviny XIX veka* (Moscow: ROSSPEN, 2012), 239–247.

[9] Tsygankov, "Universitetskii popechitel' v Rossii," 523.

During the opening of Dorpat University in 1802, a conflict flared up between those who were directly responsible for organizing the university, that is, the curators, who represented the nobility of the three Baltic provinces (Estonia, Livonia, and Courland), who had founded the university on their own initiative, and the professors who had been invited to teach there. The latter were led by Georg Friedrich Parrot. Because of his influence on Alexander I, Parrot secured the adoption of the new university charter, the Act of Foundation of an Imperial University in Dorpat (Akt postanovleniia dlia Imperatorskogo universiteta v Derpte), which was signed on 12 December 1802.[10] The significance of this document lies in the fact that it created a new supreme supervisory authority in the person of a representative of the Ministry of National Enlightenment, the member of the Commission on Schools who, in accordance with the decree of 8 September 1802, was entrusted with "special care for this university" (paragraph 3).[11]

After that, for the sake of brevity, this official begins to be called the "member-superintendent" (chlen-popechitel') (paragraph 9) in the document.[12] Thus, both curators and the "member-superintendent" figure in the act that established Dorpat University. Moreover, the power of curators and the superintendent stemmed from two opposed sources: curators were chosen by the local nobility of Estonia, Livonia, and Courland provinces but the superintendent represented the central government, that is, the Ministry of National Enlightenment. This meant, as historian Villu Tamul emphasizes, that the adoption of the Act transformed Dorpat University from a regional educational institution (Landesuniversität), as it was initially envisioned, into an "imperial" institution (Reichsuniversität), in other words, an institution controlled by the empire's central organs of governance.[13] This was also how Parrot was able to weaken the local noble curators' power (in 1803 these posts were abolished altogether), especially since in St. Petersburg he soon man-

---

[10] A. Iu. Andreev, "Imperator Aleksandr I i professor G. F. Parrot: K istorii vozniknoveniia 'universitetskoi avtonomii' v Rossii," *Otechestvennaia istoriia*, no. 6 (2006): 19–30.

[11] Indeed, the decree of 8 September 1802 on the creation of a Commission on Schools (Komissiia ob uchilishchakh) states that: "The members of this commission will divide among themselves jurisdiction over all higher and lower schools existing in the empire by regions or by provinces and on receiving statements about the condition and performance of the affairs of the schools of their department, *they are especially obligated to care for the accomplishments of all educational institutions in disseminating education*, in accordance with their understanding of each department's needs and what would be advantageous for each department, which comprises several guberniia," *Polnoe Sobranie Zakonov* (hereafter PSZ), no. 20407, vol. 27 (St. Petersburg, 1830). [The italics are mine, A. Iu. A.]

[12] PSZ, no. 20551, vol. 27.

[13] Villu Tamul, "Die Dörptsche Universität: Landes- oder Reichsuniversität? Zum Verhältnis von Deutschbalten, Stadt und Universität im 19. Jahrhundert," in *Zur Geschichte der Deutschen in Dorpat*, ed. Helmut Piirimäe and Claus Sommerhage (Tartu: Universität Tartu, 1998), 85–110.

aged to ensure that Major-General Friedrich Maximilian Klinger, who had no connection whatsoever with the Baltic nobility, became the superintendent of Dorpat University while the Livonian count Gotthard Andreas Manteuffel, who had been eager to take up this post in Dorpat, was appointed as the superintendent of the far-off Kazan educational district.[14]

Thus, semantically, the name of the position—"superintendent"—emerged in a nontrivial manner. Although at first glance one might think that "superintendent" (popechitel') was simply a translation of the word curator into Russian (lat. *curare* = rus. *imet' popechenie*/to take care of) this was by no means the case. Instead, it came from the wording of the decree of 8 September 1802, which charged the member of the ministry to whom the educational district was entrusted "to *care* for the accomplishments" of the educational institutions which fell within his remit. The formulation of "member-superintendent" (*chlen-popechitel'*, lit. "member who cares for") in the Act of Foundation of Dorpat University followed from this. From there the name of the position (simply "superintendent") was transferred into the Provisional Rules of National Enlightenment of 24 January 1803, and this wording was used right up to 1917.

In terms of the adaptation of European university concepts, the introduction of the post of superintendent instead of curator undoubtedly signified a departure from the local understanding of the university, its tasks, social status, etc., and the recognition of its significance for the empire as a whole. This entailed the creation of the single, centralized system of university education in Russia, which arose in the first half of the nineteenth century.

Although the superintendent was mentioned repeatedly in legislation on the Ministry of National Enlightenment, to a significant degree, the superintendents of the early nineteenth century inherited from the curators of the eighteenth century a vagueness about their sphere of competence, an insufficient definition of their authority and duties. The legislative basis for their activity was essentially only two acts, the Provisional Rules (1803) and the University Statute of 1804. In the first of these two documents, only one paragraph (paragraph 20) is devoted to the superintendent's tasks. This paragraph points out that "the superintendent is responsible for improving all the schools of the district entrusted to him," and in general he is supposed to bring the "university and other schools [...] into a flourishing state."[15] As for the 1804 University Statute, the superintendent's powers were generally not discussed in any detail, since the structure of the statute was subordinated to a description of the features of the "autonomous" university corporation.

In terms of the construction of this "autonomy," judging by the text of the statute, the superintendent was supposed to act above all as an intermediary

---

[14] Friedrich Gustav Bienemann Junior, *Der Dorpater Professor G. F. Parrot und Kaiser Alexander I* (Tallinn: Franz Kluge Verlag, 1902), 168–169.
[15] *PSZ*, no. 20597, vol. 27.

between the university and the ministry: he was supposed to submit for the minister's approval a list of names of the faculty members (both the junior members and the professors chosen by the university Council), the names of administrative officers (the rector and the deans), financial reports, and the reports of the university Committee on the Administration of the Lower Schools, etc. The superintendent was also mentioned frequently in the statute in conjunction with extraordinary matters which went beyond the range of issues which the university council usually dealt with: "the council will immediately inform the superintendent about every special case and cases demanding a resolution" (paragraph 52). Thus, one can agree with S. I. Posokhov's conclusion that "instead of a formal definition of the superintendent's responsibilities the 1804 University Statute made an assumption about the informal practice of the authority he exercised in relation to the professoriate, whose rights were laid out in detail, unlike the rights of the superintendent himself."[16]

In practice, therefore, the adaptation of Alexandrian university reform occurred in such a way that the superintendent's activity took place "between the law and reality." The main component of this prevailing "reality" was the fact that the superintendent was seen as the immediate "head" (nachal'nik) of the university although no such meaning was explicitly recorded in any of the documents or instructions of the first decades of the nineteenth century. However, in comparison with the earlier position of the university curator, there were features of the superintendent's job that differed from eighteenth-century administrative practice. Above all, the superintendent was no longer able to present the interests of the university directly to the emperor. Although the 1804 statute gave every Russian university "imperial" status, at the same time, the universities' unmediated connection with the throne was severed since the right to report to the sovereign about higher education now belonged to the Ministry of National Enlightenment. Therefore, the minister's higher power limited the superintendent's "headship" over the university. With time, a power vertical of "university-superintendent-minister-emperor" was constructed; when controversies arose, however, the university could also appeal directly to the minister and, thus, in this configuration, from the beginning the foundation was laid for a potential "sphere of tensions" in which the professoriate was on one side and the superintendent on the other. In his relations with the local authorities, the superintendent did not have enough of the authority of the "head:" their relationships with each other were not spelled out in any way and the superintendent had to build them up on his own, either through his high rank or his authoritative position in the organs of noble self-government; without this support, the superintendent would find himself in a complicated situation. The history of the opening of Kazan University in January 1805 is very instructive here: its superintendent, Stepan Iakovlevich Rumovskii, who was a member of the St. Petersburg Academy of

---

[16] Posokhov, "Popechiteli Khar'kovskogo uchebnogo okruga," 42.

Sciences, had neither the authority that came from being a grandee nor the authority accruing to a high position. He had not had any contact with the provincial administration beforehand, and probably for that reason he decided to carry out the opening of the new university "quietly, privately, without pomp and publicity, without leaving the closed walls of the gymnasium building."[17]

A Typology of Superintendents in the Period when the University Statute of 1804 was in Effect

During the first years after the reform, a certain insufficiency in the superintendent's formal "authoritative" status was compensated for in that representatives of the Russian Empire's highest aristocracy, men who were close to the throne and the levers of supreme power, held these posts. One such early superintendent was Mikhail Nikitich Murav'ëv, who was the superintendent of Moscow University in 1803–1807 and was simultaneously the deputy minister of national enlightenment as well as the secretary of the Petitions Commission (Komissiia po priniatiiu proshenii na vysochaishee imia, or Commission on the Acceptance of Petitions to the Highest Name). He had previously been the future emperor Alexander I's tutor for the Russian language and history.[18] Another was Count Aleksei Kirillovich Razumovskii, the son of Ukrainian hetman, patron of the arts, and art collector Count Kirill Grigorjewitsch Razumovskii. He was the superintendent of Moscow University from 1807 to 1810, and in 1810 he was appointed minister of national enlightenment.[19] A third superintendent from the highest ranks of the aristocracy was Count Seweryn Potocki, a Polish magnate who was chamberlain to Alexander I before he assumed the throne.[20] In 1803, two of the young emperor's closest friends became superintendents. Prince Adam Czartoryski was a scion of the aristocratic Polish-Lithuanian Gediminovich family, representatives of which served in the government of the Polish-Lithuanian Commonwealth during the last years of its existence and carried out a number of reforms, including reforms in education. He became the superintendent of the Vilna educational district.[21] Nikolai Nikolaevich Novosil'tsev, who was the president of the

---

[17] N. P. Zagoskin, *Istoriia Imperatorskogo Kazanskogo universiteta za pervye sto let ego sushchestvovaniia, 1804–1904*, vol. 1 (Kazan: Tip. Imperatorskogo Kazanskogo universiteta, 1902), 50 and 59.

[18] A. Iu. Andreev and D. A. Tsygankov, *Imperatorskii Moskovskii universitet, 1755–1917: Entsiklopedicheskii slovar'* (Moscow: ROSSPEN, 2010), 475–476.

[19] Ibid., 603–604.

[20] Zaitsev and Posokhov, *Popechiteli Khar'kovskogo uchebnogo okruga*, 17–21.

[21] Daniel Beauvois, *Lumières et société en Europe de l'est: L'université de Vilna et les écoles polonaises de l'Empire russe, 1803–1832*, vol. 1 (Lille: Reproduction des Theses, Université Lille, 1977).

St. Petersburg Academy of Sciences in 1803–1810, was appointed as the superintendent of the St. Petersburg educational district.[22] Sergei Semënovich Uvarov, who was Aleksei Razumovskii's son-in-law, became the superintendent of the St. Petersburg educational district in 1811, and in 1818 he also became the president of the Academy of Sciences.[23]

These men were the first generation of superintendents, the most brilliant cohort of the entire nineteenth century. Their work made it possible to create new universities, to which they invited prominent European scholars (above all from Germany) to be professors. They set up university laboratories, observatories, museums, libraries, and a whole host of other facilities for academic work. Every member of the first generation of superintendents had extensive connections in the world of European science and culture. They corresponded with many scholars, writers, and people active in the world of education. Thus, Seweryn Potocki exchanged letters with Goethe about possible faculty members for Kharkov University, Mikhail Nikitich Murav'ëv corresponded with the leading professors of Göttingen University,[24] and Sergei Semënovich Uvarov corresponded with Wilhelm von Humboldt.[25] Aleksei Kirillovich Razumovskii created a unique botanic garden at Gorenki, his estate outside Moscow, which he planned to use as the basis for a "phytological society" that several well-known European naturalists, including Johann Christian von Schreber, the president of the Leopoldina (now called the German National Academy of Sciences), and Alexander von Humboldt, agreed to join.

In this way, the first generation of superintendents helped to bring Russian universities into the European network of scholarly contacts in which they themselves were involved. In general, they treated the university as their own "offspring" and indeed, in many ways the universities did owe their existence to the superintendents. Thus, at the beginning of the nineteenth century the European-style curator was replicated in Russia. This was a liberal patron, a "father of the university," who strove to make the sciences flourish and, through his patronage, raised the university's status in society—in other

---

[22] E. N. Filippova, "Nikolai Nikolaevich Novosil'tsev," in *Vo glave pervenstvuiushchego uchebnogo sosloviia Rossii: Ocherki zhizni i deiatel'nosti prezidentov Imperatorskoi Sankt-Peterburgskoi Akademii nauk, 1725–1917*, ed. E. I. Kolchinskii (St. Petersburg: Nauka, 2000), 105–106.

[23] Cynthia H. Whittaker, *Graf S. S. Uvarov i ego vremia* (St. Petersburg: Akademicheskii Proekt, 1999).

[24] For more detail about the superintendents' scholarly correspondence about selecting professors, see A. Iu. Andreev, *Rossiiskie universitety XVIII–pervoi poloviny XIX veka v kontekste universitetskoi istorii Evropy* (Moscow: Znak, 2009), 407–426.

[25] Otdel pis'mennykh istochnikov (Department of Written Sources, hereafter OPI, f. 17, op. 1, d. 85, l. 208–218 and d. 82, l. 146, Gosudarstvennyi Istoricheskii muzei (State Historical Museum, hereafter GIM).

words, these men are comparable to German university curators such as Gerlach Adolph von Münchhausen or Daniel de Superville.[26]

There was a notable change in the collective portrait of superintendents during the second half of the 1810s and in the 1820s. The liberal "father-founder" became a thing of the past, giving way to the "bureaucrat," an executive officer of the Ministry of National Enlightenment. People who had neither connections nor authority in the academic world were now ascending to the post of superintendent. Accordingly, in their eyes improving teaching or the university's scientific development had little or no value. The priority for this new, second generation of superintendents was to scrupulously follow the ministry's new policies. When Alexander Nikolayevich Golitsyn became the minister of religious affairs and national enlightenment in 1817, the so-called "dual ministry" became a stronghold of reaction to Enlightenment ideas, which were countered with universalizing, extra-confessional propaganda of "Christian education" and a mandate to "teach the truths of the faith" as the Russian educational system's main task.[27] An example that shows how quickly superintendents took up this duty is a statement made by Zakharii Iakovlevich Karneev, who was the superintendent of Kharkov University in 1817–1822. When Alexander I visited the university, Karneev declared that his goal was to "awaken the spirit of true religion in university students and school pupils."[28]

It should be noted that this was not simply a change in educational policy in Russia in the mid-1810s and 1820s but rather an ideological about-face on a larger scale—the formation of a kind of "anti-Enlightenment" project. This project was nourished by the ideological currents of the epoch of Romanticism, which in turn grew out of ideas about establishing the nation and the national values being constructed therein.[29] Ideological currents of this kind were characteristic of all European countries in one way or another but in Russia they took on a markedly more conservative coloration, which showed itself precisely in the sphere of education and created significant difficulties in adapting Enlightenment educational concepts there.

---

[26] On the Göttingen type of curator, see Walter Buff, *Gerlach Adolph Freiherr von Münchhausen als Gründer der Universität Göttingen* (Göttingen: Gesellschaft der Wissenschaften, 1937); Charles E. McClelland, *State, Society, and University in Germany, 1700–1914* (Cambridge: Cambridge University Press, 1980).

[27] See E. A. Vishlenkova, *Zabotias' o dushakh poddannykh: Religioznaia politika v Rossii pervoi chetverti XIX veka* (Saratov: Izdatel'stvo Saratovskogo universiteta, 2002).

[28] Zaitsev and Posokhov, *Popechiteli Khar'kovskogo uchebnogo okruga*, 24.

[29] See Alexander M. Martin, *Romantics, Reformers, Reactionaries: Russian Conservative Thought and Politics in the Reign of Alexander I* (DeKalb: Northern Illinois University Press, 1997); A. L. Zorin, *Kormia dvuglavogo orla... Literatura i gosudarstvennaia ideologiia v Rossii v poslednei treti XVII–pervoi treti XIX veka* (Moscow: Novoe Literaturnoe Obozrenie, 2001).

Even the change in the social profile of the second generation of superintendents is characteristic of this development: men like Aleksandr Aleksandrovich Pisarev, the superintendent of Moscow University (1825–1830), Egor Vasil'evich Karneev (1822–1825), and Vladimir Ivanovich Filat'ev (1830–1834), who were superintendents of Kharkov University, Mikhail Leont'evich Magnitskii (1817–1826) of Kazan University, and Dmitrii Pavlovich Runich (1821–1826) at St. Petersburg University were well off but, unlike the first generation of superintendents, they did not belong to the highest levels of the nobility and often reached the helm of power only after a long career in government service.

Despite the obvious differences, both generations of superintendents shared some traits. One is the fact that the superintendents exercised power "by correspondence." In the 1800s first-generation superintendents rarely visited their university cities (with the exception of the capitals). During the ten years of his activity as superintendent, the only time Rumovskii was in Kazan was to take part in the opening of the university. Although Count Potocki visited Kharkov University several times, he lived permanently in St. Petersburg and then Warsaw; Mikhail Nikitich Murav'ëv too usually spent his time in St. Petersburg, serving as the deputy minister of national enlightenment. Among the second generation of superintendents of Kharkov University, students remembered that Egor Vasil'evich Karneev "lived in Petersburg and we did not see him."[30] Alexey Alexeyevich Perovskii, who was the university's superintendent in 1825–1830, also lived abroad for a long time. Later he lived in Moscow, St. Petersburg, and at his estate near Chernigov, and from 1826 on he was never in Kharkov.[31] The Kazan superintendent Mikhail Leont'evich Magnitskii, one of the "pillars" of the anti-Enlightenment reaction, took an active part in working out the policy of the "dual ministry." In conjunction with this work, he was always in St. Petersburg.

For "governing by correspondence" to be effective, the superintendent needed a trusted proxy at the university through whom he could transmit his instructions and monitor the way they were being carried out. According to the logic of the system of university governance introduced by the 1804 statute, this person would be the university rector. However, the relationship between the superintendent and the person elected as the rector was not always harmonious, and with this in mind, the superintendent was interested in choosing the "closest" possible person, someone he "understood well." For Mikhail Nikitich Murav'ëv this was the physicist Peter Ivanovich Strakhov, who was passionate about advancing science. For Aleksei Kirillovich Razumovskii it was Ivan Andreevich Geim, the polymath who catalogued Razumovskii's library at Gorenki. Despite frequent conflicts at Kazan University,

---

[30] S. I. Posokhov, ed., *Kharkivs'kyĭ universitet XIX–pochatku XX st. u spogadakh ĭogo vikladachiv ta vikhovantsiv*, vol. 1 (Kharkov: Saga, 2008), 90.

[31] Zaitsev and Posokhov, *Popechiteli Khar'kovskogo uchebnogo okruga*, 36.

until the end of his life, Stepan Iakovlevich Rumovskii gave his unwavering support to "professor-director" Ilya Fedorovich Iakovkin, who, as an administrator, reproduced the commanding, bureaucratic style of the chancellery of the Academy of Sciences that Rumovskii knew so well from his time at the Academy.[32]

But alongside the official "superintendent-rector" relationship, in many cases trusting ties between the superintendent and a member of the university corporation developed outside the work-related hierarchy. The latter did not have to join the university council and hold an authoritative position on it, but the superintendent valued the opportunity to receive detailed, unofficial information about what was going on at the university from someone like this. This type of a member of the corporation could also act as a kind of an "influential figure" for the superintendent at the university, bypassing the official channels prescribed by the statute. At Moscow University, when Aleksei Kirillovich Razumovskii was the superintendent, the assistant professor Mikhail Trofimovich Kachenovskii played this role. In the words of a contemporary, Kachenovskii "strictly and, in his own way, sensibly, maintained order in the university."[33] Because Kachenovskii was also Razumovskii's secretary and then became the head of his chancellery, he himself took care of all of the superintendent's correspondence with the university council and the board: Kachenovskii wrote recommendations in Razumovskii's name, which did not produce an entirely favorable impression on the older professors, who were used to the way things had formerly been done.[34]

It would, however, be a mistake to interpret the presence of the superintendent's "trusted people" among the professors and their informal influence on the policies of the superintendents as a particularly Russian practice. It was in fact generally very characteristic of universities in the Enlightenment. It suffices to remember the prominent role played by Josef von Sonnenfels, the professor of management and administration of the public revenue (cameralistics) at the University of Vienna, in Austrian educational reform. Von Sonnenfels, who worked out the legal basis for the need to reduce the number of universities in the state, enjoyed great prestige among the members of the court's school commission (Hofschulkommision).[35] A well-known example from the history of Göttingen University accurately encapsulates the relation-

---

[32] E. A. Vishlenkova, *Kazanskii universitet Aleksandrovskoi epokhi* (Kazan: Izdatel'stvo Kazanskogo Universiteta, 2003), 82.

[33] M. P. Tret'iakov, "Imperatorskii Moskovskii universitet, 1799–1830," *Russkaia starina*, no. 7 (1892): 125.

[34] I. M. Snegirev, "Vospominaniia," *Russkii arkhiv* 4, no. 5 (1866): 752.

[35] Helmut Engelbrecht, *Von der frühen Aufklärung bis zum Vormärz*, vol. 3 of *Geschichte des österreichischen Bildungswesens: Erziehung und Unterricht auf dem Boden Österreichs* (Vienna: Österreichischer Bundesverlag, 1984), 197. In 1786, Catherine II turned to Sonnenfels for information about the structure of Austrian universities after the reforms.

ship Russian superintendents had with certain professors: during the end of the eighteenth century and the early nineteenth century, curators sent their instructions to the university and gave advice about what it needed but the vehicle for this was not the formal body at the highest level (the university senate). Instead, the curators' directives and advice came via an energetic young professor whose personal experience was marked by the desire— grounded in Enlightenment ideals—to improve research and teaching. For many years, the authorities in Hannover received advice from a person of this kind—the university librarian and professor of ancient languages Christian Gottlob Heyne.[36]

Superintendents' Self-Representation

Some of the statements made by the members of the first and second generations of superintendents, in public speeches or notes, are good sources for comparing these two generations. In both generations the superintendents were united in that they felt they were representatives of the state, which alone could "bring [educational institutions] into a flourishing condition." "Public education is perhaps the most important of the work the government does," said Seweryn Potocki in his speech at the opening of Kharkov University.[37] But the two generations had differing ideas about the ways to do these things: first-generation superintendents saw "cordial guardianship" as necessary while also respecting the rights of scholars, whereas in the 1810s and 1820s their successors engaged in "vigilant supervision" of university affairs and these methods began to be perceived as the main instruments, or even the only ones, at their disposal.[38]

The most important task shouldered by the superintendents was still educating "loyal subjects." Until the mid-1820s, this task, which was clearly articulated for the first time during the era of the "dual ministry," was carried out in institutions of higher education in the spirit of the doctrine of "Christian education" by studying the Bible and the "unvarnished foundations of the faith." But with Nicholas I's ascent to the throne, this task, now freed from its previous admixture of mysticism, continued to be a priority from a purely utilitarian point of view. Superintendent Aleksandr A. Pisarev in particular argued that only sciences which were useful to the state should be taught in

---

[36] Ernst Gundelach, *Die Verfassung der Göttinger Universität in drei Jahrhunderten* (Göttingen: O. Schwartz, 1955), 30–33.
[37] D. I. Bagalei, *Opyt istorii Khar'kovskogo universiteta: Po neizdannym materialam*, ch. 1, *1802–1815 gg.*, reprinted in D. I. Bagalii, *Vibrani pratsi*, vol. 3, ed. V. V. Kravchenko et al. (Kharkov: NUA, 2005), 218.
[38] Posokhov, *Popechiteli Khar'kovskogo uchebnogo okruga*, 48.

the university: "for military men, military sciences, for civilians, the civic sciences, for industrialists, the commercial sciences."³⁹

Accordingly, in carrying out this task, the superintendents of the second generation opposed both freedom of learning and freedom of teaching in the university. Proposals for introducing a rigidly fixed curriculum from which they felt a number of subjects and scientific methods should excluded as "absolutely harmful and unnecessary" occupied no small place in their notes. Pisarev considered various philosophical systems and the field of political economics, which he described as a "newly introduced science, arbitrary and incomprehensible," subjects of this type. He thought that "from the time when pedagogues began to teach about how to govern states, states faltered; they began to seek out riches and we grew poorer."⁴⁰ At the same time, Pisarev proposed doing away with the teaching of natural, public, common, and private law and he wanted to abolish Roman law as a subject in its own right, instead converting it into various historical excursions taught as part of other legal disciplines. He thought Roman law should be preserved in a truncated form in the Department of Antiquities and Latin (although it had not been taught there earlier) but he advocated for strengthening the study of theology and church history. His proposals also touched on the way the natural sciences—physics, chemistry, and astronomy—were taught. He thought all theoretical aspects of these fields should be excluded from the curriculum, leaving only the applied parts of these subjects.

Likewise, the Kazan superintendent Mikhail Nikolaevich Musin-Pushkin devoted a great deal of attention to the teaching of technical disciplines and the "commercial sciences." As for other subjects, he saw his goal as protecting the students' morals from "pernicious influences," which they were receiving from the "bad rules and [poor] skills of their teachers."⁴¹ In Kharkov, Egor Vasil'evich Karneev wanted, in essence, to make the university closer—in terms of teaching—to the way things were done in the gymnasium: he proposed having students take one course in each department for a year; basing teaching on repetition and questions and answers in the classes; calling the roll; and publishing short summaries of the lectures. In this type of teaching, the professors would monitor the students closely, and the university leadership would do the same for the professors. Karneev also felt that special attention should be paid to having students specialize in the natural and precise sciences.⁴²

---

³⁹ Alexandr Aleksandrovich Pisarev's 1826 proposals for reorganizing teaching and the system of governance at Moscow University are quoted in F. A. Petrov, *Universitetskaia professura i podgotovka Ustava 1835*, vol. 3 of *Formirovanie sistemy universitetskogo obrazovaniia v Rossii* (Moscow: Izdatel'stvo Moskovskogo universiteta, 2003), 58.
⁴⁰ Ibid., 372.
⁴¹ P. Kh. Galiullina, "M. N. Musin-Pushkin: Popechitel' Kazanskogo i Sankt-Peterburgskogo uchebnykh okrugov," in *Sankt-Peterburgskii universitet v XVIII–XX vv.*, ed. Burkova et al., 92–93.
⁴² Posokhov, *Popechiteli Khar'kovskogo uchebnogo okruga*, 49.

The statements quoted and discussed above show how utterly alien ideas about academic freedom (based on the "Göttingen model" and their further development along the lines of the "classic" Humboldtian university) were to the second generation of Russian superintendents. On the contrary, these superintendents were committed to an opposing model—a rigid curriculum, control over the content of what was being taught, highly specialized instruction, an emphasis on applied, technical knowledge and preparing civil servants as opposed to pursuing "pure science." In other words, they were committed to the utilitarian features which were embodied in the Austrian university system at the end of the eighteenth century and during the first half of the nineteenth century. For example, in a note addressed to the emperor, Alexey Alexeyevich Perovskii, the superintendent of Kharkov University, wrote about his sympathies for the Austrian system, as opposed to the principles of the Russian statute of 1804:

Guided by the examples of other states, we should have taken the educational institutions of Austria as our model [...] but instead of that, we have made ourselves imitators of the universities of Göttingen, Jena and others like them, completely overlooking the fact that neither the systems of those states nor the needs of those people have the slightest thing in common with ours.[43]

The way superintendents were perceived in the ministry also evolved. Superintendents had been seen as the minister's colleagues and as coworkers who sat with him on the Main Administration of Schools, which is how they were depicted in the Provisional Rules for National Enlightenment of 1803. Now superintendents became simply "transmitters of opinion" from the university council to the ministry, as the minister of national enlightenment, Karl Andreevich Lieven, formulated it in 1828. The norm of collective decision-making in the ministry with all of the superintendents participating, as it was laid down in the Provisional Rules, also ceased to be observed. However, this rule already began to be violated soon after the reform,[44] although during their first years at the Main Administration of Schools superintendents indeed jointly discussed problems and took decisions affecting all the universities.[45] But if a departure from the collective order in the 1800s simply denoted that large measure of freedom and independence which each of the grandee superintendents of the first generation could use because of his high status, then later, especially in conjunction with the "settling" of superintendents in university towns, in their person the university received its "local" head. On the

---

[43] Rossiiskii gosudarstvennyi istoricheskii arkhiv (hereafter RGIA), f. 1021, op. 1, d. 18, l. 4–4 ob.
[44] For V. N. Karazin's complaints about this, see *Russkaia starina*, no. 10 (1875): 273–274.
[45] Thus, in July 1805, on the initiative of Friedrich Maximilian Klinger, the superintendent of Dorpat University, an issue of vital importance for the university's professors, the question of stipends for apartments, was decided. See K. A. Il'ina, "Informatsionnye resursy popechitelia," 42.

Between Science and Empire   113

other hand, because he was so far from the capital, the superintendent lost the opportunity to influence decisions taken in Petersburg on strategic questions involving the development of university education. This kind of a superintendent was not very different from any other official in the ministry. He acted in accordance with the minister's circulars and if he came out with a legislative initiative, as a rule he proposed projects whose main goal was to introduce disciplinary measures that would strengthen his authority as the immediate head of the university.[46]

In an 1834 note to Nicholas I that was devoted to Kharkov University, Education Minister Sergei Semënovich Uvarov subtly noted a certain kind of "provinciality" that he considered typical of the superintendents' mindset: "The optical illusion has its foundation in provincial prejudices and the custom of thinking that the place where you are is the center of the state."[47] Thus, while superintendents were considered the personification of the state, centralization, and unification when Russian university reform was first undertaken at the beginning of the nineteenth century, in the realities prevailing at the beginning of the 1830s, they responded primarily to local needs and saw the university as "their own," guarding and protecting it in every way possible from outside influences, including those from the capital, which often benefitted the faculty's corporate rights. For example, the reminiscences of men who studied at Kazan University at this time reflect this. They noted that "Musin-Pushkin did not tolerate any interference and [...] stood up for his own [faculty]."[48]

Changes in the Nature of the Superintendent's Authority at the University in the 1830s and 1840s

The new University Statute of 1835 decisively removed superintendents from the group of top officials at the Ministry of National Enlightenment, attached them to the universities, and gave them the functions of the senior echelon of university administration. Among the statute's general provisions, a particular paragraph about the superintendent (paragraph 8) and an entire section, "On the Superintendent and his Assistant" (paragraphs 47–60), in chapter five, which is entitled "On Those in Charge," provide evidence of this.[49] All these things represented a significant innovation inasmuch as in the preceding statute of 1804, as was discussed above, there were no paragraphs specifi-

---

[46] Tsygankov, "Universitetskii popechitel' v Rossii," 525–526.
[47] D. I. Bagalei, *Opyt istorii Khar'kovskogo universiteta: Po neizannym materialam*, ch. 2, *1815–1835 g.g.*, vol. 4 of D. I. Bagalii, *Vibrani pratsi*, edited by V. V. Kravchenko et al. (Kharkov: NUA, 2005), 204.
[48] V. N. Nazar'ev, "Zhizn' i liudi bylogo vremeni," *Istoricheskii vestnik*, no. 12 (1890): 716.
[49] *PSZ-II*, no. 8337, vol. 10 (St. Petersburg, 1836). There are references to specific paragraphs later in the text.

cally about the post of the superintendent. Thus, the superintendent's functions and rights were finally registered in university legislation, putting an end to the former lack of definition. The statute strengthened the superintendent's role, which had been established in the previous decades, as a state administrator, the "special authority" to whom the university was entrusted (paragraph 8).[50]

The superintendent's supervisory functions were clearly defined in paragraph 48, where it was stated that he "keeps watch carefully so that the places and people falling under his remit fulfill their obligations diligently," "[he] takes notice of the abilities, assiduousness and good behavior of the Professors, Adjuncts, teachers and officials of the University, he corrects negligence with reprimands and takes lawful measures to remove those who are unreliable."

On the whole, it cannot be said that when the new statute took effect, the superintendents' powers expanded significantly. The 1835 statute only outlined the sphere of their responsibilities, made them more specific, and translated several informal practices which the superintendents used outside the legal framework defining their functions into formal practices.

The specific experience of university governance after 1835 confirms this. While Sergei Semënovich Uvarov was the minister of national enlightenment (1833-1849), superintendents held their posts for a long time, usually for more than ten years, which ensured stability of governance and allowed a steady equilibrium to develop in their relationships with university corporation, which was advantageous for both sides. In fact, during these years there were almost no conflicts in which the superintendents took part.

Count Sergei Grigoryevich Stroganov (1835-1848) at Moscow University, Count Yurii Alexandrovich Golovkin at Kharkov University (1834-1846), and Prince Mikhail Aleksandrovich Dondukov-Korsakov (1832-1842) and Prince Grigory Petrovich Volkonskii (1842-1845) at St. Petersburg University were representatives of the third, or "Uvarov," generation of superintendents. During the Uvarov era, Kazan superintendent Mikhail Nikolaevich Musin-Pushkin gained the minister's trust and kept his post (the only member of the second generation of superintendents to do so). He stayed in Kazan until 1845, after which he continued to serve as the superintendent of St. Petersburg University (1845-1856)—something which rarely happened. It is notable that the superintendents of the third generation all came from aristocratic families, which is reminiscent of the superintendents of the first generation; evidently Uvarov deliberately gambled on the titled elite, whose presence at the

---

[50] Cf. even stronger wording: the superintendent is the "head and master [khoziain] of the university and its educational district." This formulation can be seen in the proposals of the superintendents themselves, but it was not part of the final version of the statute. See Svod predlozhenii popechitelei uchebnykh okrugov i Sovetov universitetov S. Peterburgskogo, Moskovskogo, Khar'kovskogo i Kazanskogo, 1834, RGIA, f. 737, op. 1, d. 159, l. 1.

head of the university administration was supposed to raise the university's social status.

The superintendents of the new generation pursued policies which were similar to each other in many ways, in agreement with Uvarov's general university policy of strengthening "Russian science," that is, preparing Russian scientists and other scholars who, after mastering European knowledge and research methods, could further develop teaching in the country's universities on their own.[51] In this sense, the third generation, like the first, took up the task of promoting scientific research in the universities whereas in the second generation of superintendents, in contrast, these concerns were virtually forgotten. The Uvarov-era superintendent could no longer be the faculty members' formidable "uncle" or their "commanding officer;" now he needed a certain scholarly perspective and respect for academic work. Ideally, he would also be convinced that the quest for knowledge per se was valuable, which was the distinctive feature of the Humboldtian idea of what the university was.[52]

Therefore, in choosing people to be superintendents Uvarov was quite careful. Thus, when he proposed to appoint Count Yurii Alexandrovich Golovkin to this post, in the note about Kharkov University to Nicholas I cited above, he succinctly outlined the crucial qualities he was looking for in a superintendent:

[I]t is necessary [to appoint] a person in charge who knows how to combine firmness of character with the ability to get along with people; a leader who stands fairly high in public opinion so that his subordinates will see him not only as someone who justly carries out the will of the highest [level of] government but also as their intercessor and defender, in a word, an impartial judge of the labors and weaknesses of his subordinates.[53]

Count Sergei Grigoryevich Stroganov, the superintendent of Moscow University, fulfilled these criteria far better than anyone else. Financially independent, uninterested in making a career as a civil servant, and educated in Europe, he could pass for the "ideal superintendent," and the period during which he was in charge of the Moscow educational district and the university was the "Stroganov era."[54] About Stroganov, a university student of that epoch wrote that

He had all the means to put the university on that height on which he stood then: the intelligence, the multifaceted education, and the administrative abilities of a state official. He was well acquainted with the state of science in the West, he genuinely understood the

---

[51] On the development of the state's higher education policy in the Uvarov era, see Petrov, *Universitetskaia professura*, vol. 3 of *Formirovanie sistemy universitetskogo obrazovaniia v Rossii*, 173–351.

[52] For more information about the connections between the "Uvarovian" and the "Humboldtian" university, see Andreev, *Rossiiskie universitety XVIII–pervoi poloviny XIX veka*, 524–548.

[53] Quoted in Zaitsev and Posokhov, *Popechiteli Khar'kovskogo uchebnogo okruga*, 44.

[54] Petrov, *Rossiiskie universitety i liudi 1840-kh godov: Professura*, vol. 4.1 of *Formirovanie sistemy universitetskogo obrazovaniia v Rossii*, 129.

meaning and the spirit of the modern movement and at the same time he knew what our society expects from Russian universities.[55]

Moscow University's visible success in the 1830s and 1840s, which was noted by many contemporaries, was based primarily on the fact that Stroganov was able to fill professorships with highly qualified young specialists who had, as a rule, completed their training by studying and doing research in leading European educational institutions and who were armed with advanced scholarly methodologies. Stroganov was able to solve this problem in close collaboration with Uvarov, as we see in their extensive correspondence about sending young academics abroad and appointing them to academic posts on their return. The minister and the superintendent acted in concert, discussing each position in detail, carefully selecting young professors, guided above all by an assessment of their scholarly and pedagogical potential.[56] Although Stroganov's and Uvarov's personal relationship later foundered, nonetheless, in Petersburg the superintendent's major success in revitalizing teaching at Moscow University was never in doubt.

Conclusion

To conclude we will summarize the ways in which the post of superintendent was adapted in the Russian university. In the process of the university reform of the beginning of the nineteenth century, the curator was transformed into the superintendent. Insofar as the creation of an entire system of Russian universities subordinated to the Ministry of National Enlightenment was the result of the reform, in this ministry the post of superintendent was conceptualized as crucial. It was the post on which the collective adoption of very important decisions depended. In his own university each superintendent acted as the representative of the central government; the superintendents ensured uniformity across the university system and an appropriately high level of university education as the pinnacle of the Russian Empire's educational system.

The first generation of superintendents, which was in charge of the universities and educational districts in the 1800s, tried hard to respond to the tasks posed by the university reform. For this, a patriarchal "founder"-superintendent was needed, the type of superintendent that was close to the Göttingen model. The superintendents of this generation belonged to a correspondence network with European scholars, which was an advantage that enabled them to bring up-to-date ideas about universities to Russia and to invite foreign professors who were the carriers of these ideas to work at Rus-

---

[55] N. M. Dmitriev, "Studencheskie vospominaniia o Moskovskom universitete," *Otechestvennye zapiski*, no. 8 (1858).

[56] Pis'ma S. G. Stroganova k S. S. Uvarovu [Letter from S. G. Stroganov to S. S. Uvarov], OPI, f. 17, op. 1, d. 81, 90, GIM.

sian universities. However, after that, for several reasons, not least because of the general change of Alexander I's political course in the mid-1810s, a completely different type of university administrator prevailed among the second generation of superintendents. Their ideas about the university (if they had any such ideas) included neither the principle of "freedom of teaching" nor that of "freedom of learning." In the best-case scenario, they were oriented toward the Austrian bureaucratic model. Inevitable conflicts with professors flared up and there were various kinds of excesses: on the one hand the mass dismissals of academic staff during the "routs" of the Kazan and Petersburg universities and, on the other hand, superintendents were forced to retire because of the professors' complaints.

As a result, by the early 1830s, the superintendent turned into an official of the ministry who was responsible for regulating university life, someone with a narrowly local perspective and minimal influence on the state's policy on higher education. It is also important here to note that a similar tendency was also observed during those years in German universities, where the former independence, the significant role of university curators which was characteristic of the eighteenth century, gradually changed into the modest functions of a state official who was, in essence, a supervisor. The only significant difference was that in Germany this took place in the process of the transition to the "classic" university and was accompanied by the strengthening of the professors' academic freedom.[57]

The new University Statute of 1835 fixed this position, legislatively confirming the superintendent as the "head of the university" who lived in the same city as the university and was directly involved in supervising it so that the professors could "conscientiously fulfill their obligations." But at the same time, when the statute came into effect while Sergei Semënovich Uvarov was the minister of national enlightenment, a remarkable synthesis between the superintendent as a "commander-in-chief" and as a patriarchal benefactor, a liberal patron of the arts and sciences, took place. While in the ministry they remained executive officers and were deprived of an independent role, in the universities in the 1830s and 1840s superintendents were guardians, "benefactors," who promoted the flourishing of the professors' scholarly and pedagogical activity, opened the way for an influx of students, and respected both the rights and freedom of scholars and the results of their work. In this one sees evidence that the idea of the "classic" research university (Forschungsuniversität) had been transferred to Russia.

Translated from Russian by Jacqueline Friedlander.

---

[57] Wolfgang Kahl, *Hochschule und Staat: Entwicklungsgeschichtliche Betrachtungen eines schwierigen Rechtsverhältnisses unter besonderer Berücksichtigung von Aufsichtsfragen* (Tübingen: Mohr Siebeck, 2004), 32 f.

Daniel Baric

CLASSICAL ARCHEOLOGY AND SCIENCE POLICY IN HABSBURG BOSNIA AND HERZEGOVINA
Carl Patsch in Sarajevo, 1891–1918

To the Austro-Hungarian military men and officials who went to Bosnia and Herzegovina in 1878 to occupy it, the province seemed to be a space untouched by Western influence. The groundwork had to be laid *ex nihilo* to integrate this frontier region into the Habsburg imperial holdings. In fact, cultural policy, which was developed by the provincial government in Sarajevo together with representatives of the Austro-Hungarian Ministry of Finance (which was responsible for both provinces—i.e., Bosnia and Herzegovina), rested on no preexisting foundation. While in the economic and social domains the Austro-Hungarian officials sought to maintain some sort of continuity with the earlier Ottoman regime, the dynamics in the realm of culture demanded the creation of a new sort of institutions.

One of the most important innovations in these provinces, which had belonged to the Ottoman Empire since the fifteenth century, was the establishment of a *Landesmuseum* (provincial museum) in the capital. This museum provided a framework for the development of research in the Viennese tradition, in which archeology was given a particularly prominent place. Research on ancient history was organized according to the model of German-speaking universities in a place where developments in archeology had previously been almost totally unknown.

In this province, with its Muslim, Orthodox, Catholic, and Jewish populations, historical memory had long been transmitted mostly along confessional lines. This precluded an approach to the past that was based on scholarly interpretation of the scattered traces of previous times and systematic excavations.[1] Opening up this region through archeology demanded particular fi-

---

[1] On the propagation of historical memory in Bosnia-Herzegovina in the late nineteenth century in a context shaped by Islam, see Philippe Gelez, *Safvet-beg Bašagić: Aux racines intellectuelles de la pensée nationale chez les Musulmans de Bosnie-Herzégovine* (Athens: École française d'Athènes, 2010), 113–117. In 1879 the province had around 1,150,000 inhabitants, of which 43 percent were Orthodox, 39 percent Muslim, and 18 percent Catholic. On the changing demographics in the Austro-Hungarian period, see ibid., 619. An earlier version of this article was published in French: Daniel Baric, "Archéolo-

nancial resources.² The scholarly and political dimensions of research into a distant past lend it particular significance. The four decades of Bosnia-Herzegovina's integration into the Habsburg realm were characterized by a multiplication of excavation sites as well as "public works" connected to them. Bosnian-Herzegovinian historiography (and Yugoslav historiography in general) interpreted this all-encompassing cultural policy as evidence of a colonial approach to scholarship which cared less about enlightening the population than about securing the dynasty's international prestige and preparing for further expansion of Austria-Hungary to the East.³

The question of the political authorities' instrumentalization of archeology is justified in a historiographical context that examines the nature of colonialism in the Habsburg case.⁴ It is important, however, to take into account the intellectual horizons and the motivations of those actors who implemented the policies that were set at a higher level. The extremely modest number of professional archeologists who were active in Bosnia-Herzegovina during the Habsburg period makes it possible to follow the development of archeology by reference to the central figure in this constellation.

This figure was Carl Patsch (1865–1945), who was the first curator of the Roman division of the *Landesmuseum* and remained involved with the museum almost until the end of the First World War.⁵ Reconstructing the intel-

gie classique et politique scientifique en Bosnie-Herzégovine habsbourgeoise: Carl Patsch à Sarajevo, 1891–1918," *Revue germanique internationale* 16 (2012): 73–89.

2   The context into which Austrian archeology inserted itself in the Mediterranean region is presented by Stephen L. Dyson, *In Pursuit of Ancient Pasts: A History of Classical Archaeology in the Nineteenth and Twentieth Centuries* (New Haven, CT: Yale University Press, 2006), 112–115. The history of Austrian archeology (especially on the territory of the present-day republic) is discussed in Ingomar Weiler, "Alte Geschichte, klassische Archäologie und Altertumskunde," in *Geschichte der österreichischen Humanwissenschaften*, vol. 4, *Geschichte und fremde Kulturen*, ed. Karl Acham (Vienna: Passagen, 2002), 83–126.

3   See Todor Kruševac's studies in *Sarajevo pod austro-ugarskom upravom, 1878–1918* (Sarajevo: Muzej grada Sarajeva, 1960); Todor Kruševac, *Bosansko-hercegovački listovi u XIX. veku* (Sarajevo: Veselin Masleša, 1978); Risto Besarović, *Iz kulturne i političke istorije Bosne i Hercegovine* (Sarajevo: Svjetlost, 1966), 7–25 ("Specifičnosti kulturnog razvitka u Bosni i Hercegovini 1878–1918"); Hamdija Kapidžić, *Naučne ustanove u Bosni i Hercegovini za vrijeme austrougarske uprave* (Sarajevo: Arhiv Bosne i Hercegovine, 1973). Conference proceedings that distance themselves from any ideological bias represent a turning point: Enes Omerović, ed., *Nijemci u Bosni i Hercegovini i Hrvatskoj: Nova istraživanja i perspektive/Die Deutschen in Bosnien und Herzegowina und Kroatien: Neue Forschungen und Perspektiven* (Sarajevo: Institut za historiju, 2015).

4   Several contributions in the volume edited by Johannes Feichtinger, Ursula Prutsch, and Moritz Csáky, focus on the case of Bosnia-Herzegovina from this perspective. See *Habsburg Postcolonial: Machtstrukturen und kollektives Gedächnis* (Innsbruck: Studien-Verlag, 2003).

5   I am very grateful to Konrad Clewing (IOS, Leibniz-Institut für Ost- und Südosteuropaforschung in Regensburg). He brought the rich archival collection of Carl Patsch, now in the Bavarian State Archive (Bayerisches Hauptstaatsarchiv, hereafter BHStA), to my

lectual path and geographic mobility of this pioneer of classical archeology on Bosnian soil allows us to understand the relationship between local actors and the central authorities in Vienna. Both had similar goals, but differed in their approach, such as their ideas of the proper pace for the work that was being undertaken. It is not always easy to trace the ways in which these two logics— one central, one regional—came into contact, coincided, and occasionally diverged. The motivations of each individual participant must therefore be determined. The *Landesmuseum*'s publications and Patsch's papers allow us to trace the career of one scholar who was part of the Austro-Hungarian state apparatus established specifically for cultural development in Bosnia-Herzegovina. With views on Rome and Greece that he developed in Sarajevo, Patsch highlighted particular connections between the ancient world and the Habsburg present. Through this, he revealed the entanglement of archeology and empire.

Carl Patsch, from Prague to Sarajevo: The Exemplary Career of an Archeologist

At the end of his life, at 75 years old, Carl Patsch wrote his memoirs, primarily for his descendants. These pages recounted carefully reconstructed actions and impressions, reaching far back into the past. Although he had already been living in Vienna for a good 20 years by this point, it was the period he spent in Sarajevo that he described in the greatest detail. For a quarter century, he played a pathbreaking role in regional archeology there. This account of his life, enriched by carefully preserved documents, gives insight into the archeologist's work from his own perspective.

Unsurprisingly for an archeologist, his memoirs began with the origins of his family. The origins were twofold, with Tyrolean and Bohemian roots.[6] His family chronicle contains a branch that resettled in Bohemia.[7] It was in this bilingual milieu, a site of German-Slavic contact, that Carl Patsch grew up.[8] He studied classical philology at the German university in Prague, where he

---

attention. Two stays at the IOS enabled me to delve into this documentation, including his extensive correspondence. For an introduction, see Karl Nehring, "Der im Südost-Institut aufbewahrte Nachlaß von Carl Patsch: Briefe an Carl Patsch," *Südost-Forschungen* 57 (1998): 287–294. See also Christian Marchetti, "Archäologisch-ethnographische Parallelen: Carl Patschs wissenschaftliche Karriere in Bosnien und der Herzegowina," in *Nijemci u Bosni i Hercegovini i Hrvatskoj*, ed. Omerović, 109–122.

6 Nachlass Carl Patsch, Folder 261, Beilage I, Bestand Südost-Institut (hereafter BSOI), BHStA.
7 Ibid., Beilage IV.
8 For his biography, see Gerhard Seewann, "Patsch, Carl," in *Biographisches Lexikon zur Geschichte Südosteuropas*, vol. 3 (Munich: Oldenbourg, 1979), 405 f.

excelled.[9] In 1886, during the second semester of his studies, he got a certificate of passing a lecture on the imperial period from his professor for Roman history, Julius Jung. The following semester, he passed introductory Latin epigraphy with flying colors, then a session on the history of Greece under Athenian hegemony as well as another on the Danube provinces under Roman rule. Already in the early semesters of his studies his later predilection revealed itself: Patsch discovered the subject of the Roman Empire and its relationship to the provinces, which he continued to work on, and to which, after many years of work, he dedicated a monograph.[10]

A few years later, he became an instructor at the University of Vienna, for the seminar on archeology and epigraphy. Even in his earliest historical and archeological works, he was already subjecting the scholarly consensus of recognized scholars to relentless critique. An established approach to written sources and direct contact with epigraphic materials allowed him to reflect critically on the literature. In recognition of his erudition, he was requested as an epigrapher to interpret the inscriptions used by Philipp Ballif (1847–1905), the regional government's planning director, in his work *Die Römerstraßen in Bosnien und der Herzegovina* (Roman roads in Bosnia and Herzegovina), published in 1893.

In 1891, Patsch traveled to Bosnia and reported on the shock he received from discovering the region, which he described as the "European Orient."[11] He observed the Balkan world as it presented itself to him during his journey, with its exotic nature somewhat tempered by the presence of a German-speaking population. He described in detail the cultural atmosphere that he encountered on the streets amidst which he was about to spend a quarter century, though he did not yet realize it at the time.[12]

Patsch, who was sent to gather information about the Roman presence in Bosnia, demonstrated sincere interest in the world into which he immersed himself. Already in his earliest reflections on his encounter with Bosnian-Herzegovinian reality, he could distinguish between the different cultures that were found there, without concealing the distance he felt from those that did not belong to his "family"—that is, representatives of the Habsburg Empire, military men, and officials.[13] This connection among German speakers in a region dominated by South Slavs struck him at the time as stronger than

---

[9] Meldungsbuch des Studierenden Karl Patsch aus Kovač als ordentlicher Hörer inscribirt in der philosophischen Facultät der k. k. deutschen Karl-Ferdinands-Universität zu Prag, Oktober 1885, Nachlass Carl Patsch, Folder 262, BSOI, BHStA.
[10] Carl Patsch, *Der Kampf um den Donauraum unter Domitian und Trajan* (Vienna: Hölder-Pichler-Tempsky, 1937).
[11] Nachlass Carl Patsch, Folder 261, 48, BSOI, BHStA.
[12] Ibid.
[13] "Officers and officials, a companionable German-speaking family:" this is how Patsch described the group of people in Bosnia-Herzegovina with whom he clearly and exclusively identified. Ibid.

the one he felt thanks to his knowledge of other Slavic languages, especially Czech. In his later publications, he mingled commentary and archeological considerations with ethnographical observations: each piece of jewelry, each garment that he observed a Herzegovinian peasant woman wearing struck him as evidence of a noteworthy conservatism.[14]

The dominant feeling from his nearly month-long epigraphic expedition is satisfaction with the large amount of information he was able to collect and the contacts he established as well as with the foreign environment in which he moved. He traveled through Bosnia, first on the narrow-gauge railway that was administered by the military, then with the postal service. From this first contact with Bosnia onwards, the military played a significant role. The young Patsch, who was accustomed to moving about in the empire's major cities, such as Prague and Vienna, was able to appreciate equally the comfort of a modern hotel outfitted according to European standards and the pristine beauty of a provincial capital still largely untouched by modernization.[15]

At that time, he also probably enjoyed being able to live off his own earnings. As the oldest of nine siblings, he was uncomfortable with being dependent on his parents. Several months before he embarked on his journey, he was appointed to be an instructor at the University of Vienna and could now count on a regular income.[16] Added to this was his scholarly enthusiasm, as he anticipated entering into a still-undiscovered region, which he could present to the world after taking stock of it with his intellectual arsenal. Carl Patsch could be sure that he had reached an academically and socially privileged position, which a lengthy stay in Bosnia could only solidify. Such a chance soon presented itself.

Because his work on the publication of Ballif had demonstrated that he was academically well prepared for the given task, the local authorities recognized him as an expert in "Roman things," as Governor Benjamin von Kállay wrote to Hugo Kutschera, the man responsible for civilian affairs. Thus Patsch was initially assigned to the gymnasium in Sarajevo.[17] From the beginning, this position was seen as a stepping stone for a later assignment to the *Landesmuseum*. The museum was supposed to publish a German-language journal, and the editors needed someone with sufficient knowledge of epigraphy: as the governor explained, this journal should be directly aimed at "the learned European public."[18] Thus from the very beginning it had ambi-

---

[14] Carl Patsch, "Archäologisch-epigraphische Untersuchungen zur Geschichte der römischen Provinz Dalmatien," in *Wissenschaftliche Mitteilungen aus Bosnien und der Herzegowina* 12 (1912): 121.
[15] Nachlass Carl Patsch, Folder 261, 48, BSOI, BHStA.
[16] Ibid., 49.
[17] Kapidžić, *Naučne ustanove u Bosni i Hercegovini*, 109.
[18] Ibid.

tious personnel recruitment standards. Candidates should be carefully selected so that they could hold their own on the international level in terms of the quality of their scholarly production.

This approach was supported by Governor Kállay personally. Already in 1886, he had seen to the appointment of Ćiro Truhelka (1865-1942) as curator of the future *Landesmuseum* in Sarajevo, while the latter was still working at the Museum of Applied Arts in Zagreb.[19] The prehistorian Moriz Hoernes (1852-1917), who had already conducted preliminary archeological studies in Bosnia beginning in 1879-1880, and who had built his academic career largely on the basis of the research he had initiated there, was appointed by Kállay to advise the regional government on museum affairs.

The establishment of the *Landesmuseum* was initially a private initiative. Julije Makanec, a doctor who reported to the city council of Sarajevo, had published an appeal in a Sarajevo German-language newspaper, which was then published in Bosnian in other local press outlets. The signers of the appeal saw the founding of the museum as a necessary response to the rapid disappearance of cultural goods, which were being taken out of Bosnian territory in large numbers.[20] The establishment of a museum served as a reaction to this situation from the city's elite, even transcending confessional boundaries. For the administration in Vienna such a *Landesmuseum* was a prestige project, a vital precondition for the international recognition of Austria-Hungary's efforts to modernize the province. This was all the more important as the Habsburg monarchy, which was granted a mandate for Bosnia-Herzegovina in Berlin in 1878, was accountable to the other European powers.[21]

Governor Kállay, who knew the South Slavic world exceptionally well,[22] wanted to promote a cultural mission in which archeology played a prominent role. In a letter of 17 May 1893 to the ancient historian Mommsen, accompanying the first German-language publications of the *Landesmuseum* as

---

[19] On the role of Ćiro Truhelka in the development of the provincial museum in Sarajevo, see the volume edited by Nives Majnarić Pandžić, *Ćiro Truhelka: Zbornik* (Zagreb: Matica hrvatska, 1994).

[20] Under the title "Sarajevo as Regional Capital" an appeal was published in the *Bosnian Post* (on 13 July 1884) by Julije Makanec, who intended for a museum to develop around an archeological association, which would increase the city's attractiveness. Cited in Risto Besarović, *Iz kulturnog života u Sarajevu pod austrougarskom upravom* (Sarajevo: Veselin Masleša, 1974), 69-96 ("Ostvarene i neostvarene zamisli o osnivanju muzeja").

[21] On the museum in its political and cultural context, see Oliver Bagarić, "Museum und nationale Identitäten: Eine Geschichte des Landesmuseums Sarajevo," *Südost-Forschungen* 67 (2008): 144-167. For the broader context, see Robert J. Donia, *Sarajevo: A Biography* (Ann Arbor: University of Michigan Press, 2006), 88-91.

[22] On the scholarly and political career of Benjamin von Kállay (1839-1903), see *Biographisches Lexikon zur Geschichte Südosteuropas*, vol. 2 (Munich: Oldenbourg, 1976), 322-324.

well as Ballif's book about Roman roads, he explained, "When I was entrusted with the highest management of the administration of Bosnia and Herzegovina by my sovereign in 1882, I set myself the goal of not only establishing and securing peace and order, civilized and statutory conditions in these regions, but also—insofar as it could seem possible and useful—to provide inspiration and support in the realm of scholarship."[23]

In reality, classical archeology was a domain that, because of the historical and confessional traditions (with the exception of a few Catholic clergy), was incapable of sparking any particular enthusiasm among the population in Bosnia and Herzegovina. This was also due to the fact that during the Ottoman period there was no standard schooling based on the study of Greek and Latin texts. Therefore, the administration itself had arranged for publications in this field, so that these "works bear witness to the activities that have previously unfolded here in the scholarly and particularly in the archeological-historical realm."[24] The German historian's response soon arrived in Sarajevo. The expressions used by the governor were partially adopted by the historian as well, to emphasize the intellectual and military advantages of the Austrian presence in Bosnia.[25] Mommsen greeted the German-language publication of research that had been undertaken in the museum, as "the wretched linguistic divide [...] has to date prevented the full appreciation of these results abroad and, what I regret even more, has deprived important local workers of the proper fruits of their labor."[26]

Carl Patsch's contributions to the museum's publications (in both Bosnian and German, with a German original version) became ever more numerous over the years he spent in Sarajevo. He was fully recognized for his scholarly contribution, so that the governor could only rejoice over having recruited him. Over the years Patsch rose to higher positions in the scholarly administration. After he was temporarily assigned to the gymnasium, he officially became part of the museum staff, where from 1898 he served as curator of the Roman division; in 1904 he founded a research center there in the Balkans, which was oriented far beyond the borders of Bosnia-Herzegovina. It was only years later that this initially private initiative was esteemed as a part of official state policy and was accordingly given financial support.

The rationale for the sincere interest that he nurtured throughout his entire career towards the Roman Empire in particular can be found in the marginalia of his writings, which lay bare Patsch's close, long-term relationship with Austrian officials and his identification with Austrian cultural policy. As evidenced by a number of official evaluations of Patsch's activities

---

[23] Kapidžić, *Naučne ustanove u Bosni i Hercegovini*, 97.
[24] Ibid., 98.
[25] Ibid., letter from 30 June 1893.
[26] Ibid.

(described below), Patsch's loyalty did not go unnoticed by the imperial authorities.

On the Transferability of Enthusiasm for the Traces and Values of the Greek and Roman Empires

Patsch wrote that his appointment in Sarajevo inaugurated a "meaningful chapter in [his] life."[27] Year by year, he tirelessly crisscrossed the country in search of new archeological findings. While he pondered the idea of a synthesis of the Roman history of Bosnia, he brought his collected knowledge to the museum's depot and worked it into publications. He also located regions where no evidence of a Roman presence had been found to date.[28] Patsch had no doubt that new findings could be made and that collaboration with the local population could be expanded.

Patsch's publications underscore his conviction that Bosnia's position within the Roman Empire was by and large, in comparison to later periods, satisfactory. The economic position that the excavations brought to light indicated a province that for centuries had enjoyed optimal prospects for development, which "awoke pleasure in possession and the desire for a comfortable lifestyle. The houses in the small towns and in the countryside attest to this. There we find mosaics that even the most cultivated provinces could boast of."[29] The rich mosaics with iconographic elements from the Mediterranean region[30] attest to trade relations and cultural contact, fostered by the country's multiconfessionality. Patsch believed that religious tolerance was one of Rome's greatest achievements: "Both Earth and heaven are outwardly Romanized; but under Greek-Roman Olympus the gods of the country live on, with borrowed Latin names and Greek images."[31] A "mixed culture" arose that could develop thanks to the "tolerance and security that the empire secured for its subjects."[32] In comparison to the neighboring regions, "the conditions were particularly good."[33]

Patsch enjoyed the support of the local authorities for his excavations and publications, as well as that of the regional government's council for museum and scholarly affairs. A central figure in this regard throughout Patsch's career in Bosnia was Moriz Hoernes, who insisted that Patsch should over time

---

[27] Nachlass Carl Patsch, Folder 261, 57, BSOI, BHStA.
[28] Carl Patsch, "Archäologisch-epigraphische Untersuchungen zur Geschichte der römischen Provinz Dalmatien," *Wissenschaftliche Mitteilungen aus Bosnien und der Herzegowina* 12 (1912): 162 f.
[29] Carl Patsch, *Bosnien und Herzegowina in römischer Zeit: Ein Vortrag* (Sarajevo: Selbstverlag des b.-h. Instituts für Balkanforschung, 1912), 25 f.
[30] Ibid., 23–26.
[31] Ibid., 34.
[32] Ibid.
[33] Ibid., 34 f.

be put in charge of all publications connected to Roman archeology. This privileged position was also thanks to Kállay's support. The latter died in 1903, and the museum's journal honored him on the first pages of the 1904 issue for his policies on archeology in Sarajevo. The text, signed by the head of the *Landesmuseum* in Sarajevo, Kosta Hörmann, and the curator of the natural history museum in Vienna, Hoernes, recognized his

> care for scholarship as the task of a leading modern politician [...] He saw the whole picture, he knew to support everything that served the goals of scholarship and thereby indirectly the goals of statecraft. Thus he was venerated by all who knew him, in his high shadows, the image of the enlightened administrator, endowed to the country by the skill of fate and the wisdom of the monarch at exactly the time when a worthy new construction should be built over the rubble of the past.[34]

In the eyes of these two curators, Kállay seemed to have embodied the ideal politician, as he ascribed a key role in the modern city to scholarship. The metaphor of ruins used here refers to the Ottoman past. "Rubble" (Trümmer) indicates the past in the sense of a relic that should give way to a new order; the word "ruins" (Ruinen) was not used, because if remains designated in this way were of noble origin, perhaps Roman, they would have been valued and could have been put on display.[35] This differentiation appears to clearly divide different, hierarchized pasts from each other; apropos of this, the *Landesmuseum* was accused by some contemporaries and more generally by local and Yugoslav historiography of a lack of interest in the Ottoman past.[36]

Patsch located and investigated Roman monuments. The Roman Empire that he described, to be sure, was not a simple projection of the unitary power of Rome, the *caput imperii*. It was a multiethnic and multiconfessional tapestry, which assimilated and preserved all material and spiritual forms that were found before conquest. The image of classical Greece blurred with that of Rome into one large, imperial whole. This classical horizon was a mark of the Austro-Hungarian elite's intellectual education; it also shines through in the metaphor of archeological knowledge penetrating into the territory of Bosnia and Herzegovina.[37]

Over the years, the Greek world took on ever more importance within the museum. This was the result both of acquisitions in the last decade of the nineteenth century through a deft broker of archeological objects on the antiquities market as well as of the Austro-Hungarian consuls in the countries bordering Bosnia, who were in contact with the museum. The catalog of

---

[34] "Einleitung," *Wissenschaftliche Mitteilungen aus Bosnien und der Herzegowina* 9 (1904).
[35] On the perception of material traces of the past, see Eva Kocziszky, ed., *Ruinen in der Moderne: Archäologie und die Künste* (Berlin: Reimer, 2011).
[36] See note 3.
[37] The "shadow realm" of Bosnia-Herzegovina had supposedly been shielded "from the torch of Helios and the enlightening visage of Pallas Athene" until the arrival of Austrian scholars. See "Vorwort," *Wissenschaftliche Mitteilungen aus Bosnien und der Herzegowina* 1 (1893): IV.

Greek vases soon totaled 200 items.[38] Although the study of the Hellenic world was not a priority for the museum at the start of its operations, these acquisitions gave the *Landesmuseum* a certain scholarly prestige, so that the regional government primed Patsch to publish in this field as well.[39] Patsch did so on the occasion of the first numismatic congress, which was organized in Paris within the framework of the World Exposition.[40] Even if Roman artifacts formed the core of the museum's holdings and were its centerpiece, both due to the symbolic value attributed to them and through their exhibition in strategic locations (initially the lapidarium was set up at the center of the regional government's palace, then at the heart of the *Landesmuseum* when it opened in 1913), official interest was not limited to this period. Patsch strove in his reports to connect the objects of Greek origin that had come to the regional museum with Aegean culture, which had been studied by the Austrian archeological museum in Asia Minor, especially in Ephesus. Thanks to the serendipitous expansion of the Greek collection and through an analysis of contacts with Hellenic culture, Patsch rose to meet the academic expectations of his superiors. They were interested in developing the museum into not only a research institution for the history of Bosnia and Herzegovina, but also an internationally significant knowledge center.[41] Although the priority at the beginning of the occupation was Roman classicism, the addition of a Greek dimension followed as part of enriching the collection, but also of reflecting on the role that the museum could take on in the Balkans. The authorities were preoccupied with the question of the wide-ranging connections between Greek outposts and the emanation across the entire Balkan peninsula of a pre-Roman culture in the Mediterranean region.[42] Thus all

---

[38] Edmund Bulanda, "Katalog der griechischen Vasen im Bosnisch-herzegowinischen Landesmuseum zu Sarajevo," *Wissenschaftliche Mitteilungen* 12 (1912): 254–300.

[39] See M. Hoernes's letter of 30 July 1910 in Kapidžić, *Naučne ustanove u Bosni i Hercegovini*, 317.

[40] Carl Patsch, "Contribution à la numismatique de Byllis et d'Apollonia," in *Congrès international de numismatique réuni à Paris en 1900*, ed. Comte de Castellane and Adrien Blanchet (Paris: Société française de numismatique, 1900), 104–114.

[41] See the letter from 4 January 1913 by Oskar Potiorek, head of the regional government of Bosnia-Herzegovina, to finance minister Leon Biliński: "That means, above all, that we maintain our provincial museum—which today is the most important museum on the Balkan peninsula—in its leading role, and that we form an institute for Balkan research connected to the museum, so that Sarajevo will be the uncontested center of Balkan research for the whole world." Kapidžić, *Naučne ustanove u Bosni i Hercegovini*, 427.

[42] At the same time, Hellenism should in no way be understood as paving the way for Orthodoxy, as it somewhat seemed to be in Serbia; Staša Babić, "Janus on the Bridge: A Balkan Attitude Towards Ancient Rome," in *Images of Rome: Perceptions of Ancient Rome in Europe and the United States in the Modern Age*, ed. Richard Hingley (Portsmouth, RI: Journal of Roman Archaeology, 2001), 167–182. It was much more about a universalist tropism, as was present in Viennese culture at the turn of the century;

conditions seemed to be fulfilled, both intellectual and material, to make the museum a driving force on the local level, with the goal of presenting the ancient past as a societal model for the Habsburg present. Beyond this, on the international level it was supposed to be able to meet the highest scholarly standard. For several reasons these ambitious goals were only partially realized, at least as far as its function in the province of Bosnia and Herzegovina was concerned.

Carl Patsch, Between the Dream of a Network and Scholarly Isolation

The articles in the *Bulletin* regularly boasted of the ever-more important and irreplaceable contribution of individuals who were able to step in when material reasons made it plainly impossible for the museum to tackle all the promising excavations that had been planned. The epigraphic monuments that had been discovered in 1897 were presented as follows:

> The circles in which our efforts are met with understanding are constantly expanding; this is a phenomenon that fills us with joy, and the number of our colleagues who support our scholarly work through action and power shows a pleasing increase [...] After all, despite its good will, the museum administration, given its wide-reaching field of activity, cannot always travel to check every discovery.[43]

Patsch's assertion seems to show that research was a collective endeavor. In that same year, the regional government in Sarajevo tried to convince the Finance Ministry to be generous towards those who had contributed significantly to the enrichment of the collections, like the Catholic priest Anđeo Nuić from Županjac (today Duvno in Herzegovina). The letter emphasized the important contribution of the amateur, who was able to identify a piece of a wall from a Roman building, thereby supporting Patsch "most powerfully."[44]

One of the largest excavation sites, where Patsch was active for years, was that of the Roman camp of Mogorjelo in Herzegovina, which was quickly deemed extraordinary for its size and condition. Kállay wrote to the regional government and Patsch as soon as he received news of the discovery of this site; as ever, he was very anxious to organize the Roman archeology as efficiently as possible. Thus he counseled them to get in touch with the archeologists who were working on a seemingly comparable military camp in Carnun-

---

Jacques Le Rider, *Freud, de l'Acropole au Sinaï: Le retour à l'Antique des Modernes viennois* (Paris: Presses universitaires de France, 2002).

[43] Karlo Patsch, "Nove rimske epigrafske tečevine iz Bosne i Hercegovine," *Glasnik Zemaljskog muzeja* (1900): 169.

[44] In a letter from 18 May 1897, the regional government in Sarajevo asked the Finance Ministry to recognize and reward the priest's "patriotic" activities; he was unlike his brothers from the Franciscan province of Herzegovina, who would demonstrate little understanding for the scholarly goals of the museum. The request was approved. Kapidžić, *Naučne ustanove u Bosni i Hercegovini*, 216–219.

tum on the Danube, not far from Vienna.⁴⁵ For this man, on whose shoulders the responsibility for the administration of archeology lay, Austrian—when possible, Viennese—scholars were the absolute gold standard. Patsch immediately suggested building a small house for the scholars and tourists who would come to Mogorjelo, and the government approved this.⁴⁶ By the summer of 1899, however, he had to think about how to protect the excavations from damage caused by the local population, mainly out of ignorance. More than two decades after the arrival of Austrian troops and administrators, the wish for the population to see these activities as pure science was far from realized in this corner of Herzegovina.⁴⁷

In 1911, Patsch gave a lecture at the Museum of Applied Art in Vienna under the title "Bosnia and Herzegovina in Roman Times." He gave a synthesis of his research and discoveries, in anticipation of a monograph that he could not complete without fundamental research into several questions. Here he presented the Viennese public with the initial results of his general reflections on the basis of three decades of studying Roman archeology.⁴⁸

The Roman question also resonated with a reflection on the current situation in the province, annexed in 1908 and thereby transplanted directly into the heart of the empire, to use a biological metaphor, which was subliminally present in Patsch's lecture. The comparison between the present situation and that of Roman times was made more frequently and more explicitly in this lecture than in his scholarly publications, and he bluntly presented an unfavorable comparison: "The Romans burned their corpses; today's Herzegovinians cannot afford a coffin, so they cover the bodies in an unadorned earth grave with stone slabs."⁴⁹ He characterized Herzegovina as a place where "in ancient times unevenly favorable economic conditions arose, so favorable that even exerting all possible forces they will never be attained again, because the natural foundations are already irrevocably lost."⁵⁰ Thus Patsch described a precarious economic situation that decades of Austro-Hungarian administration could not manage to fundamentally improve. But the Viennese authorities were not to blame; rather, at fault was the animal husbandry that had had a negative effect on the natural landscape and had widely destroyed the forest.

The archeologist's unconditional sense of belonging to the province in which he was active only grew over the years. He had also established a family in Sarajevo, with a Croatian woman from near Zagreb, with whom he took an

---

45 Ibid., 269.
46 Ibid., 273, letter from Patsch to the regional government from 6 July 1899.
47 See the 17 July 1899 excavation report from Mogorjelo, in which possible plundering was mentioned, "because more than anywhere else the people here have been taken by the delusion that the walls and blocks harbor treasures," ibid., 298.
48 Patsch, *Bosnien und Herzegowina in römischer Zeit*. See note 29.
49 Ibid., 9.
50 Ibid.

active part in social life. This ever-stronger identification with the fate of Bosnia and Herzegovina distanced him from the detached Viennese perspective on the province. Already at the start of his stay in Sarajevo, through the first anthropological and archeological congress that Kállay's government brought to the city in 1894, Patsch was confronted with this Viennese view of the peripheral province, which he found to be inadequate and condescending. To wit, during the conference, in an otherwise favorable framework for international recognition of Viennese science policy, doubts were expressed about Patsch's activity.

Patsch described a scene characteristic of this in his memoirs: Otto Benndorf (1838–1907), who occupied a central academic position in Vienna as a full professor (Ordinarius), expressed sharp criticism of the museum's activity.[51] Only with the perspective of time could Patsch understand why the classical archeologist found it impossible to grasp the specificities and value of the seemingly modest artifacts that were being promoted in Bosnia at the time. It took the reconstructed context of a peaceful Roman province for the appropriate worth to be attributed to them. These artifacts of provincial archeology should be accorded greater attention, Patsch pleaded in his publications, to overcome the (false) impression of monotony. What was most important was to go beyond the established estimation, according to which an aesthetic verdict could only be based on the standards of the classical epoch— that is, one should not "swiftly dismiss monuments that are certainly no feast for the eyes as 'barbarian,' 'late,' or the like."[52] Here Patsch was taking a similar tack to that of the Viennese School of art history and historic preservation, which preferred interest in a set over individual, aesthetically highly valued items.[53] Carl Patsch developed this approach over his entire career, while it was also becoming the norm in Vienna, a parallel that was not the result of institutional reciprocity.[54]

As a result of his efforts to expand his scholarly interests, in 1904 Patsch founded the Institute for Balkan Studies. But it took ten years for a long-term financial solution to be found. The institute was established within the *Landesmuseum*, which contributed to additional tensions with Truhelka, the head of the museum since 1903.[55] As Patsch was not merely implementing

---

[51] Nachlass Carl Patsch, Folder 261, 62, BSOI, BHStA.
[52] Patsch, "Archäologisch-epigraphische Untersuchungen," 121.
[53] Georg Vasold, "Entre histoire de l'art à Vienne et archéologie en Dalmatie: Alois Riegl et la question d'un nouveau rapport au patrimoine romain," *Revue germanique internationale* 16 (2012): 43–55.
[54] See the significance of the pathbreaking analysis of Alois Riegl's *Stilfragen* (1893) in Diana Reynolds Cordileone, *Alois Riegl in Vienna, 1875–1905: An Institutional Biography* (Farnham: Ashgate, 2014), 86–108.
[55] See Hoernes's reports on the internal tensions in the museum and the constant rivalry between Patsch and Truhelka: Kapidžić, *Naučne ustanove u Bosni i Hercegovini*, 112–115.

the expectations of the Austro-Hungarian authorities, but ever more frequently exceeded them, he met with difficulties. This brought a certain professional isolation, even if his publications found universal acclaim.

But the local population demonstrated no understanding for the enthusiasm with which he conducted his research, primarily on Roman Bosnia. The real recognition came from his colleagues, who were mostly outside Bosnia and Herzegovina.

Wartime Epilogue: Excavations and Publications as Occupation of the Terrain of Science

In 1912 Patsch's Viennese lecture "Bosnia and Herzegovina in Roman Times" appeared in Sarajevo in a translation intended for the local readership. The underlying tone of the booklet was, where imperial intentions in Bosnia were concerned, characterized by confidence; reading between the lines, one saw that the province was once occupied by Roman peacemakers, and now, thanks to trade, the administration, and the army, was equally successfully occupied by the Austrians. The Roman presence was established in three stages—arrival, local resistance, peace—and Patsch similarly treated the Austrian presence as permanently established with the 1908 annexation of the province.

Patsch described a period that was initially characterized by mistrust, which then led to open conflict, when a tenacious resistance broke out near Sarajevo in the year 6 AD. After fierce clashes between the two camps, the Roman Empire eventually won. The 1878 occupation of Bosnia had, as Patsch noted, a similar sort of short-lived, unexpected, and intensive guerilla warfare, also not far from Sarajevo. The *pax romana*, the result of severe fighting, did allow for the production of one of the masterworks of Roman art, the *Gemma Augustea*. This showpiece was a testament to the grandeur of the imperial capital, to the transcendence not only of its weapons, but also its art. Both army and art induced the last brave resistance fighters to surrender, regardless of their courage. The cameo could be marveled at in Vienna, in the court museum (today in the Kunsthistorisches Museum).[56]

Patsch suggested that armed resistance against the imperial powers was a laudable part of Bosnia's history. The population of Bosnia had given proof of their courage. But he thereby relativized the significance of this, by extolling the artistic achievements of Roman culture as a successful response to the quelled political revolt. Soon the memory of this episode would fade. Now, however, the fatal gunshots of June 1914 appear in retrospect as a sharp rejec-

---

[56] Carl Patsch, *Bosna i Hercegovina u rimsko doba: Predavanje Karla Patscha: Autorizirani prevod Nikole Vidakovića sa 30 slika* (Sarajevo: J. Studnička, 1912), 13 f. See note 29 for the German edition.

tion of Patsch's optimism over the alleged general satisfaction with the advantages an imperial system brought to the people of Bosnia-Herzegovina.

The outbreak of war marked the beginning of the last phase of Patsch's activity in the Balkans, during which certain elements of his early years in Bosnia seemed to repeat themselves. He was extremely active in the Austrian occupation of territories that bordered the Habsburg Empire. He initiated excavations, just as he had done in successive phases since 1891, and indeed with the logistical support of the army, in the easternmost part of the monarchy, in Bosnia up to the river Drina, then in the Sanjak of Novi Pazar, and finally in 1917 in Albania. All these arenas of his activity clearly demonstrated continuity in the object of his research and its scholarly and military aspects. Like his British contemporary Francis John Haverfield (1860–1919), the founder of Roman provincial archeology in Britain,[57] he was also a loyal subject of his ruler, and his work seems to have been inseparable from his engagement on behalf of the state, even if this was not accompanied by political engagement or a political mandate, as it was for Theodor Mommsen.

In the newly founded South Slavic state, which Bosnia and Herzegovina joined in late 1918, Patsch's institute (his sole place of work since 1917) was closed. After his dismissal and his return to Vienna, he was soon offered a teaching position at the university. In Sarajevo, even if there was no clear *damnatio memoriae*, his work sank into obscurity, as nothing was undertaken to continue it. Whereas in the Austro-Hungarian Empire classical archeology could serve as an eloquent reflection of prevailing relations, in Yugoslav Sarajevo it no longer counted among the priorities. Classical archeology initially remained relevant only to a small circle of people. The dream of eliciting collective enthusiasm for this past could not be brought to fruition in Austria-Hungary, because there was no comprehensive education policy in Bosnia and Herzegovina. Since the time of the Josephinian reforms, Vienna sought to keep the number of school graduates low. There was also no development in favor of classical archeology in the Kingdom of Yugoslavia, where national archeology—above all from the Middle Ages—had a political role to fulfill. Nevertheless, Patsch played a significant part at least in laying the scholarly groundwork for Roman archeology, which became significant again in Bosnia and Herzegovina only after the Second World War.[58]

As for some other university graduates from Bohemia, for Patsch a stay in the South Slavic provinces offered an opportunity to uncover new scholarly territory and to make a name for himself on that basis.[59] Cooperation with all Austro-Hungarian authorities, especially the army, was a significant aspect of

---

[57] Richard Hingley, *Roman Officers and English Gentlemen: The Imperial Origins of Roman Archaeology* (London: Routledge, 2000).
[58] In this regard the works of Esad Pašalić (1915–1967) deserve particular mention.
[59] Anton Gnirs (1873–1933) had done this in archeology in Habsburg Istria over the same time period.

his activity there. Individually, given his own career (as well as his family ties in several parts of the monarchy), he could give only a positive evaluation of the impact of the imperial order on Bosnia and Herzegovina. Among his contemporaries, however, a radically different attitude prevailed. When he returned to Vienna, the last phase of his life began, during which his publications continued to focus on the Balkan realm. Through this, at least in the scholarly field he could continue to depict Roman imperial history in all its Danubian breadth.

Translated from German by Kate Younger.

Matthias Golbeck

LETTERS FROM THE EDGE OF EMPIRE
Self-Descriptions of the Imperial Official and Scientific Amateur
N. F. Petrovskii from Turkestan, 1870–1895[1]

In the mid-1860s, a decade after the Crimean War and just as the official liberation of the Caucasus was first declared, the Russian expansion into Central Asia began under Tsar Alexander II (1818–1881). The capture of Shimkent in 1864 kicked off the Russian advance towards West Turkestan, which only ended in 1884 at the Merv oasis. In the midst of this twenty-year stretch, in 1867 the empire consolidated its new territories into the province of Turkestan, with Konstantin Petrovich von Kaufman (1818–1882) being appointed as its first governor. The empire strengthened its position through treaties, such as the Petersburg Agreement with China in 1881. This agreement clarified the borders with East Turkestan and permitted the Russian Empire to open consulates in Chinese territory. The struggle with Great Britain for regional hegemony gave rise to treaties like the Pamir Agreement of 1895 or the Anglo-Russian Convention of 1907.[2]

These developments gradually brought Russian military men and bureaucrats, scholars and travelers, and settlers and their relatives to the region. Among these was the finance official, later consul general, Nikolai Fëdorovich Petrovskii (1837–1908). This essay is based on an analysis of his correspondence between 1870 and 1895.[3] These letters allow us to examine the tightly

---

[1] An earlier version of this article appeared in Martin Aust and Benjamin Schenk, eds., *Imperial Subjects: Autobiographische Praktiken in den Reichen der Romanovs, Habsburger und Osmanen* (Cologne: Böhlau, 2015). I am grateful to the editor for publishing this revised translation.
[2] Yuri Bregel, *An Historical Atlas of Central Asia* (Leiden: Brill, 2003), 62–65; Andreas Kappeler, *Rußland als Vielvölkerreich: Entstehung, Geschichte, Zerfall* (Munich: C. H. Beck, 2001), 154–166; Otto Hoetzsch, *Russland in Asien: Geschichte einer Expansion* (Stuttgart: Deutsche Verlagsanstalt, 1966), 77–89; Hermann Kreutzmann, "Das Great Game: Asien als Bühne eines imperialen Machtkampfes," in *Über den Himalaya: Die Expedition der Brüder Schlagintweit nach Indien und Zentralasien 1854 bis 1858*, ed. Moritz von Brescius, Friederike Kaiser, and Stephanie Kleidt (Cologne: Böhlau, 2015), 89–95.
[3] For his correspondence, see Nikolai Petrovskii, *Turkestanskie pis'ma*, ed. Vladimir Miasnikov (Moscow: Pamiatniki istoricheskoi mysli, 2010). Henceforth letters will be cited with the author's name, their number, and page number, as they are given in this volume. The volume contains 189 letters from Petrovskii, four replies, two official reports,

entangled processes of military, economic, and scholarly conquest, exploitation, and administration from the point of view of an implicated actor. The task of men like Petrovskii was to secure, exploit, and administer the new territories. Their activities were legitimized by St. Petersburg's need for protection and stability along its Central Asian borders vis-à-vis regional actors, as the Russian foreign minister Aleksandr Mikhailovich Gorchakov (1798–1883) put it in an 1864 circular report. But Petrovskii and his colleagues were also implicated in the assertion of Russian interests in Pamir against Great Britain as part of the Great Game. At the same time, moreover, these officers and officials were tasked with carrying out the Russian civilizing mission.[4]

Parallel to their contribution to military expansion, Russian actors often also participated in the study of Turkestan as amateur scholars. They made use of their treasure trove of local experience and their direct access to the object of their research. Hence, they found themselves in tension with the academic field of Oriental studies, which was becoming institutionalized in St. Petersburg and Moscow in a variety of ways from the mid-nineteenth century onwards. Russian engagement in the Caucasus and in Central Asia demanded multifaceted specialist knowledge in the military and administrative realms. Scholars and amateurs were in communication both nationally and internationally through literature, conferences, and personal correspondence. Together they debated the Russian Empire's position among the colonial powers or the legitimacy of the Russian advance. Here the parallels to the case of Carl Patsch, as examined in Daniel Baric's contribution to this volume, are clear in terms of the actors' roles, even if the political contexts were markedly different.[5]

Historians have often considered the Russian Empire's expansion into Central Asia in the context of its overall Asia policy.[6] In addition to classical

---

and two other publications by Petrovskii written between 1870 and 1907. The documents come from the collection of the Institute for Oriental Manuscripts of the Russian Academy of Sciences in St. Petersburg and the Archive of Foreign Policy of the Russian Empire in Moscow.

[4] Hoetzsch, *Russland in Asien*, 27 and 113; Kappeler, *Vielvölkerreich*, 162 f. On the economic reasons for this expansion, see Sven Beckert, *King Cotton: Eine Globalgeschichte des Kapitalismus* (Munich: C. H. Beck, 2014), 321–334.

[5] Rostislav Rybakov, *300 Years of Oriental Studies in Russia: Imperial, Soviet and Post-Soviet Periods* (Moscow: Institute of Oriental Studies, Russian Academy of Sciences, 1997), 16–43; Vera Tolz, "Russische Orientalisten und der transnationale Imperien-Diskurs an der Wende zum zwanzigsten Jahrhundert," in *Globalisierung imperial und sozialistisch: Russland und die Sowjetunion in der Globalgeschichte, 1851–1991*, ed. Martin Aust (Frankfurt am Main: Campus, 2013), 126–159.

[6] A recent example is Manfred Hildermeier, *Geschichte Russlands: Vom Mittelalter bis zur Oktoberrevolution* (Munich: C. H. Beck, 2013).

political histories,[7] the topic has been examined in recent works on the Great Game.[8] There have also been studies of the important discourses and social, cultural, or religious changes associated with this expansion.[9] There are still relatively few recent works that focus on the actors involved, such as this essay.[10]

In recent years, there has been a growing number of studies of the Russian Empire that rely on the analysis of ego-documents. These works have dealt with the emergence of the individual in the empire; compared autobiographical practices in Eastern and Western Europe; and investigated textual strategies and their transformation in the self-images of important authors.[11] On the theoretical level, on one hand, Jochen Hellbeck and Klaus Heller have called for an integrative approach to sources through their concept of "autobiographical practices," incorporating autobiographies, memoirs, diaries, and letters. On the other hand, Volker Depkat has argued for greater consideration for literary studies approaches in historiographical engagement with autobiographies.[12] In addition, there have been numerous studies of particular cases in the nineteenth-century Russian Empire that have analyzed ego-documents from different social, confessional, or political groups.[13] There have

---

[7] In addition to Hoetzsch, see also Andrei Lobanov-Rostovsky, *Russia and Asia* (Ann Arbor: Georg Wahr, 1959); Jeff Sahadeo, *Russian Colonial Society in Tashkent, 1865–1923* (Bloomington: Indiana University Press, 2007).

[8] For example, Evgeny Sergeev, *The Great Game, 1857–1907: Russo-British Relations in Central and East Asia* (Washington: Woodrow Wilson Center Pr., 2013).

[9] Daniel Brower, *Turkestan and the Fate of the Russian Empire* (London: Routledge Curzon, 2003); Robert Crews, *For Prophet and Tsar: Islam and Empire in Russia and Central Asia* (Cambridge, MA: Harvard University Press, 2006); Alexander Morrison, *Russian Rule in Samarkand, 1868–1910: A Comparison with British India* (Oxford: Oxford University Press, 2008); Ulrich Hofmeister, *Die Bürde des Weißen Zaren: Russische Vorstellungen einer imperialen Zivilisierungsmission in Zentralasien* (Stuttgart: Steiner, 2017).

[10] Jörn Happel, *Nomadische Lebenswelten und zarische Politik: Der Aufstand in Zentralasien 1916* (Stuttgart: Steiner, 2010).

[11] Jochen Hellbeck and Klaus Heller, "Vorwort," in *Autobiographical Practices in Russia/Autobiographische Praxis in Russland*, ed. Jochen Hellbeck and Klaus Heller (Göttingen: V&R unipress, 2004), 7–10; Julia Herzberg, "Autobiographik als historische Quelle in 'Ost' und 'West,'" in *Vom Wir zum Ich: Individuum und Autobiographik im Zarenreich*, ed. Julia Herzberg and Christoph Schmidt (Cologne: Böhlau, 2007), 15–62; Ulrich Schmid, *Ichentwürfe: Die russische Autobiographie zwischen Avvakum und Gercen* (Zürich: Pano, 2000).

[12] Hellbeck and Heller, "Vorwort," 7–10; Volker Depkat, "Zum Stand und zu den Perspektiven der Autobiographieforschung in der Geschichtswissenschaft," *BIOS: Zeitschrift für Biographieforschung, Oral History und Lebensverlaufsanalysen* 23 (2010): 172–179.

[13] Besides the volume from Jochen Hellbeck and Klaus Heller, see the studies in Herzberg and Schmidt, *Vom Wir zum Ich*; Aust and Schenk, *Imperial Subjects*; Tim Buchen and Malte Rolf, eds., *Eliten im Vielvölkerreich: Imperiale Biographien in Russland und Österreich-Ungarn, 1850–1918* (Berlin: De Gruyter, 2015); Julia Herzberg, *Gegenarchive:*

still only been a few similar studies on Central Asia during the expansionary period.¹⁴

This essay is an effort towards filling this gap, putting N. F. Petrovskii, a participant in Russian expansion, at the heart of its analysis. His correspondents came from diverse social and professional groups. Among them were scholars who were well known at the time, like Fëdor Romanovich Osten-Saken (1832–1916) or Sergei Fëdorovich Ol'denburg (1863–1934). My focus is on the self-descriptions provided by this finance official, diplomat, and amateur researcher in professional and private spheres at the edge of the empire. It seeks to examine how Petrovskii oriented himself within predetermined structures, how he appropriated them, or how he sought to change them.¹⁵

N. F. Petrovskii, Russian Expansion, and Oriental Studies

Little information exists about the early years of Nikolai Fëdorovich Petrovskii's life. He was born on 30 November 1837 and had one brother, Sergei Fëdorovich. Nikolai Fëdorovich received his education at the Second Moscow Cadet Campus, graduating in 1858. Thereupon Petrovskii moved over to the 12th Astrakhan Infantry Unit as a staff sergeant. Beginning in 1859, he taught Russian at the Aleksandrinskii Cadet School for Orphans in Moscow; he left there in 1861 for unknown family reasons.¹⁶

Petrovskii's early years fell in the phase of autocratic stagnation during the reign of Nicholas I (1796–1855). As a young man, at the start of his career he experienced the early, reform-oriented reign of Alexander II. He felt through his career, however, that freedoms were strictly limited under a still autocratic regime. On 16 July 1862, now at the rank of staff captain, Petrovskii was imprisoned in St. Petersburg on the grounds of connections to the "London propagandists." Presumably this referred to contacts with the groups of exiles

---

Bäuerliche Autobiographik zwischen Zarenreich und Sowjetunion (Bielefeld: Transcript, 2013).

[14] Larisa Levteeva, *Prisoedinenie Srednei Azii k Rossii v memuarnykh istochnikakh: Istoriografiia problemy* (Tashkent: Izdat. Fan. Uzbek. SSR, 1986); Aleksandr Matveev, "Perceptions of Central Asia by Russian Society: The Conquest of Khiva as Presented by Russian Periodicals," in *Looking at the Coloniser: Cross-Cultural Perceptions in Central Asia and the Caucasus, Bengal, and Related Areas*, ed. Beate Eschment and Hans Harder (Würzburg: Ergon, 2004), 275–298.

[15] Cf. Anke Stephan, "Erinnertes Leben: Autobiographien, Memoiren und Oral-History-Interviews als historische Quellen," *Digitales Handbuch zur Geschichte und Kultur Russlands und Osteuropas* 10 (2004): 2 or, most recently, Malte Rolf, "Einführung: Imperiale Biographien. Lebenswege imperialer Akteure in Groß- und Kolonialreichen, 1815–1918," *Geschichte und Gesellschaft* 40 (2014): 7 f.

[16] Nikolai Petrovskii, "Kommentarii," in *Turkestanskie pis'ma*, ed. Vladimir Miasnikov (Moscow: Pamiatniki istoricheskoi mysli, 2010), 318; Vladimir Bukhert, "...i ego russkii portret," in *Turkestanskie pis'ma*, ed. Miasnikov, 21.

living in Western Europe; such contacts were apparently sufficient to deserve imprisonment, even during the comparatively liberal reign of Alexander II.[17]

On 9 December 1863 Petrovskii was released on bail posted by a *zemstvo* representative friend. It was only a year later that the staff captain was sentenced to a year in prison; his pretrial detention fulfilled this sentence, however, so he did not have to return to prison. Soon after, on 28 May 1865, Petrovskii married Sofia Alekseevna Sakhnovskaia, the daughter of a staff captain. In the same year, he entered the state control service, which was subordinate to the cabinet of ministers and monitored compliance with the state budget, which was established in 1862–1863. This makes it clear that Petrovskii, unlike Carl Patsch, was first and foremost engaged in military and fiscal activities, and only later in diplomatic ones.[18]

Petrovskii's scholarly interests came into view in 1866, when he joined the Imperial Society of Devotees of Natural Science, Anthropology, and Ethnography at the University of Moscow (Imperatorskoe obshchestvo liubitelei estestvoznaniia, antropologii i etnografii, henceforth OLEAE). But these interests remained part of his private life. The only aspect of this phase of Petrovskii's life that is known is the 1869 birth of his son Nikolai. In 1870, he was sent by the Ministry of Finances to the General Government of Turkestan, which was founded in 1867. Here he gathered information about the state of trade. He settled in Tashkent in 1872. Two publications from 1873—*Materials for Trade Statistics of the Region of Turkestan* and *On the Rearing of Silkworms and Silk Spinning in Central Asia*—illustrate the sorts of things Petrovskii was studying. He traveled around the region to gather information. In 1872 he visited Bukhara, publishing a travel report a year later in *Vestnik Evropy* (The Herald of Europe). In Tashkent he was involved in establishing the Turkestan branch of OLEAE. He was also the secretary for the Central Asian Scholarly Society (Sredneaziatskoe uchenoe obshchestvo). Petrovskii traveled to the Caucasus in 1878 and studied Russian trade there as well.[19]

Before Petrovskii transferred to the diplomatic service in 1882, he took part in an 1880 evaluation of the provinces of Saratov and Samara. The Treaty of St. Petersburg (1881) laid the groundwork for Petrovskii's later activity, with its provision for the creation of Russian consulates and the opening of numerous Chinese cities to Russian merchants. On 1 June 1882 Petrovskii was appointed consul to Kashgar. In mid-December he reported to the Foreign Ministry that the consulate had been set up. But by the end of 1883, Petrovskii had already departed again for an inspection of the Turkestan Gen-

---

[17] Hans-Joachim Torke, *Einführung in die Geschichte Rußlands* (Munich: C. H. Beck, 1997), 151–170.
[18] Bukhert, "Russkii portret," 21; Torke, *Einführung*, 167.
[19] Bukhert, "Russkii portret," 21–24. For the publications mentioned, see Nikolai Petrovskii, *Materialy dlia torgovoi statistiki Turkestanskogo kraia* (Tashkent, 1873) and Nikolai Petrovskii, *O Shelkovodstve i shelkomotanii v Srednei Azii* (Tashkent, 1873).

eral Government. He remained stationed in Kashgar until his retirement in 1903, and on 14 March 1895 he was promoted to consul general. In his professional life, in the late 1880s Petrovskii was involved in the developing territorial conflict between Russia, China, and British India over the Pamir region. Privately, the "specialized 'amateur'"[20] was a member of several imperial research societies and concerned himself with various branches of scholarship, mainly in Kashgar. Petrovskii's various professional functions and myriad scholarly interests make him seem remarkably flexible and capable of learning and adapting. In fact, though, he seems not to have been unique in this, as demonstrated, for example, by the biography of the merchant and commercial traveler Aleksei Danilovich Vasenev (1856–1917) and his contacts with Russian specialists from various academic fields.[21] Similar to Carl Patsch, Petrovskii's case offers detail-rich glimpses into the relationship between the center and one of its newly acquired peripheral regions.[22] Unlike Patsch, Petrovskii is not known to have left memoirs or any other cohesive personal narrative. We cannot say, therefore, what significance he placed on his years in Tashkent and Kashgar in comparison to his youth and education in western Russia. In his private and professional letters, his self-descriptions as a public servant and a contributor to the study of Central Asia are of the utmost significance.

Petrovskii's Self-Description as a Public Servant

As a revenue officer and consul, Petrovskii never withheld his criticism of state actions or of his superiors' demands. He often based such criticisms on the professional expertise he had gained on the job and emphasized his interest in the state's well-being. In his letters from the 1870s to Nikolai Andreevich Ermakov (1824–1897), a department head in the Ministry of Finance, or in his letters from the 1880s–1890s to Osten-Saken, then a department head in the Foreign Ministry, Petrovskii painted a picture of himself as a good, loyal but critically minded, and observant public servant, who for example, relaying his assessment of the General Government's investment policies or fiscal profligacy.[23]

But Petrovskii's criticism always also served to justify and exonerate himself. In imparting his knowledge, he frequently emphasized that he had already informed the relevant parties of the facts at hand, while simultaneously

---

[20] Tolz, "Orientalisten," 128.
[21] André Schmidt, *Im Auftrag der Wirtschaft – im Sinne der Wissenschaft: A. D. Vasenevs China-Reisen, 1882–1889* (Munich: unpublished). This work can be found in the library of the Ludwig-Maximilians-Universität, Munich.
[22] Sarah Pain, *Imperial Rivals: China, Russia, and their Disputed Frontier* (New York: Routledge, 1996), 161–166; Bukhert, "Russkii portret," 23–28, 31, 38, 55.
[23] Petrovskii, no. 16, 98–99; Petrovskii, no. 19, 103.

complaining that no one had gotten in touch with him. What the top ranks chose to do and the consequences of their decisions could not be blamed on him. In expressing his incomprehension of certain government orders, Petrovskii portrayed himself as someone whose local expertise enabled him to pronounce judgment. In this context, he also sought to use his knowledge for his personal professional advancement. By offering to share his knowledge with higher authorities, he highlighted his readiness to be of service to them and clearly hoped that this would serve to recommend himself.

One instance from his activity as consul, the case of the local powerbroker Safdar-Ali-Khan, illustrates several of these points. Petrovskii acted as the Russian Empire's contact person for the Khan from Kandzhut. On 17 November 1888, writing to Osten-Saken, Petrovskii widely criticized his employer's approach and positioned himself as a regional expert who was better informed on the local conditions in the border region:

If the ministry had just once, even in general terms, given me an indication of its plans vis-à-vis Badakhshan, Shutnan, Rushan, Wakhan, Kandzhut, Chitral, etc.—whether contacts should be cultivated or avoided, whether the goal was to unify them or to fragment them, etc.—I would gladly have set about doing so [...].[24]

In this letter, he openly questioned the ministry leadership's discernment in this matter and indicated that Osten-Saken should consider using him to enlighten the St. Petersburg decision-makers. At the same time, he assiduously distanced himself from any responsibility for any resulting problems through his complaint that he was not involved in the process.[25]

As an active, engaged official, once in the service of the Ministry of Finance, without prompting Petrovskii initiated discussions, voiced pressing issues with his interlocutors, discussed possible solutions, and also invited them to participate in joint projects. Notably, this even happened with higher-ranking officials, like Pëtr Nikolaevich Stremoukhov (1823–1885), who was the temporary director of the Asia department in the Foreign Ministry. In October 1872, for example, Petrovskii offered his detailed plans for a literacy campaign in Turkestan to Ermakov, who belonged to a committee working on related matters.[26] The young revenue officer used his professional contacts to implement his ideas in the realm of popular education, which fell outside his professional purview. Petrovskii thereby came across as an official who was engaged and ready to go above and beyond his required duties. He was aware of this, and in the 1870s–1880s he was more than once concerned about the way his "conscientiousness" was perceived. Another example of this dates from 1886, when he laid out a consular statute for the entire empire.[27]

---

[24] Petrovskii, no. 102, 212.
[25] Petrovskii, no. 102, 211 f.
[26] Petrovskii, no. 18, 102.
[27] Petrovskii, no. 80, 184; Petrovskii, no. 86, 191 f.; Petrovskii, no. 91, 196 f.

It was increasingly important to the consul that his service, which he himself considered to be extensive, be appreciated. During the Pamir conflict in the 1890s, he presented himself as an expert and a key local actor. Thus, for example, in October 1891 he reported to Osten-Saken with clear pride on the successful unmasking of British activities in the region and hastened to ask the baron to find out at the ministry whether this development had met with satisfaction, as he hoped.[28] In his telling, Petrovskii's actions remained focused on the good of the state, yet he was clearly concerned with his own advancement. To the end of the period under examination in this article, Petrovskii seems not to have grown discouraged by the fact that his efforts were not infrequently rebuffed.

Petrovskii's reports—which always adhered to a formula of explicating a problem, then proposing solutions—evolved in the 1890s, as he began to include reviews of his own earlier writings. A letter on the worsening situation in Pamir and possible scenarios for escalation written to Osten-Saken in April 1893 illustrates this. The consul presented himself to his colleague as an experienced and clear-sighted analyst; as he put it:

> Nearly twenty years ago (how time flies) I wrote officially to the deceased Pëtr Nikolaevich [Stremoukhov, M. G.] (my letter must be at the Asia department), that the English are turning the Hindukush into a new Caucasus for us. Whether it was a lucky guess or foresight, in any case my letter is in a certain sense a historical document.[29]

At the same time, in the midst of the crisis, this also served as a sort of self-reassurance: he had recognized the problems early on and voiced them clearly, and therefore he could not be held accountable for any negative consequences.

Petrovskii's self-image as a regional expert was also reflected in his ongoing efforts to publish articles on political, economic, or historical subjects in various journals. In a December 1879 letter to Aleksei Sergeevich Suvorin (1834–1912), then the publisher of *Novoe vremia* (New Times), for example, he presented himself as a broadly knowledgeable regional expert and a well-connected informant. These activities fell outside his professional activities, but they harkened back to the knowledge he had gained through his work in Tashkent and Kashgar. The titles of his publications, such as "The Scholarly and Trade Expedition of 1874–1875" (published in 1878), illustrate this. A May 1891 letter to Viktor Petrovich Burenin (1841–1926), an editor at *Novoe vremia*, confirms that these activities continued into the 1890s.[30]

---

[28] Petrovskii, no. 120, 232.
[29] Petrovskii, no. 121, 233 f. Pëtr Nikolaevich Stremoukhov (1823–1885) was, among other things, director of the Asia division of the Foreign Ministry.
[30] For the article see Nikolai Petrovskii, "Uchëno-torgovaia ekspeditsiia v Kitai v 1874–1875 godakh," *Russkii vestnik* 135 (1878): 101–121; Petrovskii, no. 30, 118; Petrovskii, no. 31, 119; Petrovskii, no. 110, 219.

Additionally, Petrovskii seems to have believed himself to be competent enough in the realm of Turkestan's recent political and military history that in 1873, for example, he offered Mikhail Matveevich Stasiulevich (1826–1911), the editor-in-chief of *Russkii vestnik* (The Russian Herald), a book review of a work by the Hungarian historian Arminius Vámbéry (1832–1913).[31] Until the end of the period under examination in this essay, however, he did not succeed in establishing himself as a regular contributor from East Turkestan in the imperial capital. In his Kashgar years, too, Petrovskii needed assistance from acquaintances in getting his articles published, like Dmitrii Fomich Kobeko (1837–1918), a high-ranking revenue officer in St. Petersburg.[32]

Above all, as consul, Petrovskii strove to utilize his criticism and engagement for his own professional advancement. In the early 1880s, he bemoaned disagreeable living and working conditions, contending that the ministry at the very least took no action to improve them. Later he cited the ministry's lack of support in the Pamir conflict as grounds for looking for a new position, complaining to Kobeko and Ermakov, among others.[33] Petrovskii subsequently asked colleagues and officials, including Kobeko and Osten-Saken, for their help, and expressed his desire for a new post. In his letters he comes across as a young official who recognized his own ambitions and who believed that his demands were justified by his experience and knowledge, but who was disappointed that they were not taken into consideration. One example is Petrovskii's failed effort in the mid-1880s to be appointed as the "political agent" of the empire in Bukhara.[34] Subsequently he sought to be promoted to consul general. In 1888 he wrote to Osten-Saken that his desire was justified by the extraordinary amount of work he had done and by the praise he had received from the ministry. This time his request was granted, though not until mid-March 1895.[35]

Petrovskii's Self-Description as a Contributor to the Study of Central Asia

Many of the characteristics that Petrovskii ascribed to himself in a professional context are the same ones that he harnessed as a contributor to the study of Central Asia. In his letters, both to professional acquaintances and to high-ranking scholars, he presented himself simultaneously as a translator and writer with many interests, a self-confident assistant, an equal interlocutor, and a simple collector. As early as 1872, writing to Ermakov, Petrovskii

---

[31] Petrovskii, no. 22, 107–109; Arminius Vambery, *History of Bokhara: From the Earliest Period Down to the Present* (London, 1873).
[32] Petrovskii, no. 43, 139 f.
[33] Petrovskii, no. 42, 136 f; Petrovskii, no. 49, 150 f.; Petrovskii, no. 51, 152 f.; Petrovskii, no. 52, 153.
[34] Petrovskii, no. 70, 172; Petrovskii, no. 71, 172 f.; Petrovskii, no. 77, 179 f.
[35] Petrovskii, no. 102, 211–213; Petrovskii, no. 115, 225–228.

described his knowledge of local languages and his ability to use them in gathering information: "I am literally the only person in Tashkent to whom the Sarts come, not to make their salams to the Russian chief, but to chat about religious and other matters from the heart."[36] With this, he set himself apart from the majority of Russian functionaries, who lacked these capabilities, in his view. Petrovskii also used his linguistic abilities on behalf of his private interest in regional culture and history. Both were taught to him in return for small favors by Dzhura-Bek (d. 1906), a former head of the city of Kitab. In writing to certain correspondents, such as Kobeko, Petrovskii in turn presented himself as a mentor. He procured materials and offered recommendations. These courtesies simultaneously served to document the success of his self-study.[37]

As consul, Petrovskii also took part directly in knowledge production and dissemination. In 1886 he reported to Osten-Saken that he had seen to the partial reprinting of William Mayers's book *The Chinese Government* in the regional newspaper *Turkestani News*, following the author's profitable lecture. A further example is Petrovskii's research and travel guide, *A Companion for Travelers to Little-Studied Countries*, a compilation of existing texts and his own experiences, whose publication he expected in 1888 through the general staff. Petrovskii also told Osten-Saken of his personal ethnographic records of traditional articles of daily use of the indigenous people, which he had made by means of a camera obscura. All in all, Petrovskii projected the image of a reflective and experienced researcher who desired recognition and public resonance for his work.[38]

Petrovskii also demonstrated personal initiative in the 1880s in the domain of meteorology. In September 1883, the consul reported to Ermakov that he had acquired the necessary instruments to take initial measurements. In later years, Petrovskii successfully convinced Genrikh Ivanovich Vil'd (1833–1902), a meteorologist and director of the Main Physical Observatory in St. Petersburg, to establish a permanent weather station. The consul independently conducted a series of measurements. As he reported to Osten-Saken in February 1887:

It turns out (after five months of observation) that Kashgar is not as dry as is assumed, and accordingly, given its relative elevation (four thousand feet), it is not so far removed from the climate of our southern Russia and could be cultivated with greater variety than Fergana, for example.[39]

---

[36] Petrovskii, no. 7, 82.
[37] Ibid.; Petrovskii, no. 71, 172; Petrovskii, no. 72, 174.
[38] Petrovskii, no. 77, 180; Petrovskii, no. 101, 209 f.; Petrovskii, "Kommentarii," 324; William Mayers, *The Chinese Government: A Manual of Chinese Titles, Categorically Arranged and Explained, with an Appendix* (Shanghai: Kelly & Walsh, 1886).
[39] Petrovskii, no. 91, 197.

Petrovskii demonstrated himself to be a scholar who measured, evaluated, and interpreted his results with a view to the empire's needs (here the agricultural exploitation of virgin lands).[40]

In the domain of the history of research on Central Asia, in Petrovskii's numerous letters to Kobeko, Osten-Saken, and the naturalist and traveler Nikolai Mikhailovich Przheval'skii (1839–1888) he ultimately conceptualized himself as an explorer. Between 1885 and 1892, the consul first tracked down the remnants of the equipment of the Bavarian explorer Adolf Schlagintweit (1829–1857), who was executed in Kashgar in 1857, and subsequently worked intensively and successfully towards the erection of a monument and an official expression of gratitude from Baron Rudolf von Gasser (1829–1904), the Bavarian envoy in Russia. In his letters, he informed his correspondents about his investigations and solicited their support, but then also took credit for any successes.[41]

In the field of archeology, Petrovskii's 1892 discovery of coins in the Iarkand region was met with interest, and a year later it led to the publication of an article with his own photographs of the finds in the *Proceedings of the Eastern Section of the Imperial Russian Archeological Society*. Here no support from a third party seems to have been necessary, as demonstrated by the fact that the amateur Petrovskii became quite well known in this field. As a translator, in contrast, Petrovskii was still dependent on the publication opportunities offered to him. His translation of the 1864 book *The Post and Travel Routes of the Orient* by the Austrian Orientalist Aloys Sprenger was published in 1894 in the *Turkestani News*, without the commentary he envisaged, in return for financial support promised to the newspaper by the governor general.[42]

As consul, in the 1880s and 1890s, Petrovskii supported Russian and international explorers and helped them in preparation for their journeys or during their transit through East Turkestan. For example, in 1887–1888, before and during Nikolai Mikhailovich Przheval'skii's fifth expedition to Tibet, Petrovskii was in contact with him and supported him through his wide-ranging local knowledge, while also repeatedly offering him assistance in dealing with the Chinese authorities. Here the consul made himself out to be

---

[40] Petrovskii, no. 38, 129; Petrovskii, no. 64, 166; Petrovskii, no. 77, 180; Petrovskii, no. 91, 197.

[41] Hermann Kreutzmann, "Die Brüder Schlagintweit im Great Game," in *Über den Himalaya*, ed. Brescius, Kaiser, and Kleidt, 97–111; Petrovskii, no. 62, 164; Petrovskii, no. 80, 184; Petrovskii, no. 82, 188; Petrovskii, no. 84, 189; Petrovskii, no. 92, 198; Petrovskii, no. 102, 212.

[42] Petrovskii, no. 116, 228 f.; Petrovskii, "Kommentarii," 326, 328; Nikolai Petrovskii, "Zagadochnye iarkendskie monety," *Zapiski Vostochogo otdeleniia Imperatorskogo Russkogo Arkheologicheskogo Obshchestva* 7 (1893): 307–310; Aloys Sprenger, *Die Post- und Reiserouten des Orients: Mit 16 Karten nach einheimischen Quellen* (Leipzig, 1864); Petrovskii, no. 140, 253.

comprehensively informed, influential, and emphatically ready to help.[43] In Kashgar, Petrovskii housed many travelers at the consulate. Although these duties often posed an additional burden to the consul, the benefit was often mutual, as in the case of the French forest inspector Eduard Blanc. Through Blanc, Petrovskii obtained the latest information about a Russian-French treaty.[44]

But Petrovskii also expressed criticism of those travelers who he believed were only masquerading as scholars, such as the geographer and zoologist Grigorii Efimovich Grum-Grzhimailo (1860–1936). The latter's hassles with the Chinese authorities prompted Petrovskii to express his thoughts on possible control measures the Geographic Society could implement on expeditions to Przheval'skii in 1886. In this way, the consul showed himself to feel a sense of duty towards research as such, rather than just towards individual scholars. His very general critique of English travelers, however, suggests that by the mid-1880s his support was limited by his patriotism and the Pamir conflict.[45]

In his dealings with the well-known scholars Viktor Romanovich Rozen (1849–1908), chairman of the Eastern Division of the Archeological Society, and the above-mentioned Indologist Ol'denburg, Petrovskii described himself in the 1880s and 1890s as a conduit for information and material and as an amateur researcher. He purchased archeological artifacts and sent them on to the metropole, reported on the plotting of his sites, and was in close contact with both of them about his discoveries. In addition to the above-mentioned coins, which led to Petrovskii's 1893 publication in Rozen's specialist journal, another of his most spectacular finds was eighty-seven sheets of extensive Sanskrit manuscripts, about which he reported to Rozen in September 1892. The Indologist Ol'denburg had confirmed to him that this is what they were, as Petrovskii emphasized to Osten-Saken in 1893. In 1892, Petrovskii hoped to find resonance in expert circles in St. Petersburg by sending these eighty-seven documents to Rozen. Ol'denburg's cataloguing of the discoveries turned Petrovskii into a collector of European standing.[46]

Petrovskii's many letters to Rozen from 1892 make clear both his own research contribution and his expert exchange with the metropole. For example, Petrovskii asked Rozen to connect him with an expert regarding a baffling glass nail that he discovered in a megalithic tomb not far from Kashgar. In May 1893 he informed Ol'denburg that he was not convinced by the appraisal he had received. Through this critique, Petrovskii staked the claim

---

[43] Petrovskii, no. 96, 203; Petrovskii, no. 98, 206 f.; Petrovskii, no. 99, 208.
[44] Petrovskii, no. 107, 216; Petrovskii, no. 108, 216 f.; Petrovskii, no. 109, 218; Petrovskii, "Kommentarii," 324.
[45] Petrovskii, no. 81, 186; Petrovskii, no. 129, 243; Petrovskii, no. 162, 271; Petrovskii, "Kommentarii," 322.
[46] Petrovskii, no. 119, 231 f.; Petrovskii, no. 125, 238 f.; Petrovskii, no. 122, 234.

that by around 1890 he had expertise equivalent to that of St. Petersburg professional circles. The consul appraised his findings, took notice of expert colleagues, solicited their opinion, and sometimes also expressed criticism of their judgment.[47]

The letters to Ol'denburg from the 1880s and 1890s confirm our previous findings. In the role of informant, Petrovskii collected numerous historical artifacts for the Indologist, as accompanying letters from 1894 and 1895 document. In 1893 and 1894 he also reported on the network of contacts he had built in the region. Additionally, between 1894 and 1895 Petrovskii compared notes with Ol'denburg about aspects of the history of East Turkestan or how to interpret individual archeological discoveries, thereby appearing to be well versed in the field. The letters to Rozen and Ol'denburg point to two individual relationships, but Petrovskii sending his findings to Rozen and his discussion with Ol'denburg about those same findings are a sign of his broader research connections. Ol'denburg's praise indicates that Petrovskii enjoyed recognition among his professional contacts in St. Petersburg. Through his confession to Ol'denburg in March 1895 that it was only through a reference in the journal of the Imperial Russian Archeological Society that he recognized the scholarly value of his findings, however, the "simple collector"[48] demonstrated the limits of his connoisseurship.[49]

Conclusions

Petrovskii's biography and self-descriptions exhibit certain characteristics of imperial biographies, as Malte Rolf has recently illustrated. There are several notable similarities to the case of Carl Patsch, as described in Daniel Baric's contribution to this volume. Like Patsch, Petrovskii belonged to a group of very "mobile actors." His postings at the edge of the empire and his official journeys to the metropole as well as to Turkestan make this clear. Likewise, the amateur scholar betook himself to local excavations and took part in meetings of the Eastern Division of the Archeological Society in St. Petersburg. The empire formed the framework of Petrovskii's thought and action, just as it did for Patsch. Thus, he reacted to local problems by proposing empire-wide solutions. Examples of this include his intention, without being directed to do so, to prepare a consular statute for the entire empire, or his thoughts regarding a central control system for Russian expeditions in the wake of derelict behavior from individual travelers. Both actors' varied per-

---

[47] Petrovskii, no. 117, 229; Petrovskii, no. 122, 235 f.; Petrovskii, "Kommentarii," 326.
[48] Petrovskii, no. 160, 269.
[49] Petrovskii, no. 122, 234 f.; Petrovskii, no. 134, 248; Petrovskii, no. 145, 257 f.; Petrovskii, no. 136, 249; Petrovskii, no. 149, 260 f.; Petrovskii, no. 142, 255.

sonal contacts at the center and the periphery also map onto these connections.[50]

Petrovskii's career illustrates what Rolf has called the "close interplay with the underlying patterns of order"[51] of the empire. If we look at his professional career from finance official to general consul, the empire, and especially Turkestan, served as his professional "enabling space." Privately, the scholarly expertise he developed at the periphery gained him recognition in expert circles by the mid-1880s at the latest. Patsch's career in Bosnia seems to have functioned similarly. Petrovskii's 1864 criminal conviction can serve on the one hand as a confirmation: in this reading, the empire offered peripheral niches for those people who had somehow fallen out of favor. But on the other hand, as a result of his conviction, Tashkent and Kashgar may well have been Petrovskii's only options, which means that the empire limited his professional career. His vain efforts to be transferred to another part of the empire or his protracted requests for promotion point to this, but it cannot be determined definitively.

Petrovskii's critical attitude, his self-initiative, and his efforts to further his career and to gain recognition for his accomplishments—three central characteristics of his self-descriptions—point to the room to maneuver highlighted by Anke Stephan. From the perspective of the finance official and the consul, his nonstop criticism and perpetual self-initiative indicate the greater scope for action he had in Kashgar. Petrovskii's suggestions for joint projects or discussion of politics in Pamir serve as examples of his desire to expand this scope. He presented himself as a regional expert in Kashgar, criticizing the center's decisions and proposing alternative assessments as a sort of "man on the spot"[52] during the Pamir conflict. At the same time, Petrovskii's ideas, which were rarely implemented, and his oft-ignored expertise prove that his professional opportunities to contribute were limited. Likewise, as a publicist and informant, he continually needed the help of third parties to get his articles published.[53]

Petrovskii's erroneous estimation of his own room to maneuver was made clear by his desire to be transferred, which he expressed beginning in the mid-1880s. The consul based his claim to certain positions on his qualifications. With reference to his early recognition and criticism of problems in other places, in the early 1890s the consul emphasized his perspicacity, while also harkening back to his contributions to the good of the state. These points indicate the continued existence of several elements of the "service biog-

---

[50] Rolf, "Einführung," 5, 9, 11; Bukhert, "Russkii portret," 31.
[51] Rolf, "Einführung," 9.
[52] Benedikt Stuchtey, "Kolonialismus und Imperialismus von 1450 bis 1950," *Europäische Geschichte Online* (2010): 7, accessed 30 March 2017, http://nbn-resolving.de/urn:nbn:de:0159-20101011129.
[53] Stephan, "Erinnertes Leben," 7 f.; Rolf, "Einführung," 8.

raphy," as elucidated by Ulrich Schmid.⁵⁴ It cannot be ascertained whether the relevant authorities did not believe his self-image of a qualified and loyal state servant and regional expert and therefore did not transfer him and promoted him only after several years.

Petrovskii's self-description as a contributor to the study of Central Asia seemed, in contrast, to correspond significantly more often with his true room for maneuver. His concrete contributions to specific scholarly fields were successful at a substantially higher rate. This is demonstrated by his publications, his contributions as an informant, and projects like the weather station or the monument to Schlagintweit. Furthermore, Petrovskii more frequently mentioned the recognition he had gotten. On top of this, his interactions with scholarly titans of his time underscores that he was successful in climbing hierarchy. This illustrates what Vera Tolz has depicted as the exchange between specialized scholarship and "specialized amateurs,"⁵⁵ going beyond institutionalized forums. But despite all that, Petrovskii remained a layman. He repeatedly laid bare his limits as a "simple collector," as he deemed himself.

Further research into members of other groups of actors in Turkestan, as well as in other peripheral regions of the Russian Empire, ought to expand upon our findings here. Comparing this case to Daniel Baric's contribution illustrates the new insights gained through comparisons with actors from other empires who, like Petrovskii, moved between imperial politics and scholarship.

Translated from German by Kate Younger.

---

⁵⁴ Schmid, *Ichentwürfe*, 375 f.
⁵⁵ Tolz, "Orientalisten," 128.

Johannes Feichtinger

"ORIENTALISTIK" IN THE HABSBURG MONARCHY BETWEEN
IMPERIAL PRAGMATISM AND "PURE" SCHOLARSHIP

A prominent example of the mutual influence of scholarship and the imperial order in the Habsburg Monarchy is the discipline of *Orientalistik* (Oriental studies), which was understood as both a practical political science and "pure" scholarship. Both versions were fashionable at different times. As a discipline that had imperial significance, Eastern studies had a long tradition outside the university setting in Vienna, stretching back to the time of the Turkish wars. As a purely academic subject, however, *Orientalistik* was a "child of our century,"[1] as Johann Kirste (1851-1920) characterized it in his inaugural lecture in Graz in 1892. The first oriental institute was founded in 1886 at the University of Vienna in the run-up to the Seventh International Orientalist Congress, which took place that year in the imperial capital.

This article examines the spheres of activity of *Orientalistik* in the Habsburg Monarchy between imperial pragmatism and "pure" scholarship. It focuses on Vienna and shows how the East was demarcated there, how its study was organized against the backdrop of the imperial order, and what function *Orientalistik* fulfilled at different times. The alliance between the Habsburg monarchy and the Ottoman Empire in the First World War made it clear to the Viennese Orientalist and geographer Hans von Mžik (1876-1961) that the East should be understood as a "volatile concept," the product of "constructions," in which "a teleological moment" was obvious: since the Middle Ages, the East had "amounted to nothing more than an opposition between Islam and Christendom, between the Turkish Empire and Europe." The space demarcated by this concept was presented "as a given and unalterable natural or cultural expanse." The wartime alliance rendered this definition obsolete. Under these auspices, Mžik perceived a "volatile" moment in the application of the concept of the East. The East, which was conceived of as a space, was ultimately a "historico-political" construct whose position and extent were defined by the demands of any given time. In using the term, the "solid ground of facts is left behind and the ground of dogma

---

1   J[ohann] Kirste, *Die Bedeutung der orientalischen Philologie: Eine Antritts-Vorlesung, gehalten an der Universität Graz am 5. Mai 1892* (Vienna, 1892), 3.

or emotional considerations" was entered, but in any case, "no excess of insight" was achieved.[2]

In the nineteenth century, scholarship that dealt with the East met two different demands: on the one hand *Orientalistik* expanded the contemporary knowledge base, and, on the other, it fulfilled a pragmatic political function. Through the study of the linguistic and cultural monuments of the ancient Middle East, the "fonts of wisdom of humanity" could be tapped and "the refined ethics of the first cultured peoples" could be conveyed to contemporaries. At the same time, through the mediation of present-day languages, trade with Near Eastern countries could be intensified and a basis for colonial enterprises could be created.[3] This double function of *Orientalistik*, which served both to safeguard economic resources and extend political dominance as well as to further knowledge, was recognized by scholars at the time. The Viennese Orientalist and writer Carl Ferdinand von Vincenti (1835–1917) picked up on this ambivalent role of Eastern studies in his feuilleton article "Zum Orientalisten-Congreß" (On the Orientalist Congress), which appeared in 1886 in Vienna in the *Neue Freie Presse*. *Orientalistik* sounded "so aloof and scholarly" and yet "today it has almost become a political force, which takes an important place in the cultural progress of our time." It opened "the treasure chests of the oldest literature" and also "outfitted the pioneers of trade and transport with powerful weapons. [...] Our 'mission' in the Orient is constantly under discussion."[4] The latter function, however, had only been insufficiently fulfilled by Viennese university-based *Orientalistik* in the late nineteenth century. Vienna had been "just the seventh suburb [Vorort]"—after Paris, London, St. Petersburg, Florence, Berlin, and Leiden—to make a bid to hold the international Orientalist congress. "Vienna," as Vincenti wrote, "just takes its time."[5]

In Vienna these two roles for *Orientalistik* were fulfilled by different institutions of learning: at the university, the new subject was understood as strictly scholarly and predominantly antiquarian. Here the receipt of "philological rites" was the sine qua non.[6] Within the university, the East was widely researched and mediated from the perspective of "dead" languages. In contrast, those Near Eastern studies that were traditionally conducted outside the

---

[2] Hans von Mžik, "Was ist Orient? Eine Untersuchung auf dem Gebiete der politischen Geographie," *Mitteilungen der k. k. Geographischen Gesellschaft in Wien* 61 (1918): 191–208, here 192, 199 f., 203 f., 206–208.

[3] C. von Vincenti, "Zum Orientalisten–Congreß," *Neue Freie Presse*, 25 September 1886, 1–3, here 1.

[4] Ibid.

[5] Ibid.

[6] Bert G. Fragner, "Die deutschen Orientalisten im 20. Jahrhundert und der Zeitgeist," *Der Zeitgeist und die Historie*, ed. Hermann Joseph Hiery (Dettelbach: Röll, 2001), 36–51.

university in Austria were considered a "school of praxis."[7] Their homes were the k.k. Oriental Academy, the Imperial Academy of Sciences, different state educational establishments, scientific societies, and the k.k. Oriental Museum, which opened in 1875. In these institutions, the living Middle Eastern languages were taught, contact with the East was fostered, and the region was studied from various angles: partially antiquarian, but also linguistic, geographic, ethnographic, geological, botanical, and zoological—for the benefit of the imperial order.

The Ambiguous Position of "Orientalistik"

The ambiguous position of *Orientalistik* dates back to the sixteenth century. In 1552 King Ferdinand I appointed the humanist and Orientalist Johann Albrecht Widmanstetter (1506–1557) as superintendent of the university at the Viennese Court and introduced the teaching of Greek and Hebrew as part of his reform of the faculty of arts.[8] At the time of the Reformation, teaching Hebrew took on new significance, as it provided a crucial foundation for biblical exegesis, "this eminent German science."[9] Hebrew was also taught at the universities of Prague and Graz, though the Catholic theological faculty in Vienna, as the theologian Neumann later put it cynically, believed this was "to do away with biblical philology entirely."[10] At the Vienna theological faculty, the study of Hebrew language and foundational texts was first incorporated into the plan of studies in 1752 under Maria Theresia (one hour per day for each over the course of four years).[11] *Orientalistik* was only freed from its "theological shackles" with the Thun-Hohenstein university reform in the mid-nineteenth century,[12] and beginning with the era of high liberalism it

---

[7] Wilhelm Anton Neumann, "Über die orientalischen Sprachstudien seit dem 13. Jahrhundert: Mit besonderer Rücksicht auf Wien. Inaugurationsrede, gehalten am 17. October 1899 im Festsaale der Universität," in *Die feierliche Inauguration des Rectors der Wiener Universität für das Studienjahr 1899/1900 am 17. October 1899* (Vienna, 1899), 41–133, here 53.

[8] Neumann, "Über die orientalischen Sprachstudien," 79–81; Sigmund Ritter von Riezler, "Widmanstetter, Johann Albrecht," in *Allgemeine Deutsche Biographie* 42 (1897): 357–361; Johann Fück, *Die Arabischen Studien in Europa bis in den Anfang des 20. Jahrhunderts* (Leipzig: Harrassowitz, 1955), 43. At Ferdinand I's expense, Widmanstetter commissioned an edition of the New Testament in the Syrian language, which was the first movable-type printed Syrian book and also the first Eastern publication printed in Vienna; it appeared in 1555. In the same year he also printed the first Syrian grammar.

[9] Vincenti, "Zum Orientalisten–Congreß," 1.

[10] Neumann, "Über die orientalischen Sprachstudien," 79.

[11] Ibid., 88.

[12] Vincenti, "Zum Orientalisten–Congreß," 2. Already before 1848, August Pfizmaier had for several years taught Chinese, Turkish, Arabic, and Persian language and literature at the University of Vienna. On nineteenth-century *Orientalistik* in Austria, see especially Leopold Hellmuth, "Traditionen und Schwerpunkte der österreichischen Orientalistik

experienced a considerable boom: at the universities of Prague, Graz, and Innsbruck, professors of *Orientalistik* were appointed,[13] and the philosophy faculty of the University of Vienna established professorial chairs for comparative linguistic research and Eastern philology and archeology. Thus, by the late nineteenth century, broad swathes of the East were covered by university professorships. The main focus was on the Arabic realm and Egypt, but attention was also paid to Japan, China, and India—countries which had received little notice at Austrian universities in previous centuries.[14] An antiquarian and philological orientation was predominant. Instruction in the ancient languages and cultures of the East even included Sanskrit. The first two Viennese professorships for Sanskrit and Eastern languages were held by Anton Boller and Jakob Goldenthal. Boller was an "autodidact," as the writer Robert Hamerling recalled, and Sanskrit did not prove to be a magnet for students: "But no attendees could be found. For three years I was Boller's nearly sole pupil."[15] The core fields of the ever more important discipline of *Orientalistik* were, in addition to Sanskrit and archeology (Georg Bühler), comparative linguistics (Friedrich Müller, Eduard Sachau), Semitic epigraphy (David Heinrich Müller), Arabic paleography and numismatics (Josef Karabacek), and Egyptian-Hamitic linguistics (Leo Reinisch). Arabic and Semitic studies experienced a considerable boom; the emphasis was predominantly on handwriting and inscriptions.[16]

The lead-up to the Seventh International Orientalist Congress in Vienna gave the five professors of Eastern languages the decisive occasion to come

im 19. Jahrhundert," in *Orient: Österreichische Malerei zwischen 1848 und 1914*, ed. Erika Mayr-Oehring (Salzburg: Residenzgalerie, 1997), 107–127.

[13] In Prague, Rabbi Saul Isaac Kaempf (1818–1892), who had completed his habilitation in 1850 at the Alma Mater Carolina in the field of ancient Hebrew language and literature, was given a nontenured professorship for Oriental languages and literature in 1859, which he held until 1885. In 1886 the philosophical faculty in Prague hired Max Grünert (1849–1929), an instructor of Semitic languages, as a nontenured professor. In Graz, the Indogermanist and Sanskrit scholar Johann Kirste was hired as a non-tenured professor in 1892 and in Innsbruck the theologian and Orientalist August Haffner (1869–1941) was appointed non-tenured professor of Semitic languages at the philosophical faculty.

[14] An overview of the professors and their activity can be found in *Geschichte der Universität Wien von 1848 bis 1898 als Huldigungsfestschrift zum fünfzigjährigen Regierungsjubiläum seiner k. u. k. Apostolischen Majestät des Kaisers Franz Josef I.*, ed. Akademischer Senat der Wiener Universität (Vienna, 1898), 359–362; Hermann Hunger, "Orientalistik," in *Geschichte der österreichischen Humanwissenschaften*, vol. 4, *Geschichte und fremde Kulturen*, ed. Karl Acham (Vienna: Passagen Verlag, 2002), 467–480; Wolfdieter Bihl, *Orientalistik an der Universität Wien: Forschungen zwischen Maghreb und Ost- und Südasien. Die Professoren und Dozenten* (Vienna: Böhlau, 2009).

[15] Robert Hamerling, *Stationen meiner Lebenspilgerschaft* (Hamburg, 1889), 176.

[16] On the significance of Arabian studies, especially in Vienna, cf. Andre Gingrich, "Science, Race, and Empire: Ethnography in Vienna before 1918," *East Central Europe* 43 (2016): 41–63, here 47 f.

together and to organize themselves into an institute at the University of Vienna. In March 1886 the Ministry of Religion and Education approved the establishment of the Oriental Institute, the first of its kind in the German-speaking realm.[17] The new institute was provided with an endowment, so that a library could be set up and the *Wiener Zeitschrift für die Kunde des Morgenlandes* (Viennese Journal for the Study of the Orient; vol. 1, 1887) could be published by the five professors, Bühler, Karabacek, D. H. Müller, F. Müller, and Reinisch.[18] The strictly scholarly self-image exhibited by the journal publishers, the "leaders of the Oriental Institute of the university," is striking. They even did not deem it necessary to prepend a mission statement to their journal. The first volume opened abruptly with Georg Bühler's work "Gleanings from Yâdavaprakâśa's Vaijayantî." In the late nineteenth century the treatment of purely antiquarian Eastern philology ultimately correlated with pragmatic imperial Eastern studies. At the Imperial Academy of Sciences, whose first president was Joseph Freiherr von Hammer-Purgstall (1774–1856), *Orientalistik* was treated in a more wide-ranging fashion than at the University of Vienna. Here the period between the late 1860s and the First World War was a "shining epoch."[19] The five *Orientalistik* professors named above, who were elected as members of the academy, were involved in various commissions at the Imperial Academy of Sciences. In 1898 the academy introduced a South Arabian commission and sent an expedition led by the philosophical-historical and mathematical-natural scientific class to Yemen and the island of Socotra. A key role in the academy's commission and in the South Arabian expedition was played by David Heinrich Müller (1846–1912).[20] Müller, a tenured professor of Semitic studies at the University of Vienna since 1885, had previously sent his student Eduard Glaser (1855–1908) to travel around South Arabia and copy inscriptions. On his trips to Yemen (1882–1895), Glaser collected materials on Old South Arabian language, culture, and his-

---

[17] The first Oriental seminar in Germany was established in 1894 at the University of Heidelberg. See Sabine Mangold, *Eine "weltbürgerliche Wissenschaft:" Die deutsche Orientalistik im 19. Jahrhundert* (Stuttgart: Franz Steiner, 2004), 173.

[18] Cf. *Geschichte der Universität Wien von 1848 bis 1898*, 361 f.; Gerhard J. Selz et al., "Der ganze Orient: Zur Geschichte der orientalistischen Fächer am Beispiel einer internationalen orientalistischen Zeitschrift. Die Wiener Zeitschrift für die Kunde des Morgenlandes (WZKM)," *Wiener Zeitschrift für die Kunde des Morgenlandes* 100 (2010): 9–35; Sibylle Wentker, "Orientalistik in Wiener Zeitschriften," in *Wissenschaftliche Forschung in Österreich, 1800–1900: Spezialisierung, Organisation, Praxis*, ed. Christine Ottner, Gerhard Holzer, and Petra Svatek (Göttingen: V&R unipress, 2015), 197–214.

[19] Richard Meister, *Geschichte der Akademie der Wissenschaften in Wien, 1847–1947* (Vienna: Verlag Adolf Holzhausens, 1947), 142.

[20] Cf. Gertraud Sturm, ed., *David Heinrich Müller und die südarabische Expedition der Kaiserlichen Akademie der Wissenschaften 1898/99: Eine wissenschaftsgeschichtliche Darstellung aus der Sicht der Kultur- und Sozialanthropologie* (Vienna: Verlag der Österreichischen Akademie der Wissenschaften, 2015).

tory and also undertook astronomical, cartographical, and meteorological research.[21]

The 1898–1899 South Arabian expedition also studied Yemen interdisciplinarily: while it focused primarily on linguistic goals and gathered a collection of inscriptions and phonographic recordings of the languages of Mehra and Socotra, the natural scientists among the expedition participants also recorded geographical and geological information about Socotra and they simultaneously studied the fauna and flora as well as ethnographic relationships. A "pure" scholarly interest in research was at the forefront.

This major undertaking also had political significance for the imperial household, however. The island of Socotra, which is located at the outlet of the Gulf of Eden, was in a geopolitically important position. Unclarified ownership claims had mobilized Emperor Franz Joseph and Archduke Ferdinand Maximilian to occupy the island in 1857. According to the historian Günther Hamann, it was the "site of one of the Austrian attempts in the second half of the nineteenth century to gain colonial outposts in Africa and Asia."[22] This undertaking bore no results. With the building of the Suez Canal, Socotra gained even greater geostrategic significance. In 1886 the island once again became a British possession—Britain had previously occupied Socotra but then gave it up again. The wide-ranging scholarly investigation of the island by the Imperial Academy of Sciences demonstrates that Austria-Hungary continued to keep Socotra in its sights. In this way, research of South Arabia prior to 1914[23] is a telling example of Austrian *Orientalistik*, in which "pure" scholarly research was also linked to pragmatic imperial goals.

Ottoman Studies and Turkology

As the most traditional subject of Habsburg *Orientalistik* that was also most important to the state, Turkology was not represented by any chaired professorship at the University of Vienna before the First World War. Only in 1917 was the first professor for Turkology appointed, in conjunction with the mili-

---

[21] The Sabaean inscriptions were published by 1981 by the Academy of Sciences, which commissioned his research trips, in the "Eduard Glaser collection."

[22] Günther Hamann, "Österreich-Ungarns Anteil an Reisen und Forschungen in den Ländern des Britischen Weltreiches," in *Österreich und die Angelsächsische Welt: Kulturbegegnungen und Vergleiche*, vol. 2, ed. Otto Hirsch (Vienna: Universitätsverlag Wilhelm Braumüller, 1968), 202–236, here 206. Cf. Harald Lobitzer and Karl Kadletz, *Grenzenlos: Forschungen von Mitarbeitern der Geologischen Reichsanstalt/Bundesanstalt außerhalb Europas* (Vienna: Geologische Bundesanstalt, 2005), 26.

[23] Cf. Andre Gingrich and Regina Bendix, "David Heinrich Müller und das Spannungsfeld der Wiener Südarabienforschung vor dem Ersten Weltkrieg: Eine Einführung," in *David Heinrich Müller*, ed. Sturm, 11–21; Gingrich, "Science, Race, and Empire," 47, 48, and 51.

tary alliance between the Habsburg and Ottoman empires.[24] The task facing the Turkologist Friedrich Kraelitz von Greifenhorst (1876–1932) was to process the rich Ottoman source collections in the archives of the Habsburg lands, in the pursuit of historical Turkology. For centuries, in Vienna Ottoman studies was not treated "purely" scientifically, since it had political and practical significance. Studying contemporary Turkic languages was traditionally done outside the university: in the seventeenth century in a boys' language institute in Constantinople, then in the eighteenth century at the k.k. Oriental Academy, which the Viennese court established expressly for this purpose. Franz von Mesgnien Meninski, who was appointed "imperial Turkish interpreter" (kayl. Türkhischen Dolmatsch)[25] in 1662, reprinted his pathbreaking *Arabic-Persian-Turkish Dictionary* (Thesaurus linguarum orientalium, 1680–1687), which for centuries was the most important teaching tool. In the eighteenth century this dictionary was republished by the pupils of an educational institution established in Vienna in 1754 by Maria Theresia at the urging of Minister Wenzel Anton Fürst von Kaunitz.[26] The k.k. Academy of Eastern Languages sought to educate "capable youths in the necessary languages of the East as well as the West" and "additionally in all sciences that [were important] to the protection of Austria's commercial and political interests in the East."[27] The "k.k. language boys" (Sprachknaben) were sent to the High Porte as interpreters and diplomats.[28] The most famous pupil was Joseph von Hammer-Purgstall, who not only published the first journal of

---

[24] In the 1915–1916 winter semester, the Ministry of Religion and Education granted Lecturer Friedrich Kraelitz von Greifenhorst, a "class II" curator at the imperial court library in Vienna, a paid, four-hour teaching assignment for the language, literature, and history of the Turkish and Tatar people. He was simultaneously granted the title of an adjunct university professor. His teaching assignment was renewed for 1916–1917 and 1917–1918. In 1917, at the decree of the emperor, Kraelitz was appointed adjunct professor *ad personam* with requisite compensation. See personnel file on Friedrich Kraelitz Edler von Greifenhorst, PA 2295, Vienna University Archive.

[25] Franz Babinger, "Die türkischen Studien in Europa bis zum Auftreten Hammer-Purgstalls," *Die Welt des Islams* 7/3–4 (1919): 103–129, here 115.

[26] Franciscus a Mesgnien Meninski, *Lexicon Arabico – Persico – Turcicum*, 4 vols. (Viennae, 1780–1802).

[27] Babinger, "Die türkischen Studien in Europa bis zum Auftreten Hammer-Purgstalls," 123. For more, see Ernst Dieter Petritsch, "Erziehung in 'guten Sitten, Andacht und Gehorsam': Die 1754 gegründete Orientalische Akademie in Wien," in *Das Osmanische Reich und die Habsburgermonarchie: Akten des internationalen Kongresses zum 150-jährigen Bestehen des Instituts für Österreichische Geschichtsforschung. Wien, 22.–25. September 2004*, ed. Marlene Kurz et al. (Vienna: Oldenbourg, 2005), 491–501; Ernst Dieter Petritsch, "Die Anfänge der Orientalischen Akademie," in *250 Jahre: Von der Orientalischen zur Diplomatischen Akademie in Wien*, ed. Oliver Rathkolb (Innsbruck: StudienVerlag, 2004), 47–64.

[28] For a more detailed account, see Michaela Wolf, *Die vielsprachige Seele Kakaniens: Übersetzen und Dolmetschen in der Habsburgermonarchie* (Vienna: Böhlau, 2012), 179–188.

Eastern studies, filled with observers' accounts in various languages and original sources, and filled the *Wiener Jahrbücher der Literatur* with Eastern poetry, but also produced a monumental *History of the Ottoman Empire* (10 vols., 1827–1833) at a time when this empire already had clay feet. In his famous 1852 *Akademierede* on multilingualism, Hammer-Purgstall also called for the establishment of professorships for the Arabic, Persian, and Turkish vernacular, which he hoped would promote "the flow of trade in the Eastern world."[29] Nothing came of this, however. Other institutions were entrusted with the politically practical task of teaching the main three contemporary Eastern Languages (Arabic, Persian, and Turkish); these included the Imperial Public School for Eastern languages, established in 1873 and subordinate to the Ministry of Religion and Education and the k.k. Polytechnic. Eastern languages were also particularly well tended to at the imperial court library, with its long tradition of erudition and cultivating Eastern languages.[30] It should be recalled that the University of Vienna appointed its first non-tenured professor for Turkology in conjunction with the Habsburg-Ottoman alliance during the First World War. In 1915, according to an order from the minister for religion and education, due to the "great and constantly increasing significance to Austria of the Near East and Turkey in particular," it seemed fitting "to dedicate an increased amount of care to scholarly instruction in the Turkish language, literature, and history at the University of Vienna."[31]

The ambiguous position of *Orientalistik* between imperial pragmatism and "pure" scholarship in the Habsburg Empire was fundamentally due to the university's structural organization. From 1623 to 1773 the theological and philosophical faculties at the University of Vienna were largely in Jesuit hands. Presumably at this time there was no urgent desire on the part of the university to concern itself with and impart to students the non-Biblical Eastern languages (e.g., Turkish), even if they were important for the state, diplomacy, and trade. Such instruction took place outside the university. This was also why, even in the nineteenth century, the greatest Austrian Orientalists such as Joseph von Hammer-Purgstall; the Arabist Aloys Sprenger, secretary of the Royal Asiatic Society of Bengal; the founder of Islamic studies Alfred Freiherr von Kremer; and his student Ignaz Goldziher (from 1872 adjunct

---

[29] Joseph Freiherr von Hammer-Purgstall, "Vortrag über die Vielsprachigkeit, gehalten an der kaiserlichen Akademie der Wissenschaften in Wien am 29. Mai 1852," in *Die feierliche Sitzung der kaiserlichen Akademie der Wissenschaften* (Vienna, 1852), 99.

[30] This tradition was established by Ogier Ghislain de Busbecq (1522–1592), who had brought hundreds of manuscripts, including the oldest Dioscorides manuscript, with him to Vienna from his time at the Sublime Porte. See Smail Balić, "Der Orient im österreichischen Schicksal und die Orientforschung an der Österreichischen Nationalbibliothek," in *Festschrift Josef Stummvoll*, ed. Josef Mayerhöfer and Walter Ritzer (Vienna: Verlag Brüder Hollinek, 1970), 123–135, here 125.

[31] Personnel file on Friedrich Kraelitz Edler von Greifenhorst, PA 2295, Vienna University Archive.

professor (Privatdozent) and from 1894 titular ordinary professor at the University of Budapest) did not hold university positions. Before the mid-nineteenth century, as explained above, Eastern studies at the university was broadly limited to teaching the biblical languages, which the theological faculty was responsible for. Hence it is not surprising that at the opening of the Seventh International Orientalist Congress, held in Vienna in 1886, the honorary president, Paul Freiherr Gautsch von Frankenthurn (1851–1918), the minister for religion and education, had to ruefully concede that other states had taken the lead in "exploring the Orient" and "studying the same," animated by "colonization," and had to acknowledge his "sincere admiration" for these "unrivalled models to follow." In the past, according to the minister, Austria had "first and foremost dealt with its practical and political needs," but since 1850 the universities had also caught up in the scholarly cultivation of Eastern studies. Thanks to the "attentive care and prolific support" since that time, as honorary president of the Orientalist congress, Gautsch could already point to early successes in his welcome address: "It is precisely in Austria where a new course has recently been taken, one which seeks to attain a clear overview of the cultural achievements of Islam through grasping its prevailing ideas."[32] The minister omitted the fact that Alfred Freiherr von Kremer had composed his pioneering *History of the Prevailing Ideas of Islam* (Geschichte der herrschenden Ideen des Islams, 1868) not as a university professor, but as a diplomat in the k.u.k. Ministry of Foreign Affairs.

"Orientalistik" in Vienna—Colonial and Postcolonial

In his address to open the Seventh International Orientalist Congress, held in the ceremonial hall of the University of Vienna on 27 September 1886, conference president Alfred Freiherr von Kremer (1828–1889), one of the most important scholars of Islam of his time, sketched a new model for *Orientalistik*. He grounded it in "facts that are just as unfamiliar as they are indicative of the significance of Vienna and Austria and Hungary as intermediaries between the Levant and Europe."[33] Striking a tone critical of imperialism, Kremer outlined the Habsburgs' long history of entanglement with the East: he spoke of an "exchange of intellectual and material goods at a large scale," which had not yet ebbed away. "Civilization," according to Kremer, was "not the feat of a single nation, even if a talented one: it is the result of the peaceful exchange of ideas, of intellectual as well as material goods."[34] But what had happened in the East in the past two centuries was Western arrogance:

---

[32] Anonymous, "Siebenter internationaler Orientalisten–Congreß," *Wiener Abendpost*, 27 September 1886, 1–4, here 1.
[33] Ibid., 3.
[34] Ibid., 2.

The East was brought under the influence of Western culture initially through the hubris of European weapons; large swathes of land fell under direct European rule. [...] The languages and mores of the West encroach everywhere, even sometimes threatening native traditional culture, and impose themselves often with entirely unwarranted presumptuousness.[35]

The Orientalists' task, he argued, was to counteract colonialism and shine a light on the points of commonality and intersections between the East and West: "But now, precisely in this direction and with this goal in mind, our studies and works have taken on a significance the scope of which and the ultimate fruits of which can at present scarcely be anticipated."[36]

By reference to the history of entanglements between the Habsburgs and Ottomans, Alfred Freiherr von Kremer showed the international assembly of Orientalists that the East could be meaningfully studied not only under colonial premises, but also from a postcolonial perspective *avant la lettre*. It should not be overlooked that in Austria too the voices in favor of separating the West from the East, its reductive simplification and demonization, had not been silenced. Thus, for example, the Indologist and ethnographer Michael Haberlandt (1860–1940),[37] who had founded the Association for Ethnography, a museum for ethnography, and the *Austrian Jurnal of Ethnography* (Österreichische Zeitschrift für Volkskunde) credited scholarly research with having conveyed "much more about what is unique and curious about the Eastern nations, about their spatial and temporal delimitation and separation, about the recognition of national contexts and historical entanglements than about an artificial recognition of what is shared." One of the "main results," according to Haberlandt, was that there was "undoubtedly an Orient, not just Oriental nations." Eastern studies had demonstrated that "the unity of the Orient [...] is a reality:"

Everywhere there are the same foundations for the family and society, the same level of development maintained in the petrification of custom; everywhere the despotism of the great and the servility of the masses. The same way of life everywhere under the immovable law of the climate, the same brutality of the passions and, their consequence, the indignity of woman across the entire region, the spirit witty but barren—everywhere the same inability to give voice to that spirit through writing.

He voiced the opinion "that this motionlessness of the Orient is not unlike the rigidity of a spiritual death."[38]

The following tentative conclusion can be drawn: in the Habsburg Monarchy, *Orientalistik* had imperial significance. This significance stemmed less from its traditional function for biblical hermeneutics or its antiquarian ex-

---

[35] Ibid.
[36] Ibid.
[37] On Michael Haberlandt, see Reinhard Johler, "Die Okkupation Bosnien-Herzegowinas und die Institutionalisierung der österreichischen Volkskunde als Wissenschaft," in *Bosnien-Herzegowina und Österreich-Ungarn: Annäherungen an eine Kolonie*, ed. Clemens Ruthner and Tamara Scheer (Tübingen: Francke, 2018), 335–369.
[38] Michael Haberlandt, "Der Orient," *Neue Freie Presse*, 2 Oktober 1886, 1–3, here 2 f.

pression at the University of Vienna than from the pragmatism with which it had fulfilled practical and political needs outside the university setting since the sixteenth century. This sort of involvement with the East was revived in the late nineteenth century,[39] to "inaugurate a new era in the history of Eastern studies," as the feuilleton writer Leon Kellner put it. This new era of European *Orientalistik* was meant to reflect the existing "preconceptions" and give full attention to the reciprocal influence of East and West.[40] But this ideal stood in opposition to a discursively engineered reality, in which the East provided a projection surface for processes of imperial self-empowerment: the means for this was a sort of "Orientalism" (in Edward W. Said's sense),[41] whereby the East presented an identity-forming counterexample to a modern, enlightened Europe, while also giving rise to the *mission civilisatrice*. In considering the various approaches and stances of Austrian Orientalists in the late nineteenth century, it is clear that "k.u.k. Orientalism"[42] was more ambivalent, more multifaceted, and more complex than the Orientalism of Edward W. Said, who had the classical colonial powers in mind: for him, it consisted mainly of a European sense of superiority and in latent or openly anti-Islamic sentiment.[43]

K.u.k. Orientalism

In the late nineteenth century, high politics presented Austria as a *Kulturstaat*, which sought to secure its "supremacy in the European Orient" over its main rival, Russia, through the conveyance of "true cultural conditions."[44]

---

[39] See Johann Heiss and Johannes Feichtinger, "Distant Neighbors: Uses of Orientalism in the Late Nineteenth-Century Austro-Hungarian Empire," in *Deploying Orientalism in Culture and History: From Germany to Central and Eastern Europe*, ed. James Hodkinson et al. (Rochester, NY: Camden House, 2013), 148–165.
[40] Anonymous, "Siebenter internationaler Orientalisten-Congreß," 4.
[41] Edward W. Said, *Orientalismus* (Frankfurt am Main: S. Fischer, 2009), 12.
[42] Johannes Feichtinger, "Komplexer k. u. k. Orientalismus: Akteure, Institutionen, Diskurse im 19. und 20. Jahrhundert in Österreich," in *Orientalismen in Mitteleuropa: Diskurse, Akteure und Disziplinen vom 19. Jahrhundert bis zum Zweiten Weltkrieg*, ed. Robert Born and Sarah Lemmen (Bielefeld: Transcript, 2014), 31–63.
[43] On the particular characteristics of Orientalism in the late Habsburg Empire and in international comparison, see Andre Gingrich, "Orientalismus," in *Habsburg neu denken: Vielfalt und Ambivalenz in Zentraleuropa. 30 kulturwissenschaftliche Stichworte*, ed. Johannes Feichtinger and Heidemarie Uhl (Vienna: Böhlau, 2016), 156–162; Andre Gingrich, "The Nearby Frontier: Structural Analyses of Myths of Orientalism," *Diogenes* 60, no. 2 (2015): 60–66; Bert G. Fragner, "Wir im Orient—der Orient in uns," in *Orient und Okzident: Begegnungen und Wahrnehmungen aus fünf Jahrhunderten*, ed. Barbara Haider-Wilson and Maximilian Graf (Vienna: Neue Welt, 2016), 37–52.
[44] Kronprinz Rudolf, "Politische Denkschrift 1886: Skizzen aus der Österreichischen Politik der letzten Jahre," in *Kronprinz Rudolf: "Majestät, ich warne Sie..." Geheime und private Schriften*, ed. Brigitte Hamann, 2nd ed. (Munich: Fischer, 1998), 143–177, here 159.

Crown Prince Rudolf defined this goal unambiguously in a "Political Report on the 1884 Trip to the Orient" addressed to the emperor: "We will be the masters of the European Orient!"[45] This claim to supremacy was tied to the prerogative of facilitating culture: "Russia can only bring Asia to the European Orient, because it itself is not yet sophisticated; but we work under the laws of a higher cultural historical mission, in the name of European progress."[46] The crown prince saw Austria's "important" and "true task" in its "great civilizing role,"[47] "to show the Balkan peoples a piece of Austrian order, European culture, and above all modern tolerance" and "to be the bearers of Western culture to the Orient."[48]

When the East was discussed in the late Habsburg Monarchy, a distinction was frequently drawn between an inner and an outer East. The Ottoman Empire was deemed the outer East, where so much was still—as was said then—in a "bedraggled state."[49] The outer East therefore offered a projection field for imperial civilizing visions, for which the Oriental Museum, which opened in Vienna in 1875, proved to be a central site. Arthur von Scala, the museum's director and publisher of the *Oesterreichische Monatsschrift für den Orient* (Austrian Monthly for the Orient), drew in this journal a picture of the East "as a world apart," which could not be judged by the "general gauge of the culture of the West,"[50] and eloquently expressed the civilizing mission: "Austria's mission as bearer of European culture and mores to the neighbouring East hovers before our eyes."[51] In civilizing this outer, allegedly entirely different East, *liberal* imperial activists saw a basic prerequisite for the fundamental task of the museum: paving the way for trade relations. The inner East was equated with the "European East,"[52] backwards to be sure, but capable of being civilized. It was occasionally construed as a "cross-over area"—half Eastern, half Western—lying within or outside of the monarchy. Through the construction of such in-between spaces, double borders could be drawn, namely between the entirely other East, that is, Turkey and Russia; a European East, whether Slavic or Hungarian, as an in-between space; and the West, Europe.

The civilizing visions of two different groups of actors corresponded to these in-between spaces: in one, *conservative* imperial actors, like the influential publicist, politician, and historian Alexander von Helfert (1820–1910)

---

[45] Kronprinz Rudolf, "Politischer Bericht über die Orientreise 1884," in *Kronprinz Rudolf*, ed. Hamann, 119–134, here 134.
[46] Kronprinz Rudolf, "Politische Denkschrift 1886," 160.
[47] Kronprinz Rudolf, "Politischer Bericht über die Orientreise 1884," 133.
[48] Kronprinz Rudolf, "Politische Denkschrift 1886," 153 and 160.
[49] *Oesterreichische Monatsschrift für den Orient* 1, no. 1 (1875), 3 f.
[50] *Oesterreichische Monatsschrift für den Orient* 1, no. 2 (1875), 1.
[51] *Oesterreichische Monatsschrift für den Orient* 1, no. 1, 2.
[52] Benjamin von Kállay, *Ungarn an den Grenzen des Orients und des Occidents* (Budapest: Khor, 1883), 3.

saw the need for civilizing. In relation to Bosnia-Herzegovina, Helfert differentiated between uncivilizable "national Turks"[53] and civilizable "Muhammadized South Slavs."[54] His goal was to deport the Turks and to "reoccupy landscapes that belonged to us of old."[55] In the other, the concept of civilizable in-between spaces opened up the possibility for competing civil society activists to appropriate the imperial strategy of the civilizing mission and use it for national self-empowerment processes.[56] Concepts such as "Half-Asia" (Karl Emil Franzos)[57] were developed not by imperial but by national actors. With such concepts the notion of an inner East was reinforced, and they made it possible to speak of a cultural differential within the Habsburg Monarchy and between its kingdoms and crownlands (e.g., Galicia), which national activists, as the spearheads of modernization, sought to level out through civilizing.

K. u. k. Orientalism, however, did not end with visions of inner and outer civilizing missions based on the construction of difference; it also incorporated a differentiated image of the East, which manifested itself in the imperial pragmatic approach examined here. Beyond the pejorative representations of the East, the reciprocal trade relationships between Occident and Orient were also thrown into new light in the nineteenth century. This is attested to by the emphasis on the "peaceful exchange of ideas, of intellectual as well as material goods"[58] with the Ottoman Empire, as well as the teaching of contemporary Arabic, Persian, and Turkish at the k.k. Oriental Academy since 1754 for the purposes of diplomatic exchange and the development of trade relations as well as a sustained grappling with Islam, which was traditional in Austria since the time of Hammer-Purgstall. Finally, as mentioned at the outset, in the late Habsburg Monarchy the concept of the East was unmasked as a construction by Mžik, and historical conjunctures and downturns in dividing the world into east and west were asserted critically. If we speak, therefore, of k.u.k. Orientalism, we mean the product of the derogatory construction of difference in Said's sense and the histories of entanglement in the sense of imperial pragmatism that were already being emphasized in the nineteenth century.[59]

---

[53] [Joseph] Freiherr von Helfert, *Bosnisches* (Vienna, 1879), 240 and 265.
[54] Helfert, *Bosnisches*, 251.
[55] Ibid., 172.
[56] See Johannes Feichtinger, "Modernisierung, Zivilisierung, Kolonisierung als Argument: Konkurrierende Selbstermächtigungsdiskurse in der späten Habsburgermonarchie," in *Ränder der Moderne: Neue Perspektiven auf die europäische Geschichte, 1800–1930*, ed. Christof Dejung and Martin Lengwiler (Cologne: Böhlau, 2016), 147–181.
[57] See Andrei Corbea-Hoisie, "Halb-Asien," in *Habsburg neu denken*, ed. Feichtinger and Uhl (Vienna: Böhlau, 2016), 73–81.
[58] Anonymus, "Siebenter internationaler Orientalisten–Congreß," 2.
[59] See Johannes Feichtinger, "Nach Said: Der k. u. k. Orientalismus, seine Akteure, Praktiken und Diskurse," in *Bosnien-Herzegowina und Österreich-Ungarn: Annäherungen*

The East lies in Europe

During the First World War, the East appeared in a totally new light. The "close armed alliance" between Austria and the Ottoman Empire (established in 1914) gave a mobilizing academic discipline the chance to position itself as an agent in this process of rapprochement:[60] already in the first years of the war, new Oriental institutes, societies, and periodicals were founded outside the university structure, including the Research Institute for the East and Orient (Forschungs-Institut für Osten und Orient), which was intended "to promote the revitalization of the sciences in Austria [...] on a German foundation."[61] Notably, in this phase the East was located in an entirely different place by scientists and scholars:[62] the traditional notion of a distinction between Europe and the East was maintained, but the East (and Turkey in particular) was suddenly also understood as an integral part of Europe. The Viennese cultural geographer Erwin Hanslik (1880–1940) explained it thus in 1916: "In the universal sense, Europe is constructed out of three communities: the West, the East (Osten), and the Orient."[63] The "Societal map of the Orient by Erwin Hanslik," "executed by the k.u.k. Military Geographic Institute" (see Figure 1) demonstrates that Europe comprises a "West European social body (Gesellschaftskörper)," an "East European social body," and a "European-Oriental social region," in which the "Aryan Orient" (from India to Anatolia) is embedded. The body metaphor allowed Hanslik to clearly differentiate Eastern and Western Europe from each other; a better understanding of the European East was the goal of his cultural research. Hanslik used the concept of the "European-Oriental social region" to label the Orient, in turn clearly set apart from Eastern Europe, while still conceiving of it as part of Europe, in accordance with the prevailing discourse during the First World War. The Research Institute for the East and Orient, which was founded in 1916 as an offshoot of Hanslik's Institute for Cultural Research in Vienna, also used a notion of Europe that incorporated the Ottoman Empire and the

---

an eine Kolonie, ed. Clemens Ruthner and Tamara Scheer (Tübingen: Francke, 2018), 315–333.

[60] Robert Bleichsteiner, "Die orientalistische Tätigkeit der kaiserlichen Akademie der Wissenschaften in Wien," *Österreichische Monatsschrift für den Orient* 42 (1916): 234–238, here 234.

[61] Erich Pistor, "Was haben die Mitglieder vom Institut in wissenschaftlicher und praktischer Beziehung zu erwarten?," *Berichte des Forschungs-Institutes für Osten und Orient in Wien* 1 (1917): 4–6, here 5.

[62] For a more detailed treatment of this point, see Johann Heiss and Johannes Feichtinger, "Orient als Metapher: Wie Österreichs Osten vor, während und nach dem Ersten Weltkrieg vorgestellt wurde," in *Orient und Okzident*, ed. Haider-Wilson and Graf, 53–77.

[63] Erwin Hanslik, "Die kulturellen Grundlagen der neuen europäischen Machtgemeinschaften," *Österreichische Monatsschrift für den Orient* 42 (1916): 36–41, here 36.

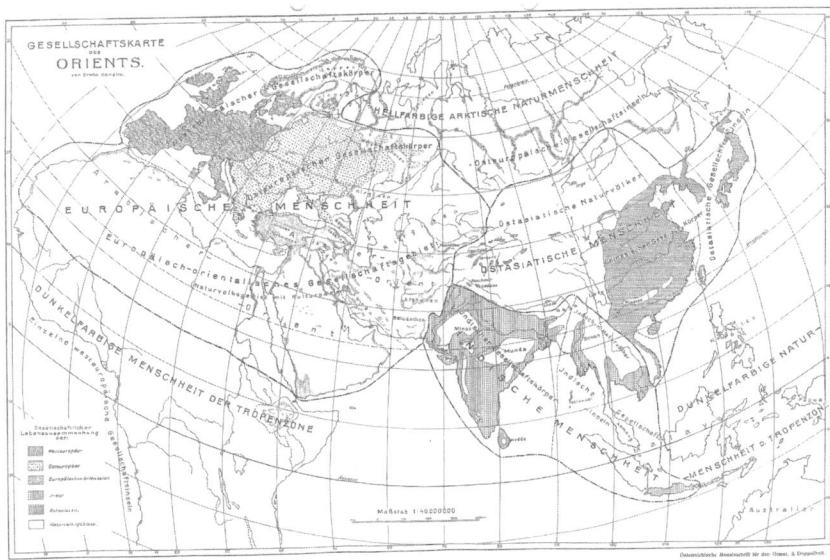

Figure 1: Erwin Hanslik, Der nahe Orient, Indien und Ostasien: Kulturstudien mit einer Kulturkarte des Orients (Vienna: Institut für Kulturforschung, [1914]). Here the Kulturkarte is called Gesellschaftskarte

"Aryan Orient."[64] The Europeanization of the East did presume the softening of central markers of difference such as religion. In this sense, in his lecture "The Question of the Future of Islam" in 1916, the Viennese Orientalist Rudolf Geyer (1861–1929) emphasized that "the total content of Islamic thought is in no way different from that of Christianity." He characterized Islam as the form "in which Christianity found a point of entry into all of Arabia."[65]

During the First World War, scholars of the East demonstrated a specific and likewise surprising form of imperial pragmatism. For example, the deciphering of Hittite cuneiform by Friedrich Hrozny, a Viennese university lecturer in Semitic languages, was understood as a demonstration that "the true and steadfast standing together of the allies so brilliantly proves itself in scholarly fields."[66] Because Hittite was counted as an Indo-Germanic language and

[64] Georg Hüsing, "Was ist Orient?," *Berichte des Forschungs-Institutes für Osten und Orient* 1 (1917): 40–45, here 43 and 45.

[65] Rudolf Geyer, "Die Zukunftsfrage des Islam: 7. und 8. Vollsitzung am 24. und 31. Mai 1916," *Berichte des Forschungs-Institutes für Osten und Orient* 1 (1917): 7–10, here 8.

[66] S[alomon] Frankfurter, "Die Lösung des Hethiterproblems," *Wiener Zeitung*, 1 January 1916, 12–14, here 14. On Friedrich/Bedřich Hrozný, see most recently Frank Hadler, "Bedřich Hrozný, 1879–1952: Pionier der Hethitologie mit transnationaler Karriere zwischen spätem Habsburgerreich und früher sozialistischer Tschechoslowakei," in *Vor-*

its similarity to Latin and Greek could be identified, it was easy to draw new integrative conclusions "about population flows [Völkerströmungen] and the blending of populations [Völkermischungen]," "the venue of which over the past millennium has been the Middle East and Eastern Europe."[67] In 1917, as mentioned above, the University of Vienna established its first professorship for Turkology.

This attempt to integrate the East into Europe, conditioned by the policies of alliance, came to an end with the collapse of the monarchy and the Ottoman Empire. The Research Institute for the East and Orient, whose very name captured the difference between enemy and ally, East and West, lost its purpose; this was not the case for *Orientalistik*, which henceforth pursued a new agenda, whether within or outside of a university setting. During the First World War, *Orientalistik* had gotten a powerful boost, and the military alliance had fostered an expectation of greater penetration in the future of new markets. It is notable that this also signaled an about-face among the Orientalists at the University of Vienna: they relinquished the pathbreaking ideal of cultivating "pure" scholarship, which they had adhered to in the late nineteenth century. Two examples serve to demonstrate this. The man appointed nontenured professor for language, literature, and history of the Turkish-Tatar peoples in 1917, Friedrich Kraelitz vom Greifenhorst, was not only a founding member of the Research Institute for the East and Orient; from 1915–1918 he also took part in studies conducted on prisoners of war in Austrian, Hungarian, and German camps,[68] and he was consulted as a censor and interpreter. During the First World War, the liberal priest, ethnographer, and Orientalist Alois Musil (1868–1944), since 1909 a tenured professor of biblical auxiliary sciences and Arabic language at the University of Vienna, served as a military scout for the dynasty in the Near East.[69]

träge und Abhandlungen zur Wissenschaftsgeschichte 2013/2014, ed. Rainer Godel et al. (Stuttgart: Wissenschaftliche Verlagsanstalt, 2016), 149–169; Frank Hadler, "Graben wie die Großen in Kleinasien: Ein frisch berufener Prager Professor umreißt mit weltpolitischen Argumenten sein archäologisches Karrierefeld," in *Kultur und Beruf in Europa*, ed. Isabella Löhr, Matthias Middell, and Hannes Siegrist (Stuttgart: Franz Steiner Verlag, 2012), http://www.europa.clio-online.de.

[67] Anonymous, "Eine wichtige Entdeckung," *Reichspost*, 27 December 1915, 5.
[68] See Britta Lange, *Die Forschungen an Kriegsgefangenen, 1915-1918: Anthropologische und ethnografische Verfahren im Lager* (Vienna: Verlag der Österreichischen Akademie der Wissenschaften, 2013), 11, 123, 354, 361, 366, 367, 388, 394, and 395.
[69] See Udo Worschech, *Alois Musil: Ein Orientalist und Priester in geheimer Mission in Arabien, 1914–1915* (Kamen: Hartmut Spenner, 2009).

Conclusion: Scholarship and Empire

The relationship between scholarship and empire in the case of the Habsburg Monarchy can be described as complex.[70] This complexity is particularly striking in the case of *Orientalistik*, thanks to its dual orientation: as a science that in the interests of imperial pragmatism produced knowledge important to the state, and as a discipline that provided the foundations for biblical hermeneutics and expanded knowledge about the East. For centuries, imperial pragmatism (diplomacy, trade relations) determined scholarly engagement with the East. In the second half of the nineteenth century, the Habsburg Monarchy's changed Eastern policy cast the space defined as the East in a new light: outside the university, new institutions of learning developed civilizing visions for the "European Orient" and for spaces beyond the "civilizational frontier" (Asia). At that same time, at the University of Vienna, *Orientalistik* developed, largely separate from politics, into a "pure" science that mainly studied the East from the philological perspective of "dead" languages. By 1900 a civilizing missionary mission in the East was hardly discussed any longer. In 1886 the Oriental Museum was renamed the k.k. Austrian Trade Museum and in 1897 its collecting activities were discontinued.[71] During the First World War the idea of a civilizing mission was entirely meaningless, as the East was presented and mediated as a part of Europe. This image of the East disappeared with the collapse of the monarchy. In the successor states, *Orientalistik* did not dodge its new task of national self-assertion. In Prague, Alois Musil, who had to give up his Viennese professorship in 1919 as a Czech, established the Oriental Institute as a "colonial institute without colonial ambitions."[72] Tomáš G. Masaryk lent him his support in this. In 1943 the Oriental Institute was incorporated into the Reinhard-Heydrich-Stiftung, Reichsstiftung für wissenschaftliche Forschung in Prag and entrusted to the leadership of Adolf Grohmann (1887–1977), an Arabist who had completed his habilitation in Vienna in 1916. Until 1915 he and Hrozny published the *Archiv orientální*. In Vienna, too, the leading Orientalists emerged from their cocoons after the Anschluss in 1938. Viktor Christian (1885–1963), a student of Hrozny's and one of the most vehement supporters and representatives of the National Socialist ideology, was appointed dean of the philosophical fac-

---

[70] See Marianne Klemun, ed., *Wissenschaft und Kolonialismus* (Innsbruck: StudienVerlag, 2009).

[71] See Johannes Wieninger, "Das Orientalische Museum in Wien, 1874–1906," in *Vienne: Porta Orientis*, ed. Dieter Hornig, Johanna Borek, and Johannes Feichtinger (Rouen: Presses universitaires de Rouen et du Havre, 2013), 143–158.

[72] Sarah Lemmen, "'Unsere Aufgaben in der Orientalistik und im Orient': Die Gründung und die erste Dekade des Prager Orientalischen Instituts in der Zwischenkriegszeit," in *Orientalismen in Mitteleuropa*, ed. Born and Lemmen, 119–143; Sarah Lemmen, *Tschechen auf Reisen: Repräsentation der außereuropäischen Welt und nationale Identität in Ostmitteleuropa, 1890–1938* (Cologne: Böhlau, 2018), 63–72.

ulty at the University of Vienna in 1938 and the university's final rector in the Nazi era in 1945.[73]

The history of *Orientalistik* in the Habsburg Monarchy is exemplary of the reciprocity between scholarship and imperial orders. In the capital of Vienna it slipped into two different roles: while outside the university it fulfilled practical political tasks in the interests of imperial pragmatism and was thus enhanced institutionally (through the founding of the k.k. Oriental Academy, of k.k. language schools and the k.k. Oriental Museum), at the university the path was paved for it to become a "pure" science. The post-imperial new East European order rendered the imperial political function of *Orientalistik* in Vienna obsolete. After the First World War it was put to national political uses, so that the discipline of *Orientalistik* and its leading representatives, including those at the university, increasingly relinquished the traditional practice of "pure" scholarly cultivation in favour of national political activities.[74]

Translated from German by Kate Younger.

---

[73] See Mitchell G. Ash, "Die Universität Wien in den politischen Umbrüchen des 19. und 20. Jahrhunderts," in *Universität – Politik – Gesellschaft*, ed. Mitchell G. Ash and Josef Ehmer (Göttingen: V&R unipress, 2015), 29–172; Gingrich and Bendix, "David Heinrich Müller," 19 f.

[74] Professors of Orientalistik Rudolf Geyer, Schülder, and the successors of David Heinrich Müller, Friedrich Kraelitz and Viktor Christian, were members of the so-called Bear's Den (Bärenhöhle), an informal antisemitic, Christian conservative, and German nationalist network at the University of Vienna, which impeded the habilitation and appointment of leftist and Jewish candidates. See Klaus Taschwer, "Geheimsache Bärenhöhle: Wie ein antisemitisches Professorenkartell der Universität Wien nach 1918 jüdische und linke Forscherinnen und Forscher vertrieb," in *Alma mater antisemitica: Akademisches Milieu, Juden und Antisemitismus an den Universitäten Europas zwischen 1918 und 1939*, ed. Regina Fritz, Grzegorz Rossoliński-Liebe, and Jana Starek (Vienna: new academic press, 2016), 221–242.

Arpine Maniero

ORIENTAL STUDIES IN THE RUSSIAN EMPIRE IN THE CONTEXT OF IMPERIAL POLITICS AND REGIONAL DISCOURSE

Introduction

Edward Said's thought-provoking theory on the connection between academic research and imperial politics[1] sparked a veritable debate among historians and other scholars about whether there was such a thing as "typical" colonial thinking, even in the Russian context. In various ways, the topic became the focus of attention, above all for Western researchers who examined Said's theory with reference to the so-called "Russian case." There were discussions on, for instance, the self-understanding of the "imperial" centre and the "colonial" periphery, their interrelationships in the context of the development of Oriental Studies, and the question of whether academic or political-administrative impulses were decisive in this process.[2] The tension between political needs and research based on the ideal of academic objectivity has led to a very ambivalent evaluation of the role of Russian Orientalists so, many questions about what their actual scope of action was and whether academic specialists had any specific influence on important political decisions still remain.

The question of whether the Russian Orientalists, as Said defined Orientalism, were in the service of the imperial state calls for a strong focus on colonial pretensions to power, or rather, on the so-called civilizing mission. To a certain extent, this question also makes it hard to examine other fields of research in a

---

[1] Edward W. Said, *Orientalism* (New York: Pantheon Books, 1978).
[2] Kerstin S. Jobst, "Where the Orient Ends? Orientalism and Its Function for Imperial Rule in the Russian Empire," in *Deploying Orientalism in Culture and History. From Germany to Central and Eastern Europe*, ed. James Hodkinson et al. (Rochester, NY: Camden House, 2013), 190–208; David Schimmelpenninck van der Oye, "Orientalizm—delo tonkoe," *Ab Imperio*, 1 (2002): 249–264; Nathaniel Knight, "Grigor'ev in Orenburg, 1851–1862. Russian Orientalism in the Service of Empire?," *Slavic Review* 59, no. 1 (2000): 74–100; Adeeb Khalid, "Russian History and the Debate over Orientalism," *Kritika. Explorations in Russian and Eurasian History* 1, no. 4 (2000): 691–699; Nathaniel Knight, "On Russian Orientalism. A Response to Adeeb Khalid," *Kritika. Explorations in Russian and Eurasian History* 1, no. 4 (2000): 701–715; Maria Todorova, "Does Russian Orientalism Have a Russian Soul? A Contribution to the Debate between Nathaniel Knight and Adeeb Khalid," *Kritika. Explorations in Russian and Eurasian History* 1, no. 4 (2000): 717–727; Charles T. Evans, "Vasilii Barthold. Orientalism in Russia?" *Russian History* 26, no. 1 (1999): 25–44.

differentiated way. Above all, the question of regional stakeholders' scope of action is still insufficiently examined. At the same time, the development of individual schools of Oriental Studies, their efforts to include regional academics in research, and the region-specific questions emanating from the peripheries against the backdrop of imperial politics represent a very exciting field of research. That is, a closer look at the peripheries makes possible the interesting observation that regional stakeholders often saw the local problems that they identified from the state's point of view, even when they themselves were not involved in imperial practices. Nevertheless, this elite, who had been educated mainly at European universities, increasingly participated in academic research with their own questions and ambitions. Indeed, they saw the adaptation of European academic concepts from national viewpoints which were not always in harmony with the standards coming from the centre.

This essay discusses the debates around the role of academic Oriental Studies in the Russian Empire. It examines the differences in ideas academics in the capital, regional academics, and political decision-makers had about the peripheries' participation in imperial academia. The discrepancy between the academic needs specified in the regions and the education policy that was actually implemented is particularly interesting. An example from Tbilisi illustrates this point: efforts to establish a university or a polytechnic in the Caucasus provided an additional stimulus for a debate on the role of the peripheries within academic Oriental Studies that had been going on since the second half of the nineteenth century. In the 1880s, this discussion expanded to include even broader circles when the urban intelligentsia and the economic elite began to participate.

Discussions about the importance of scholarly research also took place in Kazan after the department of Oriental Studies was relocated from the university in Kazan to St. Petersburg in 1854–1855. This measure was declared necessary to centralize this academic discipline but even experts in the capital such as Viktor Rosen (1849–1908), a Russian Orientalist residing in St. Petersburg, saw it as a step that would rob the region of an important cultural factor.[3] For regional stakeholders, it was more evidence of the state's ignorance of already established academic traditions and their further development.

Even though the debates on the southern and eastern peripheries of the empire were qualitatively very different, in both cases the question of better organization of local education was central to the debates. Improving local education was intended, among other things, to stimulate and promote academic research and open up these regions culturally. It did not just require well-educated specialists but also called for academic resources, which were lacking in the regions. In the framework of this discourse, Oriental Studies took on a twofold purpose: it was important for achieving purely academic objectives and to carry out the state-defined, utilitarian task of developing a reliable administrative service class. In what follows, this dual significance is considered in terms

---

[3] Vasilii V. Bartol'd, *Raboty po istorii vostokovedeniia* (Moscow: Nauka, 1977), 590.

of what it meant for the development of academia on the peripheries of the empire and what it meant in terms of how academic specialists positioned themselves between these two poles.

Oriental Studies against the Backdrop of the Imperial Claim to Power

While the Russian Empire was acquiring territories in Central Asia and the Caucasus over the course of the eighteenth and nineteenth centuries, questions about how to integrate these regions into the empire and the supposed need to "civilize" indigenous peoples became topics of discussion on both the political and academic levels. As Andreas Kappeler noted, although the Armenians and the Georgians, to take two examples, looked back "on a tradition of civilization," the peoples who had been living under Persian or Ottoman rule were considered "Asian" without any particular differentiation and, according to this point of view, "Russia had to bring the blessings of European civilization to them."[4] Thus, in 1833 the Russian State Council believed that the population in the Asian regions should be made "to speak, think, and feel Russian."[5] It was a policy that certainly impacted the organization of education in the regions. In 1869, the Ministry of Public Enlightenment published a "collection of documents and articles on the education of indigenous people" that stated that the end result of the education policy in the peripheral regions should be "complete Russification and assimilation with the Russian people."[6]

Many famous Orientalists regarded the integration of the Asian regions of the empire from a much less aggressive point of view; above all, they rejected "full cultural Russification."[7] Although tsardom's Orientalists hardly questioned the empire's role, their opinions diverged sharply from those of policymakers in state service. In particular, the former thought Russification should not necessarily come at the price of the total loss of local languages and cultures. Instead, the improved education of indigenous people as well as the involvement of the local elite in the "collection and spreading of knowledge" was, in the words of British historian Vera Tolz, an important key to fostering local identities and their link to a pan-Russian identity.[8] At the same time, in contrast to other European empires, the borders between Russia and its Asian peripheries were intentionally blurred in order to highlight the empire's special

---

[4] Andreas Kappeler, *The Russian Empire. A Multiethnic History* (Harlow: Longman, 2001), 171.
[5] Quoted in ibid., 176.
[6] Vladimir O. Bobrovnikov and Irina L. Babich, *Severnyi Kavkaz v sostave Rossiiskoi Imperii* (Moscow: Novoe Literaturnoe Obozrenie, 2007), 271.
[7] Vera Tolz, *Russia's Own Orient. The Politics of Identity and Oriental Studies in the Late Imperial and Early Soviet Periods* (Oxford: Oxford University Press, 2011), 25.
[8] Vera Tolz, "Orientalism, Nationalism, and Ethnic Diversity in Late Imperial Russia," *The Historical Journal* 48, no. 1 (2005): 137.

relationship to its "colonies." Hence many Orientalists and also political stakeholders substantiated the empire's historic obligation to acquaint the Asian regions of the empire with European civilization. In addition, they presented this obligation as a guarantee for long-term hegemony and the successful integration of these regions. Using the Armenians as an example, Georgian-Russian Orientalist Nikolai Marr (1864–1934) asked rhetorically whether the Russians had not been the ones to liberate them from Persian and Ottoman oppression and whether they had created the basis for their economic and cultural development, and, as a result, had also laid the foundations for the development of a national intelligentsia.[9]

However, there is no agreement in current research on how to evaluate the role of Russian Orientalists in imperial politics. This is even more true because their rhetoric—in which the dual significance of Oriental Studies as an academic discipline and as an instrument of the imperial claim to power emerges prominently—indicates some ambivalence. Specialists, such as St. Petersburg Orientalist Vasilii V. Grigor'ev (1816–1881), Nikolai Marr, and Azerbaijani-Russian Aleksandr K. Kazem-Bek (1802–1870), presented imperial rule as the saviour of the indigenous peoples of Asia, who were "straying in the darkness of superstition and ignorance."[10] They believed that the empire's purpose was to introduce its Asian citizens to education, progress, and "genuine" faith—and they felt this should not be done through violence and bloodshed but through the dissemination of knowledge and increased access to education.[11]

In precisely this regard, Vera Tolz ascribes great importance to the role of the capital's Orientalists in forming regional elites. However, at the same time she exposes the tensions that arose from the need to integrate national minorities into the empire without expediting the loss of their national and cultural characteristics. Consequently, she develops the theory that Russian Orientalists played a decisive role in the emergence of national movements on the peripheries of the empire. Nevertheless, their efforts contributed to better education and hence to the emergence of an educated but also nationally minded elite. In contrast to the theory that this was more of an unintended by-product of the work of the capital's Orientalists, and by no means their primary intention, Tolz links the imperial academics' collaboration with the educated elite in the regions to the nation-building processes of the early Soviet era.[12]

On the other hand, other researchers see the development of Oriental Studies as the actions of individual Orientalists, most certainly carried out in con-

---

[9] Nikolai Marr, "Kavkazskii kul'turnyi mir i Armeniia," *Zhurnal Ministerstva Narodnogo Prosveshcheniia*, new series LVII (1915): 329.
[10] Vasilii V. Grigor'ev, *Ob otnoshenii Rossii k vostoku* (Odessa: n. p. 1840), 12.
[11] Ibid., 9.
[12] Vera Tolz, "Imperial Scholars and Minority Nationalisms in Late Imperial and Early Soviet Russia," *Kritika. Explorations in Russian and Eurasian History* 10, no. 2 (2009): 263.

junction with imperialist claims to power. The professionalization and institutionalization of this discipline are likewise shown to be important instruments for the administration of the regions and the organization of education. Therefore, Viennese historian Kerstin Jobst also presents the Lazarev Institute, founded in 1815 in Moscow by the Armenian Larzarev (Lazarian) family, as the equivalent of the École des langues orientales[13] in Paris and thus as an early example of the creation of a research institution in the Russian Empire that served state interests. According to Jobst, this institute was nevertheless "just one indication that the professionalization of ethnography and orientalistic studies was closely connected with Russian imperial expansion."[14] However, precisely this private school, which was originally established to educate young Armenians, can only very conditionally be considered an early example of an institution serving imperial claims to power because initially it barely took account of the centre's efforts— or rather, it was not founded with the objective of providing academic support for imperial expansion. Only in the following decades was such significance explicitly ascribed to this academy—which tellingly was not converted to the (Armenian) Lazarev Institute for Oriental Languages until after the Russo-Persian War of 1828 and the integration of the South Caucasian region into the empire. In the curriculum, this significance for the state became more pronounced with the empire's growing interest in the Caucasian region during the second half of the nineteenth century.

It is true that knowledge about the indigenous populations of the newly conquered regions—and the academics who were responsible for developing this knowledge—played an important role in imperial policy. However, when it came down to participation in the "conceptual and practical creation of power," Russian Orientalists' real possibilities for taking action were relatively limited—which was the case although they were certainly in demand as experts in specific areas.[15] In turn, their interaction with the representatives of national minorities was essentially intended to "strengthen the organization of the academic potential in the Asian regions of the empire."[16] In particular, from the second half of the nineteenth century onwards, they increasingly focused on the need to put academic research ahead of state needs. This trend was closely linked to the development of academic Oriental Studies and reflected the need to develop the peripheries of the empire academically.

---

[13] This school was founded in 1795 in Paris and initially specialized in teaching Arabic, Turkish, and Crimean Tatar. Later it also taught other Asian languages. Kerstin Jobst highlights inter alia the close connection of this school with the work of French Arabist Silvestre de Sacy and his "academic support" for Napoleon's expedition in 1798. Cf. Jobst, "Where the Orient Ends?," 194.
[14] Ibid., 195.
[15] Knight, "On Russian Orientalism," 705.
[16] Bartol'd, *Raboty po istorii vostokovedeniia*, 590.

Oriental Studies as a Part of the "Historical-Philological Discipline"

In the course of the nineteenth century, those institutions that had specialized early on in teaching Oriental languages played a special role in the development of academic Oriental Studies. To name just a few significant examples, these were the aforementioned Lazarev Institute, the department for teaching Oriental languages established in the Asian section of the Ministry of Foreign Affairs, the school for Oriental languages established in 1837 at the Richelieu-Lyceum in Odessa, and the faculty for Oriental Studies established in 1852 at the university in St. Petersburg. In addition, the academic school that developed around Viktor Rosen, including among others Valentin A. Zhukovskii (1858–1918), Nikolai Marr, and Vasilii V. Bartol'd (1869–1930), was established in St. Petersburg. In Moscow, the Lazarev Institute united such specialists as Fëdor E. Korsh (1843–1915), Vsevolod F. Miller (1848–1913), and others. The teaching of Oriental languages in the university environment was initially regulated by the Russian University Statute of 1804, which provided for the establishment of corresponding chairs at the universities in Moscow, Kharkov, and Kazan; the University Statute of 1835 made the same provisions. Nevertheless, Bartol'd explained that language teaching could not be provided to the extent stipulated in the statute, with the exception of the university in Kazan, because Oriental Studies was being completely "sacrificed" to the "real and imaginary" needs of the state.[17]

The development of Oriental Studies was indeed closely linked to the state's need for qualified specialists in economics and politics, while, from the point of view of many Orientalists, this subject was a part of the "historical-philological discipline" and should adopt its academic methodology. An 1891 editorial written by Russian Slavicist and historian Vladimir I. Lamanskii (1833–1914), which appeared in the first issue of the magazine *Zhivaia starina* (Living Antiquity), illustrates this. In it he argued for expanding the teaching of Oriental languages at Russian universities. He specifically suggested opening a faculty for Oriental languages at Moscow University, which would also benefit the faculty for Oriental Studies at St. Petersburg because having a second faculty would initiate productive competition and critical examination of the work that colleagues were doing. Lamanskii ascribed great significance to this faculty, not only for developing the discipline of Oriental Studies but also in terms of its potential to help expand and stimulate trade with Asian countries.[18]

Viktor Rosen, who responded to Lamanskii in an article in the *Zhurnal ministerstva narodnogo prosveshcheniia* (Journal of the Ministry of Public Enlightenment), welcomed the fact that even people outside the circle of experts

---

[17] Vasilii V. Bartol'd, "Obzor deiatel'nosti fakul'teta vostochnykh iazykov," in *Sochineniia*, vol. 9, ed. Iurii E. Bregel' et al. (Moscow: Nauka, 1977), 48.
[18] Viktor R. Rosen, "O vostochnom fakul'tete i vostochnykh kafedrakh," *Zhurnal Ministerstva Narodnogo Prosveshcheniia* CCLXXIII (1891): 160.

were advocating the expansion of Oriental Studies in Russia. However, he criticized what he believed to be their lack of motivation. According to Rosen, Oriental Studies in Russia would only be on solid ground if chairs for individual Oriental languages were opened at all the Russian universities. These chairs should not be independent faculties, but part of the history or philology faculties and they should have the same rights as chairs for Romance, Germanic, or even Slavic philology. He argued that the decisive factor should not be the "great advantage for academia and trade" which these chairs were expected to provide but rather, they should be created solely for the benefit of the "historical-philological discipline".[19]

Thus, Rosen, and later Bartol'd, criticized the role that various parties were increasingly ascribing to this academic discipline, that is, primarily to educate specialists either for administrative tasks in the Asian regions of the country or to facilitate trade and diplomatic relations with Asian countries. According to Rosen, Russia had maintained trade relations with the countries of the Orient for centuries, and the extent to which this trade had developed had nothing to do with Oriental Studies. He claimed the faculty of Oriental Studies at St. Petersburg and the Lazarev Institute in Moscow were fully capable of meeting this need,[20] adding that what Russian academia lacked was the capacity to educate specialists who would dedicate themselves solely to academic tasks. Therefore, the type of knowledge produced by the men holding the chairs for Oriental languages established at the faculties of history and philology was crucial, and, even if at a later date such Orientalists did not have any students, their academic preparation and valuable writings alone would do Russian Oriental Studies a great service.[21] According to Rosen, such specialists were needed not only in Moscow but at every Russian university. He believed that the impulse for developing Oriental Studies in this way should come from the universities of the empire's two centres and he argued that the holdings of the Rumiantsev Museum, the Imperial Public Library, and the Asiatic Museum already provided the necessary basis for teaching and academic research. Contemporary and future stakeholders should always remember that

Oriental Studies is simply a branch of the historical-philological discipline and that it only has academic meaning and significance when the existing standards and methods, which have been developed over many generations for the aforementioned discipline, are strictly adhered to.[22]

Bartol'd also used the necessary integration of Russian Oriental Studies in European academic life as an argument when Russian Orientalist Sergei F. Ol'denburg (1863–1934) announced plans to establish an institute of higher

---

[19] Ibid., 161.
[20] Ibid., 162.
[21] Ibid., 164.
[22] Ibid., 160.

education with a focus on Oriental Studies in Tashkent.[23] In terms of the multifaceted academic preparation required of academics, which could only be guaranteed in the empire's two centres, Bartol'd did not think that creating such an institute, far from European academic centres and not part of a university, would be a good idea. He also argued that while the work of local specialists was important, from an academic point of view, future Orientalists should enjoy a many-sided education. That would give them the skills to refrain from blindly trusting their sources and instead to evaluate them critically, in accordance with European academic standards. He claimed that the development of academic Oriental Studies was therefore only possible in centres where "the correct view of the requirements which each academic should fulfil have been firmly established."[24] The importance he ascribed to the European schools and their impact on the upcoming generation of Orientalists is shown in his intention to encourage students who sought an academic career to go abroad after university and spend some time under the guidance of Western researchers.[25]

Russian Orientalists' academic perspective on the regions which were the foci of their academic interests was essentially shaped by the idea that their subject should be integrated into the disciplines of history and philology. Therefore, in the context of imperial policy, it is important to remember that the majority of the academics who made significant contributions to the emergence of academic Oriental Studies ultimately were interested primarily in academic research. According to Schimmelpenninck van der Oye, although many academics were certainly involved in state service and also contributed to the drafting of laws, the majority of them performed research mainly outside the political sphere.[26]

The Debates in the Caucasus

Although the education of local elites and their inclusion in administrative and pedagogical fields was a firmly anchored component of imperial education policy, from the point of view of the local elites this was not sufficient to meet regional needs. What is more, they regarded the stimulus of academic research as a necessary requirement for a comprehensive societal modernization. At the same time, Russian intervention was the best option for bringing the achievements of (Western) European academia and culture to the notice of local elites. Even stakeholders like Crimean Tatar intellectual Ismail Gasprinskii (1851–

---

[23] Bartol'd, *Raboty po istorii vostokovedeniia*, 493.
[24] Ibid., 494.
[25] Ibid.
[26] David Schimmelpenninck van der Oye, *Russian Orientalism. Asia in the Russian Mind from Peter the Great to the Emigration* (New Haven, CT: Yale University Press, 2010), 258.

1914), who advocated for the nation-building process under the banner of Europeanization, often urging that European academic thinking be disseminated "among us," emphasized the role of the empire in nation building. The empire functioned as the natural facilitator between Europe and Asia, as Gasprinskii put it, between "academia and ignorance, between progress and stagnation."[27]

However, the opening of Russian gymnasiums as well as the targeted education of specialists in certain fields initially served the objectives of giving citizens in the peripheral regions of the empire an understanding of the Russian language and creating political loyalties. It meant that the administration of these regions could be placed in the hands of the local elite who had been educated at Russian universities. Thus, in 1845, for the first time, a regulation stipulated the number of imperial subjects from Transcaucasia who could study at St. Petersburg University on scholarships from the state. However, this regulation contained so many limitations that even the curator of the Caucasian education district,[28] Baron Aleksandr Nikolai (1821–1899), criticized it harshly. He was especially critical of the very small number of faculties which were tasked with educating certain kinds of specialists to prepare them for state/civil service while, for example, the Aleksandrovskii Lyceum or the Imperial Academy of Arts, which were not part of this state-sanctioned programme, were capable of educating specialists who were equally useful for the Caucasus.[29] Moreover, he considered it obstructive that, with the exception of the Oriental Studies Department at St. Petersburg University, the Imperial School of Jurisprudence, and the Lazarev Institute, no other institutes of higher education offered teaching given in the local languages. Baron Nikolai believed it was not possible to educate useful specialists for the regions without teaching in their mother tongue.[30] In addition, in 1898, a subsequent curator of the education district, Kirill P. Ianovskii (1822–1902), criticized the programme of education for being based solely on the needs of the civil service, which, in his opinion, only inculcated in the students a life-long superficiality and resulted in a noticeable inability to carry out independent academic work.[31]

In subsequent years, higher civil servants repeatedly highlighted the lack of an efficient education policy in the Caucasian region as a trend that went

---

[27] Ismail Gasprinskii, *Polnoe sobranie sochinenii*, vol. 2 (Kazan: Institut istorii im. Sh. Mardzhani AN RT, 2017), 86.
[28] On the initiative of Count Mikhail S. Vorontsov, in 1848 the Caucasian education district was formed as an official unit for the administration of the schools located in it; it comprised inter alia the governorates of Baku, Elisabethpol, Kutaisi, Tbilisi, the Black Sea, Erivan, and Stavropol.
[29] Rossiiskii Gosudarstvennyi Istoricheskii Arkhiv (Russian State Historical Archive, hereafter RGIA), f. 733 (1849–1850), op. 89, d. 184, 3.
[30] Ibid.
[31] Kirill P. Ianovskii, *Otchet popechitel'ia Kavkazskogo uchebnogo okruga o sostoianii uchebnykh zavedenii za 1898 god* (Tbilisi: Tip. Kantseliarii Glavnonachal'stvuiushchego grazhdanskoi chast'iu na Kavkaze, 1899), 69.

against imperial interests. According to Nikolai Marr, the reason not only the Russian public but also academics (the latter being very different from their colleagues in Europe) knew so little about the Caucasus was the lack of a general education in Oriental Studies at universities. In contrast, French, English, German, and Italian academics could already present significant results in the study of Transcaucasia. Even the school textbooks for Armenian and Georgian language and literature were written by foreigners, while the Russian state continued to maintain its passivity and indifference towards this region.[32]

In the eyes of the local elite, the solution to these problems was to create a university in the Caucasian education district. In the 1880s the discussion about this reached a high point in Tbilisi. In the course of the nineteenth century, this city had asserted itself as the regional centre of Transcaucasia and as the intersection between the Ottoman Empire, Persia, and the Russian Empire and it laid a claim to becoming an additional centre of Oriental Studies. The city administration had even decided to reserve 10,000 roubles annually from the city budget, which was to be provided along with a parcel of land to establish the university.[33] The decision to open polytechnics in Kiev and Warsaw gave an additional stimulus to the debates in the Caucasus. In Tbilisi donations were collected for the future university; these collections were sometimes spontaneous and sometimes were initiated by the cities of Tbilisi, Batumi, and Kutaisi. According to various sources, the congress of oil industrialists in Baku, the city credit society, the merchant bank in Tbilisi, and also private donors collected up to 230,000 roubles.[34] However, state officials did not support the idea of establishing a university and even collecting donations was quickly prohibited.[35]

In September and October 1906, discussions about opening a university in the Caucasus took place in Lesno and St. Petersburg. The initiator of the talks was Georgii M. Tumanov (1854–1920), a delegate of the Tbilisi city duma in St. Petersburg and a member of a committee founded for this purpose at the Caucasian governorship. Aleksandr S. Posnikov (1846–1922), the dean of the Economics Department at the Polytechnic Institute chaired the discussion in St. Petersburg. Other professors and representatives of the Tbilisi city administration and influential large businesses in the Caucasus were also in attendance. Above all, the discussion focused on whether the need for a university or a polytechnic was more urgent. On the one hand, those in attendance felt that this region ought to contribute to imperial scholarship. On the other hand, it was feared that wealthy donors would prefer that a polytechnic be established

---

[32] Marr, "Kavkazskii kul'turnyi mir," 243.
[33] K. Djalalian, "Bardzraguyn usumnaran T'iflisum," *Murch* 1 (1907): 67.
[34] Georgii M. Tumanov, *Zametki o gorodskom samoupravlenii na Kavkaze* (Tbilisi: Druck Liberman, 1902), 69.
[35] Djalalian, "Bardzraguyn usumnaran," 67.

to educate the professionals needed for industry, trade, and agriculture. However, according to Sergei I. Druzhinin (1872–1935), a professor at the St. Petersburg Polytechnic, a region with a population of nine million people could not be denied the opportunity to acquire an institution of higher education in their home region, no matter which project gained the upper hand.[36] Ultimately, such an educational institution, a place where representatives of all Caucasian nationalities could study, would not only contribute to the emergence of an educated elite but would also combat inter-ethnic hatred.[37]

Grigorii F. Tigranov (1859–1930), the head of the supervisory section of the Mining Department, also advocated placing the needs of "Caucasian Studies" at the forefront. He believed that in this peripheral region both a university and a polytechnic were absolutely necessary but if one had to choose, it should be a university.[38] He thought a university would not only contribute to general education and development in the Caucasus but, as a body of academics, it would also promote "Caucasian Studies," which at the time were at the centre of Russian Orientalists' attention. He thought that although almost a century had passed since the unification of the Caucasus with the Russian Empire, there had been no adequate research into the history of any of the peoples in this region, whose fate had been linked for centuries with the histories of larger empires and who were influenced by the cultures of these empires. He added that the best specialists in the languages of this region were still the British, the French, and the Germans, despite the existence of the Lazarev Institute:

For what purpose does the faculty for Oriental languages at the university in St. Petersburg exist? And is it not remarkable that the most well-known professors for Caucasian research are either from the autochthonous population, or were at least born here?[39]

If there had been a university in the Caucasus half a century earlier, the Russian public and government would have gotten to know this region better, and the local population would also know its own country better.[40] However, according to Tumanov, the state underestimated the need for a university and its role in the development of Oriental Studies, and likewise, private donors could not be convinced of the importance of the purely academic role of an institution of higher education.[41]

Despite a definite upsurge of what the local elite called "Caucasology" in the course of the nineteenth century, state opinion regarding establishing a university in this education district remained unchanged in the first decade of the twentieth century. In the eyes of the responsible ministries, the academic ac-

---

[36] Ibid., 77.
[37] Ibid.
[38] Ibid., 75 f.
[39] Ibid., 76.
[40] Ibid.
[41] Ibid., 68 f.

tivity of regional academics was not as important as the creation of a loyal administrative service class. From a political point of view, there was a real danger that the university would become a tinderbox for political conflict in light of the widespread separatism in the region. It was feared that on the southern and the eastern peripheries of the empire better education as well as knowledge of one's own—national—history and culture would lead to separatist outbreaks— and this fear was a serious hurdle both for education policy and academic research.

In many respects, the debate around founding an institution of higher education in the Caucasus had features similar to Ol'denburg's aforementioned suggestion about establishing an institution of higher education devoted to Oriental Studies in Tashkent. In the 1890s, Ianovskii in particular also developed a similar plan to establish a faculty of Oriental Studies in the Caucasus based on the one in St. Petersburg.[42] The idea that Oriental languages should be taught in the region where they were spoken came to the fore in both cases but another important argument was that establishing tertiary education in the Caucasus would enable young people to acquire an education in their home region, thereby avoiding the hardships connected with a university education in distant cities. The regions would also profit because they would gain valuable specialists. Upon finishing school, the young people who acquired a higher education in their own region could be deployed locally in many sectors whereas highly educated specialists from the regions often did not return after their studies in the imperial centres.[43]

Oriental Studies in Kazan

There were also discussions in Kazan about the role and significance of the peripheries in imperial academia. However, there was an essential, qualitative difference from the debate in Tbilisi because in Kazan there was already a university with a long-established Oriental school that provided a solid basis for efforts to play an important role within imperial academia. However, it did not mean that in this regard academic Oriental Studies was not in conflict with the state needs.[44]

---

[42] Irina V. Cherkaz'ianova, "K. P. Ianovskii–prosvetitel' i organizator narodnogo obrazovaniia na Kavkaze," *Vestnik Stavropol'skogo gosudarstvennogo universiteta. Istoricheskie nauki* 50 (2007): 13.
[43] Bartol'd, *Raboty po istorii vostokovedeniia*, 499–500 ff.
[44] Therefore, for example, Mikhail Magnitskii, who arrived in Kazan in 1819 as an inspector and curator of the education district, demanded that Oriental languages be taught only to the extent to which it was necessary for trade and diplomacy. Cf. Schimmelpenninck van der Oye, *Russian Orientalism*, 100.

In Kazan, the groundwork for Oriental Studies was laid in 1769, when classes on the Tatar language began in the gymnasium that had been founded a decade earlier. However, the chair in Oriental languages, which was established at Kazan University, was the basis for not only teaching but also researching the history and culture of both Asian regions of the empire and neighbouring countries. Therefore, it was no accident that in the debates on centralizing Oriental Studies the university in Kazan figured alongside the one in St. Petersburg.

In fact, the expertise that predated the debates already existed in Kazan, which made the university an important centre for Oriental Studies in Russia. In 1806, when the College of External Affairs (Kolegiia inostrannykh del) proposed that schools for several Oriental languages be established in Kazan, Irkutsk, Orenburg, and Tbilisi, Stepan Ia. Rumovskii (1734–1812), the curator of the Kazan education district,[45] made a counterproposal. He advocated for establishing a school for Oriental languages at the gymnasium in Kazan that would specialize in Tatar, Arabic, Turkish, Persian, Georgian, Armenian, Japanese, Chinese, Manchurian, and Kalmyk. He envisioned this school as a test case: once it had become well established and the desired results achieved, similar schools in the aforementioned cities should be founded to produce the specialists needed by the College of External Affairs.[46] Rumovskii's proposal received strong support from German Arabist Christian Martin Frähn (1782–1851), who at that time held the chair for Persian and Arabic languages at the university in Kazan. Frähn thought that this university, as the centre of Russian Oriental Studies, was better suited for this task than any of the empire's other universities.[47]

Oriental Studies reached its high point in Kazan in the 1830s and 1840s. This was due as much to efforts made by Mikhail N. Musin-Pushkin (1795–1862), the curator of the Kazan education district, as it was to the work of several professors, the majority of whom had been invited to come to Kazan from Germany.[48] This period was distinguished by the opening of additional chairs for Oriental languages and by the work of specialists such as Kazem-Bek, Nikolai F. Katanov (1862–1922), Nikolai I. Il'minskii (1822–1891), and others. These developments led Musin-Pushkin to declare in an 1814 report to the ed-

---

[45] The Kazan education district was established in 1803 and covered the governorates of Vyatka, Kazan, Samara, Saratov, Simbirsk, and Astrakhan as well as Siberia and Central Asia.
[46] Ramil M. Valeev, *Kazanskoe vostokovedenie. Istochniki i razvitie. XIX v.–20-e gg. XX v.* (Kazan: Izdatel'stvo Kazanskogo universiteta, 1998), 161.
[47] Ibid.
[48] Miraksym A. Usmanov, "The Struggle for the Reestablishment of Oriental Studies in Twentieth-Century Kazan," in *The Heritage of Soviet Oriental Studies*, ed. Michael Kemper and Stephan Conermann (London: Routledge, 2011), 169–202.

ucation minister that the variety of Oriental languages being taught at the university, and the in-depth nature of the instruction, rivalled or surpassed what was available at any European educational institution.[49]

Despite this upturn, the education of the autochthonous population remained unsatisfactory, because, at the university in Kazan, to a greater extent than at any other educational institution in the Russian Empire, the teaching of Asian languages was intended to educate qualified civil servants for the Asian region.[50] Producing qualified civil servants occurred at the expense of academia, as Frähn complained in a report compiled in 1835. In his estimation, teaching Asian languages at both the gymnasium and the university would provide the Interior, Foreign, and Finance Ministries with the necessary specialists. Although Frähn generally thought this policy was correct, he believed that the academic organization of Oriental Studies was completely lacking. He advocated that at least those students who, from the beginning, showed special talents and wanted to dedicate themselves to academia should be integrated into the Ministry of Education. University education should not just meet the state's administrative needs but should also benefit academia.[51]

The situation changed for the worse when the Department of Oriental Studies was relocated from Kazan to St. Petersburg in 1854–1855. It is true that the university in Kazan had functioned since its foundation as an educational institution for citizens in Kazan and those from the regions of Nizhnii Novgorod, Penza, Saratov, Siberia, and the Caucasus.[52] But from the second half of the nineteenth century on, it was intended that many of those receiving state scholarships would attend the universities in Moscow and St. Petersburg, a fact that played a fundamental role in the decision to centralize Oriental Studies in St. Petersburg. Relocating this discipline to one of the leading centres, that is, either to St. Petersburg or Kazan, and concentrating it in a central location had already been the topic of discussion early on at the highest political level. Although Musin-Pushkin and Frähn had initially spoken out in favour of the university in Kazan, the initiative of these two stakeholders was ultimately the decisive factor in relocating the Department of Oriental Studies from Kazan to St. Petersburg. Many contemporary stakeholders considered the relocation a devaluation of the role that the regions could play in imperial academia, but they also saw it as an attempt to increase control over academics.[53]

---

[49] In 1828, chairs for Turkish-Tatar and Arabic-Persian languages were established there, while in the 1830s and 1840s additional chairs for Mongolian, Chinese, Armenian, Manchurian, and Sanskrit were added. Schimmelpenninck van der Oye, *Russian Orientalism*, 104.
[50] Ibid., 96.
[51] Ramil M. Valeev and Dmitrii E. Martynov, *Rossiiskoe universitetskoe vostokovedenie v arkhivnykh dokumentakh. Tsentry, sobytiia i nasledie. XIX – nach. XX vv.* (Kazan: Kazanskii gosudarstvennyi universitet, 2009), 16.
[52] Schimmelpenninck van der Oye, *Russian Orientalism*, 96.
[53] Usmanov, "The Struggle for the Reestablishment of Oriental Studies," 170.

The relocation significantly limited the teaching of Oriental languages as well as academic research in Kazan, especially because a great deal of the teaching materials had to be handed over to the new faculty. Incidentally, the Richelieu Lyceum in Odessa was also impacted by this decision. However, according to Bartol'd, it was particularly painful for the university in Kazan because a considerable amount of money had been spent on collecting Oriental manuscripts. In the following years, the curator of the education district, Vladimir P. Molostvov (1794–1863), sought a refund for at least some of these expenses, without any success.[54]

Both the lack of the necessary teaching materials and the departure of specialists had a significant impact on the subsequent attempt to maintain the teaching of at least some Oriental languages in Kazan. The teaching of the Turkish-Tatar language was an urgent necessity at the university in Kazan and could not be discontinued without considerable damage to the region. By contrast, Katanov and the others thought that teaching languages such as Chinese or Armenian was no longer sustainable due to a lack of resources.[55]

With the 1884 statute, the loss of the Oriental Studies Department was compensated for with two new chairs, one for Turkish-Tatar and the other for Hungarian-Finnish. The two extraordinary professors who were appointed, however, had to deal with considerable difficulties from the very beginning. These subjects were not obligatory for the students and were—in the words of Jan Baudouin de Courtenay (1845–1929), a Polish linguist and Slavicist in Kazan—situated "between heaven and earth."[56] He nevertheless regarded the foundation of these chairs as an attempt to highlight Kazan University's special role as a link between the centre and the Asian peripheries. Baudouin de Courtenay asked what university other than the empire's easternmost university would have been able to better fulfil this role, adding that it was no accident that in Kazan there had previously been an entire faculty for Oriental Studies, which had only been relocated to St. Petersburg by coincidence and through misunderstanding.[57] Although Baudouin de Courtenay saw the establishment of the two chairs as an attempt to maintain Kazan's "Orientalist mission," the new chairs by themselves were not enough to match the earlier level at which the languages and customs of the autochthonous population of the Orient had been taught. He added that, against the backdrop of the most recent academic developments in ethnography, linguistics, and other areas, this gap would be felt particularly painfully and the local researchers were "condemned to be eternal autodidacts" without an institute or academic support.[58] Meanwhile, the Asian region was being studied by foreign researchers or scholars in the

---

[54] Bartol'd, *Raboty po istorii vostokovedeniia*, 104.
[55] Valeev, *Rossiiskoe universitetskoe vostokovedenie*, 54.
[56] Quoted in Valeev, 46.
[57] Ibid.
[58] Ibid., 47.

imperial centres who themselves were only seldom able to travel to the Far East. Instead, they were occupied with processing data that had been collected by others. Baudouin de Courtenay energetically countered the argument that the departure of the faculty might impair education in Kazan, but Oriental Studies would not suffer. According to this logic, any subject at the university could be closed down. Meanwhile, the existence of, for example, Finnish and Hungarian philology in Kazan would allow the university to take on a leading role in linguistic and philological research on the Orient.[59]

Both the teaching of Oriental languages and academic research were considerably weakened with the relocation of the Oriental Studies department. Although Orientalists such as Vasilii V. Radlov, Nikolai F. Katanov, Il'minskii, and many more continued their work in Kazan, they worked under considerably more restricted conditions. The participation of local stakeholders in imperial academia was for the most part limited to founding various amateur groups dedicated to the study of subjects like natural history, anthropology and ethnography, Caucasian archaeology, and the exploration of western Siberia and Central Asia. Furthermore, many military and administrative civil servants in these regions participated in scholarly research, above all ethnographical studies, which led, among other things, to the emergence of a certain kind of academic infrastructure, which took the form of libraries and museums.

Conclusion

The establishment and development of academic Oriental Studies in the Russian Empire was closely interwoven with the domestic and foreign policy interests of the empire and therefore was always under the suspicion of serving colonial policy. Even when the role of Oriental Studies as an aid to developing trade and diplomatic relations with the countries of the Orient was acknowledged, Orientalists such as Rosen or Bartol'd fought for their discipline to be understood as an inalienable part of the "historical-philological discipline." Their focus, first and foremost, was on academic objectives that could be achieved solely by adapting an academic methodology which had been developed over centuries by historians and philologists.[60] The relationship of Oriental Studies to the local elite was shaped significantly by this idea and was determined by the mission that it set itself—to spread "European academic thinking" among the indigenous population of the Russian Empire.

Regional stakeholders also understood that their task was to create an awareness that it was necessary for academia to develop on the peripheries of the empire. What is more, while exploration of the Orient was presented as the Russian Empire's historic mission, even by imperial Orientalists, regional

---

[59] Ibid., 46.
[60] Bartol'd, "Raboty po istorii vostokovedeniia," 494.

stakeholders were convinced that it could be best carried out in the regions themselves. In terms of the necessary interaction between the centre and the periphery, the participation of local powers in academic research was considered essential, even by leading Orientalists. Although the academics from the capital were better equipped in terms of the academic literature at their disposal in the capital's libraries, they could nevertheless benefit from local research in situ. Conversely, intensive exchange would allow representatives of local academia to become acquainted with European academic methodology, which was often lacking in their works and led to flawed results.[61]

In academic circles, there was certainly interest in having regional specialists participate in academic research; competent specialists who were also native speakers of the respective languages and representatives of regional cultures were needed for both teaching and research. The state authorities were also convinced of the value of statistical and ethnographic knowledge, which was imperative for the administration of the Asian regions of the empire and the Caucasus. Until the end of tsarist rule, however, promoting scholarly research in the regions—including establishing universities—was not considered necessary for the smooth functioning of the military or civil administration or for fostering cultural integration.

Translated from German by Catherine Venner.

---

[61] Ibid., 491.

Borbála Zsuzsanna Török

EXPLORING THE K. U. K. PROVINCE
"Landeskunde" and "Honismeret" in Nineteenth-Century Transylvania

The political context of scientific practice has been a prominent subject of historical inquiries during the past decades. A considerable number of innovative studies have demonstrated how science's claims to objectivity and truth were put in the service of the modern nation-state all over the nineteenth century.[1] Another body of work explored the scientific production and communication in modern colonial empires; these studies analyzed the inequalities between corpuses of knowledge produced in metropolitan and colonial areas and asked how the transportation of knowledge across large distances and social milieus sustained, legitimated, or challenged colonial regimes.[2] That the two literatures developed simultaneously but largely without taking each other's results into consideration comes hardly as a surprise. While the original geographic focus of the former was the (Western) European nation-state, the latter had its roots in post-colonial studies that inquired into the Asian, African, and American societies and their interactions with the European colonizers. Perusing the literature published in the 1990s and early 2000s in these distinct sub-fields, one gets the impression of entirely different political dynamics shaping the institutionalization of "Western" science in the nineteenth century and in other regions of the world.

[1] A few examples: Ralph Jessen and Jakob Vogel, eds., *Wissenschaft und Nation in der europäischen Geschichte* (Frankfurt am Main: Campus, 2002); Theodore M. Porter and Dorothy Ross, eds., *The Cambridge History of Science*, vol. 7, *The Modern Social Sciences* (Cambridge: Cambridge University Press, 2008).

[2] Again, some examples from an impressive amount of scholarship Roy McLeod and Philip F. Rehbock, eds., *Nature in Its Greatest Extent: Western Science in the Pacific* (Honolulu: University of Hawaii Press, 1988); Patrick Petitjean, Catherine Jami, and Anne-Marie Moulin, eds., *Science and Empires: Historical Studies about Scientific Development and European Expansion* (Dordrecht: Kluwer, 1992); David Philip Miller and Peter Hans Reill, eds., *Visions of Empire: Voyages, Botany and Representations of Nature* (Cambridge: Cambridge University Press, 1996); Benedikt Stuchtey, ed., *Science Across the European Empires, 1800–1950* (Oxford: Oxford University Press, 2005); Kapil Raj, *Relocating Modern Science: Circulation and Construction of Scientific Knowledge in South Asia and Europe, Seventeenth to Nineteenth Centuries* (Delhi: Permanent Black, 2006); James Delbourgo and Nicholas Dew, eds., *Science and Empire in the Atlantic World* (New York: Routledge, 2008); Marianne Klemun, *Wissenschaft und Kolonialismus* (Innsbruck: StudienVerlag, 2009).

In recent years the situation has changed. Research on nation-state building and the dynamics of empires has become methodologically and topically more intertwined. The New imperial history has created a broad typology of empires that no longer focuses on the classical colonial-maritime systems of rule like Spain, Britain, or France only, but accommodates continental polities without colonies proper, such as Russia, the Ottoman Empire, and the Habsburg Monarchy. In contrast to traditional research, this newer strand of exploration has also pointed to the coexistence of European nation-state formation and imperial politics within the same polity.[3] A case in point is the Habsburg Monarchy, where the study of scientific culture has targeted the rivalries and cooperations emerging from the multiethnic social structure in relation to the inequalities resulting from the monarchic political system.[4] Here a contextual study of science and scholarship has addressed the double dynamics of nationalism and the politics of an imperial context, although there is no consensus yet either about the dynamics and contours of this imperial context nor about an adequate methodological toolkit for studying it.[5]

---

[3] Jürgen Osterhammel, "Imperien und Nationalstaaten: Die Beharrungskraft der Reiche," in *Die Verwandlung der Welt: Eine Geschichte des 19. Jahrhunderts*, ed. Jürgen Osterhammel (Munich: C. H. Beck, 2009), 565–672, 1361–1369, here 603; Karen Barkey, *Empire of Difference: The Ottomans in a Comparative Perspective* (Cambridge: Cambridge University Press, 2008); Stefan Berger and Alexei Miller, eds., *Nationalizing Empires* (Budapest: Central European University Press, 2014); Alfred J. Rieber, *The Struggle for the Eurasian Borderlands: The Rise of Early Modern Empires to the End of the First World War* (Cambridge: Cambridge University Press, 2014); Pieter M. Judson, *The Habsburg Empire: A New History* (Cambridge: Belknap Press of Harvard University Press, 2016); Jana Osterkamp, ed., *Kooperatives Imperium: Politische Zusammenarbeit in der späten Habsburgermonarchie* (Göttingen: Vandenhoeck & Ruprecht, 2018).

[4] Richard Georg Plaschka, Horst Haselsteiner, and Anna Maria Drabek, eds., *Mitteleuropa-Idee, Wissenschaft und Kultur im 19. und 20. Jahrhundert: Beiträge aus österreichischer und ungarischer Sicht* (Vienna: Verlag der Österreichischen Akademie der Wissenschaften, 1997); Moritz Csáky, "Zentraleuropa im Spannungsfeld kultureller Kommunikationsräume," in *Die Ost-West-Problematik in den europäischen Kulturen und Literaturen*, ed. Siegfried Ulbrecht (Prague: Neisse Verlag, 2009), 53–74; Johannes Feichtinger and Ursula Prutsch, eds., *Habsburg postcolonial: Machtstrukturen und kollektives Gedächtnis* (Innsbruck: StudienVerlag, 2003); Veronika Wendland, "Imperiale, koloniale und postkoloniale Blicke auf die Peripherien des Habsburgerreichs," in *Kolonialgeschichten: Regionale Perspektiven auf ein globales Phänomen*, ed. Claudia Kraft, Alf Lüdtke, and Jürgen Martschukat (Frankfurt: Campus, 2010), 211-235; Klemens Kaps and Jan Surman, "Postcolonial or Post-Colonial? Post(-)Colonial Perspectives on Habsburg Galicia," *Historyka: Studia metodologiczne* 42 (2012): 7–35.

[5] For a variety of stances, see Deborah R. Coen, *Climate in Motion: Science, Empire, and the Problem of Scale* (Chicago: University of Chicago Press, 2018); Wolfgang Göderle, *Zensus und Ethnizität: Zur Herstellung von Wissen über soziale Wirklichkeiten im Habsburgerreich zwischen 1848 und 1910* (Göttingen: Wallstein, 2016); Jan Surman and Mitchell Ash, eds., *The Nationalization of Scientific Knowledge in the Habsburg Empire, 1848–1918* (Basingstoke: Palgrave Macmillan, 2012); Daniela Sechel and S. A. King, "Circulation of Medical Knowledge in East Central Europe, 18th–20th Centuries," *East Central Europe*

Exploring the k. u. k. Province 189

The same considerations apply when studying the dynamics of scholarly communication in the multilingual cultural settings of East Central Europe, fractured as they were by language, social and symbolic capital, and access to political resources. My study has its focus in Transylvania, one of the eastern multiethnic provinces of the Transleithanian half of the Habsburg Monarchy. This region found itself since the end of the seventeenth century under Habsburg and, between 1867 and 1918, Hungarian rule and has been considered a "locus classicus" of entangled and rival nationalisms in the modern era.[6] Since its canonization in the nineteenth century, provincial scholarship has divided the past of the region into Romanian, Hungarian, and Saxon (Transylvanian Lutheran German) compartments, which not only marginalized each other's presence but ignored other ethnic and denominational histories, such as that of Jews, Armenians, Greeks, and Roma.[7] Neither the historical nor the societal representations of Transylvania have been "faithful" depictions but reveal "conflicting perceptions and desires of the [ethnic] communities that generated them."[8] The challenge of writing its history, beyond mastering the linguistic diversity, consists of uncovering the common grounds of the variety of the political stances and of distinct cultural identifications that took shape in these milieus.

My inquiry ventures exactly in this direction. The essay focuses on the earliest Transylvanian learned societies, the Transylvanian Saxon Verein für Siebenbürgische Landeskunde (Landeskundeverein, the Association for Transylvanian Landeskunde, 1842–1947) and the Transylvanian Museum Society (Erdélyi Múzeum Egyesület, 1859–1950, reestablished 1990). These learned societies grew out of the intellectual milieus of the late eighteenth century,

---

40, no. 3 (2013): 268–295; Tatjana Buklijas and Emese Lafferton, "Science, Medicine and Nationalism in the Habsburg Empire from the 1840s to 1918," *Studies in History and Philosophy of Biological and Biomedical Sciences* 38 (2007): 679–686.

6   Rogers Brubaker, *Nationalism Reframed: Nationhood and the National Question in the New Europe* (Cambridge: Cambridge University Press, 1996), 4. About newer analyses of entangled Transylvanian nationalisms, see Sorin Mitu, *Geneza identității naționale la românii ardeleni* (Bucharest: Humanitas, 1997); Sorin Mitu, *National Identity of Romanians in Transylvania* (Budapest: Central European University Press, 2001); Zoltán I. Tóth, *Az erdélyi román nacionalizmus első százada, 1690–1792* (Csíkszereda: Pro-Print Könyvkiadó, 1998); David Prodan, *Supplex Libellus Valachorum* (Bucharest: Editura Științifică, 1967); Keith Hitchins, *A Nation Discovered: Romanian Intellectuals in Transylvania and the Idea of Nation, 1700–1848* (Bucharest: Encyclopaedic Publishing House, 1999); Keith Hitchins, *A Nation Affirmed: The Romanian National Movement in Transylvania, 1860–1914* (Bucharest: Encyclopaedic Publishing House, 1999).

7   Sorin Mitu, "Introduction: The Argument," in Sorin Mitu, *National Identity*, 1–13, here 1 f.

8   John Neubauer et al., "Transylvania's Literary Cultures: Rivalry and Interaction," in *History of the Literary Cultures of East-Central Europe: Junctures and Disjunctures in the 19th and 20th Centuries*, ed. Marcel Cornis-Pope and John Neubauer (Amsterdam: John Benjamins, 2006), 245–282, here 256.

they were established in the mid-nineteenth century, and became the regional loci of modern Hungarian and German scholarship on Transylvania before World War I. The provincial cultural elites that created these societies attempted to define what was Transylvania(n) by creating "networks of exchange" beyond the province, addressing audiences both at home and abroad.[9] Their scholarly mapping of the territory had its origins in a knowledge field that was called *Landeskunde* in German and *honismeret* in Hungarian. Having no equivalent in English, it can only be circumscribed as the encyclopedic and systematic description of the land or the "fatherland."[10] My study captures moments in the institutionalization of *Landeskunde* and *honismeret*, while looking at their political horizon in relation to the nation-state and the multiethnic context. A prominent stimulation came from Romanian scholarship, which was simultaneously being established in Transylvania.

The study cuts across different administrative and political regimes. It starts in the 1790s, when Transylvania was a separately governed administrative territory of the Monarchy. After a brief period of unification with Hungary in 1848, the status quo ante was restored by the neo-absolutist administration, to end in 1867. The ensuing Austro-Hungarian Compromise marked the division of Habsburg rule into its Cis- and Transleithanian components and the incorporation of Transylvania into the latter. The ensuing half century witnessed the creation of a modern educational system sponsored by the state and the churches, in correspondence to the European trend of scientific modernization in a national framework. How did the reality of a multiethnic society and its hierarchies shape "national science" and "national humanities"? How is the imperial context helpful for modeling this process?

To answer these questions, I will discuss the meaning and practice of provincial scholarship, *Landeskunde*, this composite Enlightenment knowledge that adapted itself to its changing intellectual and institutional environments over the long nineteenth century. Although marginalized by the current historiography of the modern sciences, throughout the long nineteenth century *Landeskunde* was widely practiced by patriotic circles in the Habsburg Monarchy, in Germany and in Russia, among others. It belonged to *Statistik* or *Staatenkunde* (state description, descriptive statistics) which sought the systematic collection of data relevant for state administration. *Landeskunde* envisaged the description of a political territory and its inhabitants—*Land und Leute*—along a host of knowledge fields. The geographic, natural historical,

---

[9] Borbála Zsuzsanna Török, *Exploring Transylvania: Geographies of Knowledge and Entangled Histories in a Multiethnic Province, 1790–1918* (Leiden: Brill Publishers, 2015).

[10] There has still been no exhaustive definition of *Landeskunde* in English. For its interpretation as a form of scientific travel "at home," while the gathering of empirical data served the goal of improvement, see Henry E. Lowood, *Patriotism, Profit, and the Promotion of Science in the German Enlightenment* (New York: Garland Publishing, 1991), 205–261.

ethnographic, historical, topographic, and philological inquiries into the administrative territory required the collective action of many individuals, leading to the establishment of learned societies dedicated to this task in the late eighteenth—early nineteenth centuries.[11]

The two aforementioned learned societies in Transylvania were founded half a century after the first journals in this field had been published, but they consciously built on these earlier initiatives. The institutionalization of Romanian humanities in the province, on the other hand, was linked to a newer, more modern scholarly movement. While *Landeskunde* had a territorial breadth, the third Transylvanian scholarly institution, the Transylvanian Association for Romanian Literature and the Culture of the Romanian People (Asociația Transilvană Pentru Literatura Română și Cultura Poporului Român or ASTRA, 1861–1950, re-founded in 1990), dedicated itself to exploring the "nation." This was a major difference, which is at the base of my hypothesis, according to which *Landeskunde*, particularly in its origins was a technique of imperial governance, an intellectual tool of the province's mostly German and Hungarian state bureaucrats. Due to the socio-political fragmentation of the local society, these circles remained closed for the Romanian-speaking, Greek Catholic and Greek Orthodox population.[12]

My essay thus proposes that the scholarly exploration of the fatherland—and this was by no means a singular case that characterized only Transylvania but applied to other provinces of the Habsburg Monarchy as well—had an imperial dimension throughout the "long" nineteenth century. It was essentially a science of state management, practiced by members of the bureaucracy. Furthermore, *Landeskunde* focused on territories which involved the mapping of all the peoples of the province in ways that legitimized the claims of the state elites to domination, assigning them a superior status. This became manifest in the self-assigned civilizing mission of the German linguistic strata over the speakers of other vernaculars. The self-ascribed status of superiority was emulated by the regional Hungarian elite, which is the second part of my analysis. The post-1849 Austrian government backed Transylvanian Saxon initiatives, particularly in the neo-absolutist decades, while the new Hungarian state massively supported the Hungarian learned society, which will be briefly discussed in the following sections.

A "Nationalizing Sub-empire"

For a long time, the reputation of the Habsburg Monarchy during the long nineteenth century was that of a living anachronism, an autocratic and une-

---

[11] See Lowood's chapter "Science for the Fatherland," in Lowood, *Patriotism, Profit*, 205–290.
[12] Török, *Exploring*, 72–74, 126, 144–149.

ven polity wrought by nationalist conflicts amidst emerging modern European states. More recent research has dusted off the image of the Habsburg Empire by pointing to its consolidation as a polity with the flexibility to accommodate change and to strike a compromise with the political and educated elites from its various lands. While politics and bureaucracy increasingly valued expertise over mere rank, the early modern composite monarchy did not fully transform into a modern state under the rule of law.[13] It has been duly portrayed as a

> weakly integrated multiethnic entity, a collection of territories with often ancient historical identities of their own [...] the Danube Monarchy did not develop into a federation; the whole state actually became more heterogeneous after 1867. [...] Right to the end, the various components of the Habsburg Monarchy were integrated in the imperial manner: a shared imperial culture and identity took shape to some extent, without being politically enforced, while horizontal social integration continued to be restricted.[14]

Instead, as Andrea Komlosy argues, the cohesion of the empire emerged from the legal and administrative unification of the provinces, from the formation of a single market area, a single army, and from general laws of education. The Austro-Hungarian Compromise institutionalized the incomplete transformation into two halves, of which the Transleithanian half actively pursued Hungarian nation-building within a liberal framework, while the Cisleithanian half "maintained" the de facto dominance of German elites.[15]

Another feature of the Habsburg polity was the existence of mutual dependencies between overlapping military, political, and economic cores and peripheries,[16] which were being sustained by the fact that the economic and political weight of the individual lands did not correspond. Thus, Hungary was a political core region throughout the time covered by my study, while economically it played a more peripheral role.[17] The economic interdependencies precluded internal homogenization and reinforced, instead, the cohesion of unequal regions.[18] The Austro-Hungarian Compromise safeguarded

---

[13] Gerald Stourzh, *From Vienna to Chicago and Back: Essays on Intellectual History and Political Thought in Europe and America* (Chicago: University of Chicago Press, 2007); Judson, *Habsburg Empire*; Peter Becker, "Der Staat—eine österreichische Geschichte," *Mitteilungen des Instituts für Österreichische Geschichtsforschung* 126 (2018): 317–340; Jürgen Osterhammel, "Die Habsburgermonarchie," in *Die Verwandlung der Welt*, ed. Osterhammel, 624–627, 1366 f.; Andrea Komlosy, "The Habsburg Monarchy, 1804–1918: Imperial Cohesion, Nation-Building and Regional Integration," in *Nationalizing Empires*, ed. Miller and Berger, 369–427.
[14] Jürgen Osterhammel, *The Transformation of the World: A Global History of the Nineteenth Century* (Princeton, NJ: Princeton University Press, 2014), 434 f.
[15] Komlosy, "The Habsburg Monarchy," 406.
[16] Ibid., 375.
[17] Ibid., 375 f.
[18] Ibid., 407.

the economic interests of the Hungarian landed nobility, which pursued a politics of Magyarization in Transleithania.

Transylvania figures as a terrain of imperial intervention par excellence, where Vienna had weakened Hungarian estate dominance by reinforcing the rights of the Transylvanian Saxon estates (numerically a small fraction both of the provincial diet and also proportionally among the population) in the course of the eighteenth century. The confessionalization of the Greek Orthodox, Romanian-speaking elite was another attempt to forge new Habsburg alliances against the provincial Hungarian elites.[19] It is telling that Transylvanian Romanian nationalists during the Hungarian *Vormärz* of the 1830s demanded corporate and not individual or territorial rights.[20]

The entanglement of territorial-linguistic constituencies as a cohesive force became the legitimizing ideology of the Habsburg Empire after the Hungarian war of independence. Carl von Czoernig, director of the Austrian Statistical Bureau, argued on the basis of his intricate ethnographic map, accompanied by a monograph published in 1857, that national dissent was not only a political crime, it was social nonsense in the intertwined multilingual, multinational, and multiconfessional reality of the Habsburg Monarchy.[21] However, the assertion of diversity at the central governmental level did not lead to equalization measures in the Lands. In contrast to the more homogeneous legal system of national citizenship, the Habsburg "imperial rights regime" was characterized in terms of the "nonequivalence" of its multiple population, which partly took a legal expression, partly remained a matter of administrative practice.[22] Informal dominance came also from the hierarchy of languages (oral-written, rural-urban), German being the most prestigious and most widely used in trade, administration, and scholarship. By 1867 the Hungarians achieved a dominant status in Transleithania, where the Nationality Law of 1868 guaranteed vernacular language use, albeit the latter did not become a constitutional right and did not prevent the politics of Magyarization, which became particularly virulent in the 1880s. The pursuit of Magyar supremacy in the subsequent decades until World War I made Komlosy refer to the "Hungarian sub-Empire" and not simply to a national state.[23]

---

[19] Ibid., 394 f.
[20] Kinga-Koretta Sata, "The People Incorporated: Constructions of the Nation in Transylvanian Romanian Liberalism, 1838–1848," in *We, the People: The Politics of National Peculiarity in Southeastern Europe*, ed. Diana Mishkova (Budapest: Central European University Press, 2009), 79–105.
[21] Karl von Czoernig, *Ethnographie der österreichischen Monarchie* (Vienna, 1857), VII.
[22] Jane Burbank and Frederick Cooper, *Empires in World History: Power and the Politics of Difference* (Princeton, NJ: Princeton University Press, 2010); Osterhammel, *Verwandlung*.
[23] Komlosy, "The Habsburg Monarchy," 377.

Provincial Scholarship—Selective State Support

The Hungarian liberal legislation of 1867–1868 guaranteed the expression of cultural identity for the national minorities in the framework of church-run educational systems and within the framework of a grassroots civil movement, based on voluntary associations. From the 1880s onwards, however, growing national populism and an imperialist rhetoric accompanied the attempt of the authorities to control education and the cultural life of non-Magyars. These developments prompted a tone of defensive nationalism in the Transylvanian Saxon and Romanian public spheres.

The Landeskundeverein, fashioning itself as the centre of the Saxon civic movement, culture, and education, led by the Lutheran bishop, became the dominant agent in shaping the regional German identity discourse. It supplied its themes and tonality and besame increasingly inclusive towards the countryside.[24] In the 1880s the minister of education encouraged municipalities, local authorities, and other legal bodies to support cultural and educational organizations, and the Landeskundeverein successfully contacted a number of these with an invitation to join its ranks. The most significant new partners were the local chapters of the Hermannstädter Sparkassa which, together with the Bodenkreditanstalt in Hermannstadt, became an important sponsor of the activities of the association.[25] The Landeskundeverein remained a civic institution that was financed by private money and by cooperatives.

The Museum Society, founded by the provincial Hungarian social elite, was centered in Cluj (Hungarian: Kolozsvár/German: Klausenburg). Its scientific work included sections in the humanities, the natural sciences, medicine, law, and the social sciences. The main purpose of the society was, however, to support a public museum, including a library and a manuscript collection (and later on, an archive), several natural historical collections, archeological collections (numismatics, lapidarium, archeological finds) as well as other, smaller collections (art, ethnography). The collections were assembled during the first wave of Central European musealization, the time when the local Hungarian nobility turned their private collections into public ones, expecting thereby to gain the financial support of the state. Initiated as a civic institution in 1859, the Museum Society later became a state-supported institution. After the Austro-Hungarian Compromise of 1867, a contract with the Hungarian Ministry of Education enabled the Museum Society to establish a liaison with the newly established Transylvanian University in 1872, which warranted the desired state support for the museum's holdings. It also launched a process of modernization of the society's scientific sections, com-

---

[24] "Statuten des Vereins für Siebenbürgische Landeskunde vom Jahre 1870," in Heinz Herbert, "Geschichte des Vereines für Siebenbürgische Landeskunde," *Archiv für siebenbürgische Landeskunde* 28 (1898): 139–236, here 187–189.

[25] See the list of members and also Herbert, "Geschichte des Vereins," 168 f.

bining the activities of an academy with those of a museum, a financially ambitious plan. This double agenda constitutes a significant structural difference to the Landeskundeverein which did not have a museum of its own. Also, the Landeskundeverein kept close contact with Transylvanian Saxon grammar school education under the guidance of the Lutheran Church, while the Museum Society targeted academic research.

The newly created Hungarian state supported the scientific activity of the institutions run by the "titular nation." The foundation of the first modern Transylvanian university was but the most spectacular change in this regard, opening new prospects of regional development in education and research and expressing a policy that addressed the needs primarily of the regional Magyar-speaking titular nation. The university declared itself responsible for the maintenance of the collections, and the Museum Society was granted a yearly rent (5,000 Ft/10,000 Crowns) for the personal costs and the maintenance of the collections. The rector did not spare efforts to turn the Museum Society into the university's research institution, and by the turn of the century around one-third of the ordinary association members were recruited from the academic staff—an indication that the two institutions were merging scientifically. The natural sciences section was founded in 1879, formally separated from the humanities. It was to be joined by the section of medical science in that same year, and later by the section for legal and social science (1906). Pressure towards specialization turned the sections into quasi-independent institutions with separate budgets and with the right to elect internal members who had no obligations towards the society. This came along with the further infrastructural development of the society.[26]

The impact of the innovation was so significant that it raises the question of comparability with the Landeskundeverein. However, the Museum Society was not transformed into a research appendage of the university. In particular, its historical-archeological collections and library were crucial for developing a distinct regional research program for Transylvania, which indeed enables the comparison with its counterpart, the Saxon association. By contrast, the Landeskundeverein maintained, as mentioned, its ties with secondary education, and its scholarly profile developed at a steadier if less spectacular pace. While the Transylvanian German-speaking Lutherans made up ca. 9,5 percent of the population compared to ca. 33 percent of Magyars,[27] the two institutions became nearly identical in size and also with regard to their research agendas in Transylvania.

---

[26] See Török, *Exploring*, 163–230.
[27] For a rough orientation, the census of 1900 counted ca. 233,000 Germans, 815,000 Magyars, 1,397,000 Romanians, and 32,000 members of other nationalities on the territory of Transylvania (57,804 km²). Irina Livezeanu, *Cultural Politics in Greater Romania: Regionalism, Nation Building and Ethnic Struggle, 1918–1930* (Ithaca, NY: Cornell University Press, 1995).

National Appropriations of Scholarship

Rooted in early modern descriptions of states,[28] eighteenth-century *Landeskunde* was understood as an intellectual engagement with the "fatherland." Its purpose was the use of acquired knowledge for the betterment of the overall conditions and therefore a contribution to the common good. The territorial-administrative scope of the earliest *Landeskunde* journal, the *Siebenbürgische Quartalschrift* (1790–1801), run by a reading society in Hermannstadt, is well illustrated by its topics. These included the history of Saxon settlements (six articles); Transylvanian political history (two); the Saxon Lutheran Church (six); the Roman Catholic and Unitarian Churches in Transylvania (two); Sources of Transylvanian history (five); chorography of towns (twelve); political arithmetics and descriptive statistics (eleven); politics (one); agriculture (three); weather (three); geography (five); physics (one); botany (one); medicine and medical *Polizey* in Transylvania (15); the ethnography of the peoples of the province: the Saxon dialect (five); Romanians (one); Serbians (one); *historia litteraria*: the status of Transylvanian scholarship (three); Transylvanian Saxon scholars (two); Transylvanian Hungarian scholars (seven); Saxon schools (three); politicians; and statesmen from Transylvania (eighteen). The task was "making the Fatherland acquainted with itself, turning its attention to important truths [facts]" concerning its "moral," "political," "scholarly," and "economic" aspects via writings on geography, natural history, the "morality of its [Transylvanian] inhabitants once and now," "pragmatic perusals of history," and also literary reviews and news of "important events of our times." Knowledge was said to be the basis for "private and public happiness."[29] The holistic disciplinary structure and the regional focus were typical of the methods of *Landeskunde*, revealing an interest in practical knowledge based on the natural history, *historia litteraria*, and in the local and the useful, which could be registered and communicated to the larger home public. These learned practices made the circle around the *Quartalschrift* largely similar to many other scholarly associations in the European urban zones, characterized by the collaborative effort of collecting *Staatsmerkwürdigkeiten* (state peculiarities).

The editors of the *Quartalschrift* did not advance either cosmopolitan or democratic projects. Being aware of their elite status in the province, the Saxon authors argued for region-wide communication, acting as members of overlapping German, Austrian, and Transylvanian cultural networks. Not without gestures of fatherly condescendence towards the two other "nations," this stance committed itself to dynasty-loyal Austrian patriotism. Their de-

---

[28] Mohammed Rassem and Justin Stagl, eds., *Statistik und Staatsbeschreibung in der Neuzeit, vornehmlich im 16.–18. Jahrhundert: Bericht über ein interdisziplinäres Symposion in Wolfenbüttel, 25.–27. September 1978* (Paderborn: Schöningh, 1980).

[29] Daniel Neugeboren, "Ueber die Lage und Hindernisse der Schriftstellerei in Siebenbürgen," *Siebenbürgische Quartalschrift* 1, no. 1 (1790): 1–27, here 25.

mand for the use of German in public communication indicates that the advocators of *Landeskunde* understood their role as legitimate initiators of provincial improvement, which was also to the benefit of the Viennese government. This is the reason why the scholarly focus as well as the addressees of the *Quartalschrift* transcended narrow ethno-cultural boundaries. The journal also translated the writings of established Hungarian authors of *honismeret* and reported on meetings of the Transylvanian Society for the Cultivation of the Hungarian Language (Erdélyi Magyar Nyelvmívelő Társaság, 1793–1806) as well as several learned societies abroad.

The focus remained on Saxon history, religion, and education, but there was an interest in the Transylvanian Hungarian culture too.[30] The articles on history centered on the feudal privileges of the Saxon estate, respectively the satisfaction of and hindrances to corporate freedom under various kings and governors of the past. The history of the corporate nation was a subject of chief political importance among the contemporary Saxon public as well. On the other hand, the practitioners of *Landeskunde* construed the provincial society in terms of ethnic and civilizational differences. They promoted a hierarchical vision of the provincial population, championing German culture and its scholarly advancements to the benefit of the Hungarian, Romanian, Roma, and Jewish inhabitants and classified the Transylvanian "nations" according to stages of education and improvement. The Romanians were situated at the bottom of this cultural ladder, presented as a population of noble (Roman) descent, although uncultivated because of their social and political status.[31] Magyars were seen as people with a glorious intellectual past, but the centuries-long Turkish wars and religious skirmishes had eroded "taste and scholarship," according to this view.[32] The *Quartalschrift* recommended the use of German, as the Hungarian language seemed too backwards to serve as a cross-cultural language of scholarly communication. The Saxons, the authors held, had the best prospects for "enlightenment and the refinement of taste."[33]

The insufficiency of Hungarian as a language of scholarly exchange, mentioned also by Neugeboren in the *Quartalschrift*, became a chronic concern of Transylvanian Hungarian scholars. The emerging Hungarian *honismeret* championed the cultural emancipation of all Magyars across the Transylvanian-Hungarian border. This, they hoped, would be achieved by cultivating the language and adapting it to scientific needs:

To introduce the Hungarian language, it is necessary already at the beginning to found a society of learned patriots in the Hungarian country, which should 1) translate into Hun-

---

[30] The journal published the biographies of Transylvanian Hungarian and Székely scholars among the German-speaking Transylvanian Saxon ones.
[31] Neugeboren, "Ueber die Lage und Hindernisse," 6.
[32] Ibid., 9–10.
[33] Ibid., 11.

garian all kinds of books written about our patria, so that all the sons and daughters of the patria have the opportunity to read them in their own language without painfully learning foreign languages; and that would be the path to Hungarian national Enlightenment. [...] 2) This society should translate all Greek and Latin authors into Hungarian, so that one could become acquainted more easily with these authors.[34]

The Transylvanian Society for the Cultivation of the Hungarian Language and the related Manuscript Editing Society (Kéziratok Társasága) became the intellectual centre of Magyar *honismeret* well beyond the Transylvanian border. A systematic collection of the political-legal historical evidence of the Transylvanian estates, scattered in the official and ecclesiastical archives as well as private libraries of the province, became a comprehensive scholarly enterprise which could not have been achieved without the Transylvanian German counterparts, not to mention assistance from Pest.[35] The Hungarian and Saxon learned circles knew each other well and worked together, despite some issues of permanent dispute, such as the legitimacy of Hungarian-language policy or the presumed inferiority of Hungarian culture. Looking at the list of active members of the historical section of the Hungarian Language Society, one finds, besides Hungarian savants, the names of Martin Bolla, a Catholic professor of history at the legal academy in Kolozsvár, Johann Michael Ballmann, a Lutheran grammar school professor in Mediasch, György Kovachich, the well-known historical scholar from Buda, and the Transylvanian German grammar school professor and historian Josef Karl Eder.[36]

It was the social elitism of the provincial Magyar and German circles that distanced them from Romanian scholarship. This became evident after the death of Joseph II, when Greek Catholic ecclesiastic elites drew up a petition to Emperor Leopold, the "Supplex Libellus Valachorum," vindicating the rights of a fourth nation, including privileges similar to the other Transylvanian estates. The "Supplex" made up part of a campaign to obtain corporate rights for Transylvanian Romanians, which can be traced back to the 1740s. The historical and language studies of the time supported the claim that the Romanians were the oldest and most numerous people in Transylvania, who had belonged to the Transylvanian political leadership before the Saxon, Székely, and Hungarian estates had formed a political union to protect their privileges against peasant revolts in 1437. The core of the "Supplex" was the assertion of Roman continuity in Transylvania, a theory that supported the confessional and political struggle, proclaiming the right to collective exist-

---

[34] Cited by Sándor Enyedi, "Introduction," in *Aranka György Erdélyi társaságai* (Budapest: Szépirodalmi Könyvkiadó, 1988), 9–39, here 12 f.
[35] See József Szinnyei, *Magyar írók élete és munkái* [The lives and works of Hungarian intellectuals], online bibliographical lexicon, http://mek.oszk.hu/03600/03630/html/index.htm.
[36] Jakab Elek, *Aranka György és az Erdélyi Nyelvművelő és Kéziratkiadó Társaság* (Budapest, 1884), 36.

ence and emancipation of Transylvanian Romanians.[37] The practitioners of *Landeskunde* had little appreciation for the intellectual achievements of emerging Romanian scholarship, and the "Supplex" earned but contemptuous half-sentences about the "uncivilized Romanians" and their "foolish political attempts" from both the Hungarian and the Saxon side.[38]

Who was Where First? National Pedigrees and Claims to National Superiority

How did the territorial, inter-ethnic, and national scope of scientific practice change during the *Ausgleich*? *Landeskunde* and *honismeret* had been absorbing new disciplines in the humanities and the natural sciences already since the mid-nineteenth century. The differentiation of the former field and the emergence of modern ethnography, historical philology, and non-classical archeology rejuvenated the two associations' scholarly activities. At the heart of the attention of Transylvanian *Landeskunde* and *honismeret* was the history and putative ethnic origins of the Transylvanian Hungarian and Saxon "nations" respectively. These "national antiquities"[39] included the collection of historical data, now joined by philological and archaeological inquiries. At the same time, however, attention continued to be paid to the polyglot provincial society.

The "national antiquities" acquired a specific political twist in Transylvania, as they came to play a role, since the "Supplex Libellus Valachorum," in the ongoing strife that centered on historical corporate rights as the basis for participating in provincial rule.[40] Yet, as in the case of the Romanians, the ethnic and geographic origins of the Magyars and Saxons were controversial, constituting a common concern of scholarship across administrative borders.

---

[37] Éva H. Balázs, *Hungary and the Habsburgs, 1765–1800: An Experiment in Enlightened Absolutism*, trans. Tim Wilkinson (Budapest: Central European University Press, 1997), 98; Hitchins, *Nation Discovered*, 133–143; Katherine Verdery, *Transylvanian Villagers: Three Centuries of Political, Economic, and Ethnic Change* (Berkeley: University of California Press, 1983), 120; Béla Köpeczi, ed., *Erdély története 1830-tól napjainkig* (Budapest: Akadémiai Kiadó, 1986), 1112.

[38] Lucas Joseph Marienburg, *Geographie des Grossfürtstenthums Siebenbürgen* (Hermanstadt, 1813, facsimile Vienna: Böhlau, 1987), 95. See also August Ludwig Schlözer, *Kritische Sammlungen zur Geschichte der Deutschen in Siebenbürgen* (Göttingen, 1795–1797), 662 f.; György Aranka, *Anglus s Magyar Igazgatásnak egyben vetése* [Comparison of the English and Hungarian Administration] (Kolozsvár, 1790) cited in Győző Concha, "Az angolos irány politikai irodalmunkban a múlt század végén [The English stream in our political literature at the end of the last century]," in Győző Concha, *Hatvan év tudományos mozgalmai között* [Amidst the scientific movement of sixty years] (Budapest: Tudományos Akadémia, 1928), 213–227, esp. 221 f.

[39] Monika Baár, *Historians and Nationalism: East-Central Europe in the Nineteenth Century* (Oxford: Oxford University Press, 2010), 167–195.

[40] See Sata, "People Incorporated," passim.

The debates touched issues such as the history and archeology of ancient Dacian and Roman Transylvania, the philological analysis of toponyms, and the comparative ethnographic exploration of folk customs. Fin-de-siècle theories of the Balkan descent of Transylvanian Romanians, which were gaining plausibility at the time, argued against the idea of local Latin-Dacian origins. These debates were influenced by the political rivalries between Hungary and the Romanian Kingdom. The Museum Society took part with great élan in the unfolding strife, while the Landeskundeverein increasingly assumed the role of a mediator between the adversaries. The journal of the Transylvanian Hungarian institute, Erdélyi Muzeum, published a number of contributions on Romanian language, ethnography, and history.[41] The chief contributor on Romanian ethnogenesis was Gheorghe Moldovan (1845–1930), a professor of Romanian philology at the University of Kolozsvár and its principal in the years 1906–1907. Moldovan rejected the theory of Roman continuity on Transylvanian soil, thus joining the Budapest-based mainstream:

> While we find no Dacian traditions of any kind among the Romanians, the preponderance of Balkan traditions and influences is decisive regarding the determination of their place of origin. [...] The ethnographic facts of my book demonstrate that the [theory of] Dacian origins is untenable. If the Romanians buried the [theory of] Dacian continuity in the Hurul Chronicle and the linguistic bravados in the Testament and Dictionary by [Treboniu] Laurian without any impair, they will also bury this Dacian continuity as well, which will not affect by any means their vitality, their cultural efficiency, and their promising future.[42]

Moldovan criticized the "tendentious intentions of Romanian historiography" and the "tale of Dacian continuity and its falsifications." In the same breath, he condemned the political movement of Transylvanian Romanians for more autonomy as being irredentist and in effect "dangerous for the Hungarian state," a fact which indicated how strongly intellectual and political claims were interlinked in contemporary thinking. Against the pan-Romanian agitation of the Bucharest-based Romanian Cultural League (Liga Culturală Română), he praised the Magyarizing efforts of the nationalist Transylvanian

---

[41] A few examples: Endre Veress, "Erdély—és magyarországi régi oláh könyvek és nyomtatványok [Old Romanian books and prints from Transylvania and Hungary]," *Erdélyi Múzeum* (1910): 143–176 and 313–383; Endre Veress, "Documentele lui Stefan cel Mare [The documents of Stefan cel Mare]," *Erdélyi Múzeum* (1914): 279–284; Lajos Kropf, "A lipcsei rumén szeminárium [The Romanian seminar in Leipzig]," *Erdélyi Múzeum* (1900): 61–70. For a topical and disciplinary overview, see the chapters on Romanian philology, Romance languages and literatures, and Romanian and Transylvanian history, see Antal Valentiny and Géza Entz, *Az Erdélyi Múzeum név-és szakmutatója, 1874–1917, 1930–1937* [Systematic catalogue of *Erdélyi Múzeum*] (Cluj: EME, 1942), 41, 66 f., and 82, 88–90.

[42] Gergely Moldován, "A románság balkáni eredetéhez [About the Balkan origins of the Romanians]," *Erdélyi Múzeum* (1899): 61–71. See also Gergely Moldován, "Székelyek-e a mócok? [Are the Moti of Székely origin?]," *Erdélyi Múzeum* (1894): 343–358 and 403–416.

Hungarian Cultural Association (Erdélyi Magyar Kulturális Egyesület) as "useful and excellent work."[43]
The Landeskundeverein took the position of a critical observer but formulated no theory of its own. The journal was nevertheless interested predominantly in the Hungarian and German-Austrian side of the debate, and it featured most prominently the works by the German Romanicist Eduard Robert Rößler and his Austrian critic Julius Jung,[44] as well as the Hungarian Pál Hunfalvy,[45] while the Romanian and Transylvanian contributions appeared merely briefly as bibliographical references.[46] All in all, the Landeskundeverein adopted the theory of a Balkan descent of Transylvanian Romanians,[47] but also stressed close linguistic interaction with the Transylvanian Hungarian and Saxon dialects.[48] Yet, they keenly noticed whenever the academic debate drifted towards political polemics. Typically, a review of a book by Hunfalvy stated:

> For those familiar with the writings of Hunfalvy, most of what they find here is already known. The book follows not exclusively scientific objectives but also political ones, and it places itself into the service of Hungarian nationality politics. He will reap [among Romanian scholars] what he sows, and that will be no friendly reply. As long as he deals with the history of the Romanians, we can ascertain that he argues in a transparent way and is right in the details. The way he applies the linguistic evidence for the internal and external history of the Romanians is valuable in many respects but also shows many weaknesses. As al-

---

[43] Gergely Moldován, "A román nemzetiségi törekvések [The Romanian national political attempts]," *Erdélyi Múzeum* (1896): 392–394; Gergely Moldován, "Nyílt levelek a bukaresti román kulturális liga elnökéhez [Open letters to the president of the Romanian cultural league in Bucharest]," *Erdélyi Múzeum* (1895): 38 f.; Gergely Moldovan, "Román kérdés—magyar nemzetpolitika [Romanian question—Hungarian national politics]," *Erdélyi Múzeum* (1895): 40–44, 450, 512, 561; Gergely Moldován, "Magyar-szász szövetség [Hungarian-Saxon alliance]," *Erdélyi Múzeum* (1904): 431. About the pro-Magyar nationalist stance of Moldovan see Tímea Berki, "From Grigore Moldovan to Moldován Gergely: A Career in Homeland," *Acta Universitatis Sapientiae: Philologica* 3, no. 2 (2011): 156–166.
[44] Julius Jung, *Römer und Romanen in den Donauländern: Historiographisch-ethnographische Studien* (Innsbruck, 1877), cited in Karl Gooß, "Die neuste Literatur über die Frage der Herkunft der Romänen," *Korrespondenzblatt des Vereins für siebenbürgische Landeskunde* (hereafter Korrespondenzblatt) 1, no. 2 (1878): 17–22 and no. 3 (1878): 28–39.
[45] Karl Gooß, "Zur Rumänen-Frage," *Korrespondenzblatt* 2, no. 3 (1879): 26–31; N. N., "Die Landsnahmen Siebenbürgens: Von Paul Hunfalvy," *Korrespondenzblatt* 10, no. 4 (1887): 37–40; N. N., *Korrespondenzblatt* 10, no. 5 (1887): 49–52.
[46] Gooß, "Zur Rumänen-Frage," 26–31.
[47] Fr. W. Seraphin, "Zur Geschichte der siebenbürger Bulgaren," *Korrespondenzblatt* 19, no. 19 (1896): 143–146; D. [Gustav Weigand], "Die Aromunen," *Korrespondenzblatt* 20, no. 4 (1897): 54 f.
[48] A[dolf] Schullerus, "Brenndörfer János: Román (Oláh) elemek az erdélyi szász nyelvben/Rumänische Elemente in der siebenb.-sächs. Sprache. Budapest, 1902 [Brenndörfer János: Romanian (Wallach) elements in the Transylvanian Saxon language. Budapest, 1902]," *Korrespondenzblatt* 26, no. 2–3 (1903): 36–45.

ways, H. puts forward his usual etymologies of the names of land, rivers, and settlements. Should these be his best weapons, he will be soon defeated.[49]

The intransigence of the writings by Moldovan on the pages of the Erdélyi Múzeum becomes particularly noticeable if read in contrast with the more nuanced, more informed, and comparative approach which was, despite all mentioned biases, adopted by the authors of the *Korrespondenzblatt*. One finds in the Hungarian case an attitude that embraced the nationalist attitude of the state when it came to its most painful issue, namely the governance of the unassimilated and putatively *irredenta* Romanian minority. Here the debates about the primacy of ethnic descent on Transylvanian soil reached back to the claim made by "Supplex" for political empowerment of Transylvania's Romanian population, so the scholarly debates had a high political stake, clearly visible in the articles.

Concluding Remarks

Long before the end of the nineteenth century, the encyclopedic exploration of the fatherland gave way to the institutionalization of national canons in the regional framework, with a double effect. On the one hand, scholarly professionalization enhanced the differentiation of separate national interpretations of Transylvanian German, Hungarian, and Romanian cultures, enshrined by the distinct institutional framework. On the other hand, the emerging new disciplines of the humanities called for a new exchange between Saxon, Romanian, and Hungarian experts, whether it came to source edition or the evaluation of archeological discoveries.[50] This "invisible university" was maintained by the modern scientific ethos and consisted of the creation of new pathways of communication via the systematic review and evaluation of each other's work.

Yet *Landeskunde* and *honismeret* served goals of political supremacy too, illustrated in this chapter by the century-long controversies on the historical origins of the regional non-elites. The long-lasting antagonisms based on the hierarchical concept of the nations of the province did not disappear but were refashioned in accordance with historical, archeological, linguistic, and ethnographic narratives and the material and immaterial evidence of putatively glorious national antiquities. The reaction Romanian narratives of ethnic descent caused among Magyar and Transylvanian Saxon scholars indicates how politically charged *Landeskunde* and *honismeret* were at the end of the nineteenth century. While the strife was exasperating in its Romanian-Hungarian dimension, Saxon scholars increasingly took the role of critical yet less emo-

---

[49] N. N., "Paul Hunfalvy: Die Rumänen und ihre Ansprüche. Wien und Teschen, Prohaska, 1883," *Korrespondenzblatt* 7, no. 12 (1884): 143.

[50] For more about the Saxon-Hungarian exchange, see Török, *Exploring*, passim.

tional arbiters. The intensity of the conflict between the Hungarian scholars and their Romanian peers indicates the new titular nation's claim to domination.

Landeskunde and its Hungarian adaptation conveyed hierarchic visions of the patria, though at different historical conjunctures. Both of them centered on the respective "national" imaginary, while also mapping other ethnic constituencies as well as general cultural and historical topics that were relevant for the administrative and political elite. In the period before 1848, Transylvania meant for Saxon scholars both an administrative unit and the historical home of the Saxon estates. Hungarian *honismeret* focused on national improvement and sought the cultural integration of the province with the Kingdom of Hungary. After the Austro-Hungarian Compromise of 1867, a socially more inclusive perspective emerged in Hungarian scholarship, which emulated the Saxon cultural initiatives and developed a paternalistic attitude towards the most vocal and potentially secessionist minority, the Transylvanian Romanians. The scholarship of the titular nation of *Ausgleich* Hungary, backed by general state patronage, now gave itself the task of putting its nationalities in their place.

Peter Haslinger

NATIONAL GEOPOLITICS IN HABSBURG CENTRAL EUROPE
Imperial and Post-Imperial Perspectives on Hungary and Poland
1890–1930[1]

This essay examines the various logics of late and post-imperial geopolitics,[2] choosing the Habsburg Monarchy together with two of its successor states as its sample cases.[3] The argument focuses on leading scientists and academics who played a prominent role in spatially related disciplines before the war and who either worked as specialists or performed official functions in Hungary and Poland after 1918. Before the outbreak of the First World War, they had a clear understanding of their role as an avant-garde within their own community (which they understood mainly in national rather than in professional terms). They saw the First World War as trigger, catalyst, and driving force for national projects of emancipation: for them, the last two years of the First World War also created previously unidentified opportunities to trans-

---

[1] I would like to take the opportunity to thank the organizers and participants of the conference as well as all of my colleagues from the DFG Collaborative Research Centre/ Transregio 138 "Dynamics of Security: Types of Securitization from a Historical Perspective" for their extremely useful feedback.
[2] For the historical context, let me list just a few works: Jörn Leonhardt, *Der überforderte Frieden: Versailles und die Welt, 1918–1923* (Munich: C. H. Beck, 2018); Robert Gerwarth, *The Vanquished: Why the First World War Failed to End, 1917–1923* (London: Allen Lane, 2016); Jochen Böhler, Wlodzimierz Borodziej, and Joachim von Puttkamer, eds., *Legacies of Violence: Eastern Europe's First World War* (Munich: Walter de Gruyter, 2014); Omer Bartov and Eric D. Weitz, eds., *Shatterzone of Empires: Coexistence and Violence in the German, Habsburg, Russian, and Ottoman Borderlands* (Bloomington: Indiana University Press, 2013); Julia Eichenberg and John Paul Newman, "Aftershocks: Violence in Dissolving Empires after the First World War," *Contemporary European History* 19, no. 3 (2010): 183–194; Peter Gatrell, "War after the War: Conflicts, 1918–1922," in *A Companion to World War I*, ed. John Horne (Chichester: Wiley-Blackwell, 2010), 558–575; Piotr Wróbel, "The Seeds of Violence: The Brutalization of an Eastern European Region, 1917–1921," *Journal of Modern European History* 1, no. 1 (2003): 125–149.
[3] Unfortunately, it is impossible to undertake an analysis of the very extensive cartographic output of the period within the bounds of this essay; the author is currently preparing a separate study on that topic. On this topic see (among others) Peter Haslinger and Vadim Oswalt, "Raumkonzepte, Wahrnehmungsdispositionen und die Karte als Medium von Politik und Geschichtskultur," in *Kampf der Karten: Propaganda- und Geschichtskarten als politische Instrumente und Identitätstexte*, ed. Peter Haslinger and Vadim Oswalt (Marburg: Verlag Herder-Institut, 2012), 1–12.

late messages of national emancipation into concrete territorial designs for the immediate future. During the process of formation of independent nation states, these stakeholders found themselves in positions that carried political, administrative, or institutional responsibilities—at the head of governments, ministries, state offices and institutes, as members of commissions and policy advisory bodies, or as university professors with influence over organizational issues.

During the last two decades of Austria-Hungary's existence, the contexts in which the discipline of geography developed were quite different in the cases of Hungary and Poland, just as they were compared with the later situation in the various other successor states.[4] Although no co-national state existed outside the Habsburg Monarchy in either case, the idea of them having spaces of their own at their disposal was a concept characteristic of both Hungarian and Polish debate even before 1914. In both cases, discourses on spaces and borders aimed at shaping economic and demographic realities and conveyed explicit cultural hierarchies and civilizing missions that carried clear political implications.

The legal and institutional conditions for the development of an institutionalized geography under the umbrella of a nation-state-building project were particularly favourable in Hungary. After the Austro-Hungarian Compromise of 1867, Hungarian political elites were able to pursue a policy of anchoring Hungarian as a language of science and culture and actively sponsoring national-patriotic academic careers. No such level of institutionalization underpinned by a centralized state existed to support geography as practiced in the regions into which Poland was partitioned at the time. Nevertheless, a system of Polish-speaking higher education began to develop under the conditions provided by the autonomy attained for the crown land of Galicia-Lodomeria in 1868, centered around the two universities at Cracow and L'viv respectively. The increased scope for academic advancement that this situation offered—as it contrasted starkly with the opportunities for academic activities in other Polish regions—meant that Galicia developed into a place of retreat for an epistemic Polish national geography that remained closely networked with the other partitioned lands of former Poland-Lithuania via academic career trajectories and cross-border debates.

In Western Europe too, the establishment of geography as a discipline had had a close connection with state and foreign policy initiatives from the outset. Past defeats had the effect of focussing on lost territories and, in the context of state building and colonialism, geography provided an academic resource through which one could mediate the new spatial orders and power

---

[4] On the Bohemian lands and on Czechoslovakia in the interwar years, cf. Peter Haslinger, *Nation und Territorium im tschechischen politischen Diskurs, 1880–1938* (Munich: Oldenbourg, 2010).

relations.⁵ Competition between the various different national geographies emerging at that time in Austria-Hungary was lent more dynamism by the shift in paradigms that was occurring within the discipline, pushing it in the direction of political geography and its more radical variant: geopolitics.⁶ The most influential of the protagonists in this process was Friedrich Ratzel. In his 1897 work *Politische Geographie* (Political geography), the German geographer conceived a theory of state according to which—as Marcus Sandl put it—"space as a fundamental category determined both the prerequisites for and the opportunities and goals of political action."⁷ Ratzel radicalized traditional spatial concepts by placing culture, space, and state in a normative relationship with one another,⁸ introducing a new link between territoriality and the form of the organization of the state, thus legitimizing the politics of accumulation of new space on the basis of national interests.⁹ The term "geopolitics" was coined in 1899 in the journal of the Swedish Society for Anthropology and Geography by a student of Ratzel's, Swedish political scientist Rudolf Kjellén.¹⁰ Picking up on the theories of Charles Darwin, he regarded states as organisms engaged in a constant struggle for position and resources.¹¹

In response to such interpretations, the geographers of the Habsburg Monarchy had a number of counter-models available. Much intensive discussion centered around the theses of British geographer Halford Mackinder, who had developed his geopolitical "heartland" theory as an analytical model in Oxford in 1904, based on the idea of Russian dominance of the Eurasian land mass as a threat scenario.¹² But it was mainly French geographers, above

---

5 Thus, a second professorial chair was established in the German Empire after the foundation of the empire in 1871 (the first had been created in 1820), while in France the institutionalization of geography as a discipline in its own right was a direct consequence of the loss of Alsace and Lorraine—a chair in history formerly located in Strasbourg was established in 1872 in Nancy as a professorship in history and geography and was held by Paul Vidal de la Blache. Virginie Mamadour, "Geography and War—Geographers and Peace," in *The Geography of War and Peace: From Death Camps to Diplomats*, ed. Colin Flint (Oxford: Oxford University Press, 2005), 26–60, here 28.
6 On the following, cf. most importantly Ulrike Jureit, *Das Ordnen von Räumen: Territorium und Lebensraum im 19. und 20. Jahrhundert* (Hamburg: Hamburger Ed., 2012) and for Eastern Central Europe, Maciej Górny, *Kreślarze ojczyzn: Geografowie i granice międzywojennej Europy* (Warsaw: Instytut Historii PAN, 2017), 16–20.
7 Marcus Sandl, "Geschichtswissenschaften," in *Raumwissenschaften*, ed. Stephan Günzel (Frankfurt am Main: Suhrkamp, 2009), 159–174, here 164.
8 Stephan Günzel, "Einleitung," in *Raumwissenschaften*, ed. Günzel, 7–13, here 10.
9 Jörg Dünne, "Politisch-geographische Räume: Einleitung," in *Raumtheorie: Grundlagentexte aus Philosophie und Kulturwissenschaften*, ed. Jörg Dünne and Stephan Günzel (Frankfurt am Main: Suhrkamp, 2006), 371–385, here 376.
10 Rudolf Kjellén, "Studier öfver Sveriges politiska gränser," *Ymer* 19, no. 3 (1899): 283–331.
11 Joe Painter and Alex Jeffrey, *Political Geography: An Introduction to Space and Power* (Los Angeles: SAGE Publications, 2009), 199.
12 Ibid., 201.

all Paul Vidal de la Blache and his son-in-law Emmanuel de Martonne, who attracted the attention of their colleagues in the Habsburg Monarchy. Both were deeply sceptical[13] of the approach taken by German-led political geography and of the geographical determinism associated with it. Instead, they placed greater emphasis on the active role played by human beings in shaping their spatial environment.

It was primarily the works of the five above-mentioned authors that were to fundamentally transform the scientific approach taken by the spatial sciences in Austria-Hungary at the turn of the century. It is simply amazing how quickly and yet at the same time with how much controversy the new approaches were received by an entire generation of geographers. In addition to their studies at "national language" universities in Prague, Budapest, Cracow, L'viv, or Vienna, geographers of that generation had already spent part of their study time abroad (e.g., in Paris, London, Berlin, Halle, Zurich, and Kiev). At the same time, the University of Vienna counted among the global centres for the study of geography, which had established itself as an independent discipline since the 1880s. Until his appointment to Berlin in 1906, Albrecht Penck was the central figure in Vienna and was widely recognized internationally as an expert on geomorphology, climate research, and on the landscapes that had emerged as a result of Ice Ages and glaciation processes.[14] He attracted a large number of enthusiastic students to the *Residenzstadt*. Through their involvement in excursions to the nearby Central and Eastern Alps, they began to adapt the research approaches they learned to the conditions prevailing in their own regions of origin. The best-known example of these students is Jovan Cvijić, who arived in Vienna in 1889 with a Serbian scholarship and the aim of dedicating himself to the intensive study of the Dinaric Karst region. In his dissertation, which he defended in Vienna in 1893, Cvijić was the first to develop the terminology for the study of karst landscapes that was to become widely recognized internationally. In the very same year, he was called back to Serbia to take up his appointment as a full professor at the Velika Škola, the predecessor of the University of Belgrade.[15]

---

[13] Mamadour, "Geography and War," 31.

[14] After the First World War, Penck was to become one of the founding figures of the Stiftung für deutsche Volks- und Kulturbodenforschung (Foundation for German People's and Cultural Soil Research) and, together with Karl Haushofer, coined the concept of *Volks- und Kulturboden* (People's and Cultural Soil), which was to prove a very powerful political idea during the Weimar Republic. On this, cf. (among others) Guntram Henrik Herb, *Under the Map of Germany: Nationalism and Propaganda 1918–1945* (London: Routledge, 1997); Kristin Leigh Kopp, *Germany's Wild East: Constructing Poland as Colonial Space* (Ann Arbor: University of Michigan Press, 2012); Agnes Laba, *Die Grenze im Blick: Der Ostgrenzendiskurs der Weimarer Republik* (Marburg: Verlag Herder-Institut, 2019).

[15] Maciej Górny, "Der Krieg der Karten: Geografen und Grenzziehungen in Ostmittel- und Südosteuropa, 1914–1920," *Střed* 5, no. 1 (2013): 9–39, here 11 ff.

Aside from Cvijić and his Ukrainian colleague Stepan Rudnyc'kyi,[16] Polish geographer Eugeniusz Romer can also be counted among the internationally renowned experts of the discipline.[17] While completing his studies in Vienna, Romer had also begun his career in the study of mountain and landscape forms. In the course of research conducted in the Alps and the Carpathian Mountains, Romer developed his own anthropogeographical and politico-historical perspectives. He studied the river valleys in the east of the old Polish-Lithuanian Commonwealth as well as investigating the "geopolitical function of the steppes," which he regarded as relics of prehistoric times that would disappear in the course of the cultivation of the area (in a process that he contended had already occurred in Belgium and the United States).[18] During his 1895–1896 academic year in Vienna, Romer attended the lectures of Penck, who, according to an account by Marian Morczko, "combined solid geographical knowledge with nationalistic prejudices against Slavs and Poles in particular." Penck—in common with the overwhelming majority of his German colleagues—understood no Polish and was therefore ill-equipped to appreciate the works published in that language.[19] He considered the Vistula, for example, to be the natural border of German Central Europe and of a German ethno-cultural zone of influence (the German *Volks- und Kulturboden*). According to Penck, Poland was permanently bound to Germanic Europe as a consequence of its location and geographical traits.[20]

After his habilitation in 1899 at his new anchorage in L'viv, Romer devoted himself to studying the forms of the earth's surface and the various ways in which glaciers transform landscapes.[21] What bestowed international fame and recognition upon him was a pioneering geographical work that he completed under the umbrella of the Polish national cause: the *Atlas geograficzny* (Geographical atlas) published in 1908 as the first contemporary cartographic survey of all the lands of the partitioned Polish-Lithuanian Commonwealth.[22] On the one hand, his stays in Vienna, Halle, and Berlin provided a substantial

---

16 Guido Hausmann, "Das Territorium der Ukraine: Stepan Rudnyc'kyjs Beitrag zur Geschichte räumlich-territorialen Denkens über die Ukraine," in *Die Ukraine: Prozesse der Nationsbildung*, ed. Andreas Kappeler (Cologne: Böhlau, 2011), 145–157, here 147.
17 Maciej Górny offers a particularly informative overview of these and other contemporary geographers in *Kreślarze ojczyzn*, 11–16, 20–30.
18 Eugeniusz Romer, *Ziemia i państwo: Kilka zagadnień geopolitycznych* (Lwów: Atlas, 1939), 31–33.
19 Steven Seegel, *Map Men: Transnational Dives and Deaths of Geographers in the Making of East Central Europe* (Chicago: University of Chicago Press, 2018), 60.
20 Marian Mroczko, *Eugeniusz Romer, 1871–1954: Biografia polityczna*, 2nd ed. (Słupsk: Wydawnictwo Naukowe Akademii Pomorskiej, 2010), 37 f. Cf. also Stefan Misiniec, *Eugeniusz Romer: Człowiek nauki i wiary* (Kraków: Wydawnictwo Petrus, 2014).
21 Cf. the selection in Eugeniusz Romer, *Wybór prac*, 3 vols. (Warsaw: Państwowe Wydawn Naukowe, 1960–1962).
22 Alexandra Schweiger, *Polens Zukunft liegt im Osten: Polnische Ostkonzepte der späten Teilungszeit, 1890–1918* (Marburg: Verlag Herder-Institut, 2014), 134.

boost to his professional standing and reputation, and gave him the opportunity to acquire considerable methodological and conceptual knowledge. On the other, though, he also found himself confronted with teaching content based on a spatial policy connoted very much in German national terms, geared towards penetration and the struggle to achieve dominance, assimilation, and expansion. Just as in the case of other colleagues of non-German background, this almost inevitably inspired a tendency to self-nationalize academically and to self-activate politically.[23] Wincenty Pol, for example, held the works of Karl Ritter and Alexander von Humboldt in high regard from a methodological point of view. Yet at the same time —as the first professor at a Polish chair of geography at the Jagiellonian University in Cracow—Pol polemicized in his work *Historyczny obraz Polski* (A historical image of Poland) against "Prussian geography, which, on the topic of the European lowlands, is only familiar with the lowlands of France, Germania and Sarmatia [...]. Poland [is] swept off the map of Europe as a geographical term, as a natural entity."[24]

Geographers of Polish, Czech, or southern Slavic origin, who saw the axioms of German national attitudes as both a stimulus and a threat, worked intensively to develop alternatives to German-oriented political geography and geopolitics. To this end they adopted the models offered by French geographers, with a preference for the strongest possible anti-German approach. They hoped to attract international support for their own interpretations and to gain permanent academic credibility in opposition to German and Russian imperial geographies. International forums, networks, and academic media also helped accelerate this transfer of ideas.

This new "Polish" geography functioned as some kind of a "repair geography" with the aim of deepening the cohesion between all the lands of partitioned Poland-Lithuania, thus ensuring the survival of the Polish spatial idea in all regions of Poland. And yet, the diversity of the voices to emerge from the three regions of partition was to remain a characteristic trait throughout

---

[23] During the Russo-Japanese War, for example, Romer often made trips to Congress Poland and actively participated in popular science courses for teachers, and in 1908 gave lectures in Lublin and Warsaw to audiences numbering several thousands on the entitlement of future Poland to the district of Chełm. Later, on 25 and 26 August 1912, Romer took part in a meeting of representatives of the independence movement in Zakopane in Galicia and—against the background of the First Balkan War—in Vienna on 10 November 1912 in a meeting held between representatives of Galicia and the Russian Empire. Mroczko, *Eugeniusz Romer*, 37 ff. Cf. also Misiniec, *Eugeniusz Romer*, 26.

[24] Quoted from Bronisław Kortus, "Der polnische Westgedanke und die Geographie," in *Deutsche Ostforschung und polnische Westforschung im Spannungsfeld von Wissenschaft und Politik: Disziplinen im Vergleich*, eds. Jan Piskorski, Jörg Hackmann, and Rudolf Jaworski (Osnabrück: Fibre, 2002), 239–259, here 240. Cf. Gernot Briesewitz, *Raum und Nation in der polnischen Westforschung, 1918–1949: Wissenschaftsdiskurse, Raumdeutungen und geopolitische Visionen im Kontext der deutschpolnischen Beziehungsgeschichte* (Osnabrück: Fibre, 2014).

the entire period. In contrast to Hungarian and Czech geography, which were based on clearly pre-existing ideas about borders, the Polish scholars could not reach a consensus on the question of the territorial shape of a future Poland. Triggered by a series of articles by poet, literary critic, and publicist Czesław Jankowski, which appeared together as a book in 1914, the Polish-language journalism commenced a fierce controversy in June of the same year on the ideal frontiers of any possible Polish state.[25]

The process of consolidation of national epistemic positions in Polish and Hungarian geography went hand in hand with a perception that the nation's "own" science still lagged far behind the major scientific nations in terms of methods, resources, and senior research figures. It was hoped that this reputational deficit could be at least partly compensated by the regions' scientists' own proximity to the specific subject matter and their knowledge of all the languages necessary. Researchers aimed to demonstrate this by identifying and correcting any inaccuracies, errors, and misunderstandings that could be found in research from other European countries based on a lack of detailed knowledge or simply on ignorance.[26] One example of this can be found in the translation into Hungarian of a multivolume *General Description of the Earth* (Atlas ethnographique du globe) by Italian geographer Adriano Balbi. The work was translated from the original French by Géza Czirbusz, now considered the founder of modern Hungarian geography. Czirbusz, actually a priest by vocation, had adopted the approaches of Vidal de la Blache and Jean de Brunhes into his own works, *A nemzeti művelődés geográfiája és a geográfiai fatalisták* (The Geography of national education and the geographical fatalists) and *Anthropogeográfia* (Anthropogeography). He did not, however, translate Balbi's work word for word, but rather made corrections and extensive improvements to the passages on Hungary, changes that, according to Czirbusz, had been made necessary by his Italian colleague's lack of expertise on Hungary.[27]

Especially in Hungary, the development of the expanding higher education system went hand in hand with the implementation of "scientific nation-

---

[25] Schweiger, *Polens Zukunft*, 64–68.
[26] On the historical sciences Markus Krzoska remarked aptly: "Because one knew the 'foreign interpretation' of one's own history by the partitioning powers, one strove to construct an independent image of Polish history, a national master narrative. However, this phenomenon was not associated with any democratization in the picture of one's own social role. On the contrary, one still felt that one was part of the national elite and had serious doubts as to whether one's own results would not be misused for foreign political goals." Markus Krzoska, *Für ein Polen an Oder und Ostsee: Zygmunt Wojciechowski (1900–1955) als Historiker und Publizist* (Osnabrück: Fibre, 2003), 37.
[27] Balázs Ablonczy, *Pál Teleki, 1874–1941: The Life of a Controversial Hungarian Politician* (Boulder, CO: East European, 2006), 23.

alism."[28] The tasks of recording the country's statistics and of developing its infrastructure now required an independent and increasingly professional cartographic output, one that was no longer dependent on the Habsburg imperial perspective. Aside from this, both state and academic elites in Hungary were keen to be seen on the international stage as independent actors capable of representing themselves at the cutting edge of the zeitgeist in terms of their methodologies and conceptual axioms. This resulted in the emergence of a separate Hungarian cartography of emancipation in the face of competing interpretations (including those of Habsburg imperial dominance, the Croatian claim to sovereignty, and alternative concepts stemming from the claims of national minorities and neighbouring states). One important consequence of this was that Hungarian geography increasingly began to undermine the spatial imaginary of Austro-Hungarian dualism.

Against this background, Pál Teleki, then a young geographer, increasingly became the hub around which Hungarian geography was reorganized and internationalized. Teleki's extensive language skills provided one of the foundations of his international recognition (he spoke German, French, and English, as well as some Dutch, Italian, and Romanian).[29] As with Romer, the first international recognition for Teleki resulted from a project to produce an atlas. In the latter's case, his highly praised maps depicted not Hungary, but the islands of Japan. Teleki, who knew no Japanese and had never been to Japan, launched the atlas consisting of 20 maps, which was published in Leipzig in German and Hungarian in 1909. In recognition of this achievement, he was elected to the Old Map and Chart Committee of the Geneva International Geographical Society and in 1911 to the Paris Geographical Society. Also, he received the highly prestigious Jomard Award.[30]

In parallel with his international reputation, his picture postcard career was moving along nicely in Hungary too, where he made academic and political advances from the outset. In 1909 Teleki was made chief of the Geographical Institute of the University of Budapest and was named General Secretary of the Hungarian Geographical Society (Magyar Földrajzi Társaság) in 1910. As the chairman of various other learned societies, Teleki combined the progress he made in his professional standing with networking activities in the political space, in which he was also active from early in his career. He even served as a member of the Hungarian parliament from 1905 to 1910. Consequently, he was at the centre of a number of initiatives supported by the Hungarian government. Among enterprises worth mentioning that he was

---

[28] Gábor Palló, "Scientific Nationalism: A Historical Approach to Nature in Late Nineteenth-Century Hungary," in *The Nationalization of Scientific Knowledge in the Habsburg Empire, 1848–1918*, ed. Mitchell G. Ash and Jan Surman (Basingstoke: Palgrave Macmillan, 2012), 102–112.
[29] Seegel, *Map Men*, 35.
[30] Ibid., 39.

involved in is the foundation of the Turanian Society on 3 December 1910, which, in terms of its composition, structure, and aims, would correspond to what we now call a think tank. Balázs Ablonczy distinguishes between three groups of members of this learned society: The first consisted of scientists—geographers (e.g., Lajos Lóczy, Jenő Cholnoky, and Pál Teleki), ethnologists, Orientalists, and Turkologists. The second group, aside from aristocrats, politicians, and financiers, also included István Tisza, who served two terms as prime minister, Minister of Culture Albert Berzeviczy, former minister of agriculture Ignác Darányi as well as the mayor of Budapest, Géza Teleki. The third group was made up of intellectuals who approved of the Turanian project, and from among whom in the 1920s extreme right-wing groups were to emerge.[31]

Thus far, we can come to the provisional conclusion that both Polish and Hungarian geography had already generated independent geopolitical discourses of a national character before the First World War—despite the fact that the legal and institutional background conditions differed considerably between the two cases. The outbreak of the First World War had the effect of dynamizing all these spatial ideas and political initiatives. A decisive role in this process was played by the fact that the new quality of warfare brought by the conflict came accompanied by a new way of perceiving geographical space. In this context, cultural sociologist Stefan Kaufmann, for example, pointed out that "space is no longer thought of as linear, but as a net, the front is no longer located on the front line, but is conceived as a planar structure [...]." According to Kaufmann, this modified view already bears traces of modern, twentieth-century mass society. "What was being mobilized was no longer jingoistic patriotism, but universal pathos of progress, modern avant-gardism and projects of radical anthropological optimization."[32]

Thus, the war also brought new opportunities for geographers. In its initial stages, the sciences that occupied themselves with spatial matters benefited from the boost in demand for knowledge of military geography and prognostic planning skills. This demand was also a consequence of the fact that both the progress of the war and war goal policies—which on all sides were to crystalize into clearer concepts only gradually over the course of the war—brought previously unfamiliar regions into the focus of decision-makers. Not just geographers, but also other scientists with an interest in spatial matters, including historians, ethnographers, linguists, anthropologists, statisticians,

---

[31] Ablonczy, *Pál Teleki*, 29 ff. One particularly outstanding personality in this group was a former expert advisor at the US Department of Agriculture, Alajos Paikert, who upon returning to Hungary became one of the prime movers of the Agricultural Museum. Ibid., 30.

[32] Stefan Kaufmann, "Raumrevolutionen: Die militärische Raumauffassung zwischen dem Ersten und dem Zweiten Weltkrieg," in *Der Weltkrieg 1914–1918: Ereignis und Erinnerung*, ed. Rainer Rother (Berlin: Deutsches Historisches Museum, 2004), 42–49, here 42 and 49.

sociologists—and sometimes even art historians and botanists—now took on new roles and entered new contexts of action through their contributions to the conduct of war and to the administration of occupied territories. One special field of activity was their work in developing an ethnographic cartography of corroboration and delegitimization (depending on the specific political context in question), through which one was expected to accept claims to dominance over particular towns, villages, and rural areas or to cast doubt upon the enemy's sovereignty over parts of its territory.[33]

Over the course of the war, Hungarian policy also began to diversify when it came to the question of possible territorial gains: as part of the debate on Austria-Hungary's war objectives, Prime Minister István Tisza, speaking on the future of the Kingdom of Serbia, expressed strong opposition to its annexation, yet he pursued a plan to incorporate the fertile plains to the north of the occupied land. As the war went on he suggested some territorial expansions in the west of Romania in order to gain control of the strategic Iron Gates gorge on the Danube and to establish a common border between Hungary and Bulgaria.[34] In parallel with this, circles close to government in Hungary developed a substantially more diffuse program of expansion—one open to being substantiated in various directions at the same time—directed mainly at Dalmatia, Bosnia-Herzegovina, and the western Balkans. In this connection, Pál Teleki was part of a three-member working group which was to provide a scientific basis for independent Hungarian research on the East and the Orient. The Hungarian Geographical Society[35] (due to a lack of resources and personal rivalries) was unable to provide an adequate forum for this discussion: it was the Adriatic Society, and, once more, the Turanian Society, along with the Turanian Association-Hungarian Eastern Cultural Centre, which was formally inaugurated in the great hall of the Hungarian Academy of Science on 2 May 1916, that was to take on the task of conducting scientific research on southeastern Europe and the Ottoman Empire.[36] In the case of Hungarian spatial studies—just as in the person of Ukrainian geographer

---

[33] This is described very impressively in Vejas Gabriel Liulevicius, *War Land on the Eastern Front: Culture, National Identity and Occupation in World War I* (Cambridge: Cambridge University Press, 2000). Cf. also, among other works, Vejas Gabriel Liulevicius, "Der Osten als apokalyptischer Raum: Deutsche Frontwahrnehmungen im und nach dem Ersten Weltkrieg," in *Traumland Osten: Deutsche Bilder vom östlichen Europa im 20. Jahrhundert*, ed. Gregor Thum (Göttingen: Vandenhoeck & Ruprecht, 2006), 47–65; Annemarie Sammartino, *The Impossible Border: Germany and the East, 1914–1922* (Ithaca, NY: Cornell University Press, 2010). Further relevant references can be found in Górny, *Kreślarze ojczyzn*, 49–57.

[34] Marvin Benjamin Fried, *Austro-Hungarian War Aims in the Balkans during World War I* (Basingstoke: Palgrave Macmillan, 2014), 127, 200, 222.

[35] Lajos Lóczi Lóczy, *A Magyar Szent Korona országainak földrajzi, társadalomtudomanyi, közművelödési és közgazdasági leírása* (Budapest: Kilián, 1918), III.

[36] Ablonczy, *Pál Teleki*, 36 f.

Stepan Rudnyc'kyi—one may speak of a cartographic sub-imperialism under the aegis of the German *Kaiserreich*.

However, from the moment Romania launched its surprise attack on Transylvania at the end of August 1916, anxiety began to seep into the Hungarian political positions. Hungarian-speaking scientists now doubled down on the credo that had applied even before 1914, according to which the unchanged borders of the thousand-year-old territory of Hungary represented a natural and logical unit in terms of its politico-economic and historical development. Nevertheless, the first voices began to be raised setting out possible unfavourable scenarios. The most influential publications discussing such ideas were authored by János Karácsonyi, a church historian and ordinary member of the Hungarian Academy of Sciences. In a work published in 1916 (and—unsurprisingly—reprinted in 1920), Karácsonyi was convinced that the Allied powers had already cold-bloodedly decided to split up Hungary and had drawn paid insurgents from among the country's non-Hungarian-speaking citizens over to their side. It was, however—he claimed—a "historical truth that every truth-loving man is convinced that the Hungarian nation has a historical right to the region stretching from the Carpathians to the Adriatic."[37]

In contrast to the case of Hungary, Polish territorial discourses during the war completely lacked any institutional grounding from state officials. In their case, two statements of the belligerents—issued on 9 August by the Central Powers and on 14 August by Russia—commenced an argumentative contest between the competing territorial plans of the empires in relation to Poland.[38] Although an Austro-Polish scenario always remained under discussion in the camp of the Central Powers, it was not the Habsburg Monarchy but the German Empire that proved to be the driving force behind such plans. After the failure of its offensives in the west, Germany shifted its interest in designing hegemonic futures toward the east, quickly making Congress Poland a key object in the disputes it had with Austria-Hungary over the post-war order.[39]

Against the backdrop of open or indirect competition between all powers involved in the partition of the Polish-Lithuanian Commonwealth, and despite continued major differences in regard to the shape of a future Polish state territory, the divergent range of opinions present in the Polish national camp now turned out to give them a strategic advantage. The internationali-

---

[37] János Karácsonyi, *A magyar nemzet történeti joga hazánk területéhez a Kárpátoktól le az Adriáig* (Nagyvárad: A Szent-László-Nyomda, 1916), 5 ff.

[38] Jan Karski, *The Great Powers and Poland 1919–1945: From Versailles to Yalta* (Lanham, MD: University Press of America, 1985), 4.

[39] Ursula Prutsch, "Historisches Gedächtnis in kulturpolitischer Machtstrategie: Deutschland, Österreich-Ungarn und die polnische Frage, 1915–1918," in *Ambivalenz des kulturellen Erbes: Vielfachcodierung des historischen Gedächtnisses*, ed. Moritz Csáky and Klaus Zeyringer (Innsbruck: StudienVerlag, 2000), 69–91, here 71.

zation process affecting the Polish question opened up new opportunities for Polish scientists: there was some limited space available for Polish politicians to introduce arguments into the discussions that aimed at developing a territory for the Polish state in one direction or another, depending on how the war progressed. The attempt by the German occupation authorities to influence the Polish public through a series of lectures held at the newly founded University of Warsaw (in which Albrecht Penck, among other luminaries, took part) was greeted with the customary vehement criticism for factual errors, and the effort met with little empathy from Polish colleagues.[40] A separate event held at the University of Cracow deliberately countered the German efforts: through their examination of the reasons behind the failure of the Polish-Lithuanian Commonwealth, experts like Eugeniusz Romer, Oskar Halecki, Franciszek Bujak, and Stanisław Kutrzeba also provided a programmatic contribution in advance of the expected post-war situation.[41]

During the war years, Romer in particular came to public attention through a number of publications designed to provide arguments in favour of post-war geopolitical changes. The story of the making of the *Geograficzno-statystyczny atlas Polski* (Geographical-Statistical Atlas of Poland), which he compiled in his Viennese apartment after his escape from L'viv and which was published in 1916, has since become legendary among researchers. In his foreword to the atlas, Romer claimed the following motive for his work: "[...] to serve the needs of those who wish to liberate this country and for those who wish to rule over it, the Atlas illustrates the national, social and economic conditions in the period before the Great War."[42] The explosive character of its content is also revealed in the fact that Penck not only vehemently polemicized against the work of his former pupil—accusing him, for example, of using falsified statistics for the maps in his atlas. He even went so far as to demand that the Austrian authorities confiscate Romer's materials, ban the export of the printed atlas, and arrest Romer. These attempts notwithstanding, a copy was smuggled abroad to a neutral country, from where it was sent on to the United States, where it was possible to produce an English-language version.[43]

In the last two years of the war, the opportunities for and the need to participate in the simulation of political territories for use in post-war scenarios

---

[40] Górny, "Der Krieg der Karten," 16 f.
[41] Andrzej Wierzbicki, *Wschód-Zachód w koncepcjach dziejów Polski: Z dziejów polskiej myśli historycznej w dobie porozbiorowej* (Warsaw: Państwowy Instytut Wydawniczy, 1984), 327. These lectures were published in *Przyczyny upadku Polski* (Warsaw: Gebethner i Wolff, 1918).
[42] Quoted from Kortus, "Der polnische Westgedanke," 241.
[43] Steven Seegel, "Remapping the Geo-body: Transnational Dimensions of Stepan Rudnyts'kyi and his Contemporaries," in *The Future of the Past: New Perspectives on Ukrainian History*, ed. Serhii Plokhy (Cambridge, MA: Harvard University Press for the Ukrainian Research Institute, Harvard University, 2016), 205–229, here 212.

almost automatically led to a politicization of the academic world. It had already become clear through the peace negotiations at Brest-Litovsk in February and March 1918 that maps and atlases were going to play an important role in future peace talks. Starting in the summer of 1918, academic currents emerged along national lines following a new, multidisciplinary model. Representatives of these currents made their academic knowledge available to political actors or themselves became active in politics in a contest over the best arguments: geographers, historians, statisticians, ethnologists, and economists, as well as delegations from the military and the press (representing national news agencies and newspapers of record), provided the core of such circles of expertise.

Inside the collapsing Austria-Hungary, this inevitably led to intensified efforts to create competing national cartographies, which focussed especially on regions and border sectors where controversies were to be expected. An academic *habitus* provided just one basis for their self-image as a part of the national avant-garde within their own national community. Each national camp was therefore both willing and able to prepare political messages in the mode of scientific statements and to resort to cartographic tricks similar to the ones used by their opponents. The consequences of this politicization of science were serious: it was to lead to permanent alienation between academics who had formerly been colleagues and to the disintegration of scientific circles that had previously been closely networked through shared career paths, scientific societies, and publication forums.

By the end of the war, this politicization of academia was also in full swing in Hungary. In November 1918, the new bourgeois-democratic government under Mihály Károlyi founded the Országos Propaganda Bizottság (National Committee for Propaganda), whose role was to win back the sympathies of the nationalities for the new Hungary and thereby preserve as far as possible the territorial integrity of the country. Of the nearly nine million leaflets and brochures it produced, 6.4 million were aimed at populations living in overwhelmingly non-Magyar areas.[44] The famous *carte rouge*, which was to play such an important role in the 1920 peace negotiations in Trianon, was also produced in English, French, and Hungarian by Pál Teleki as early as December 1918. His methodology and choice of colours overemphasized the Hungarian-speaking minority population in Transylvania at the expense of the Romanian-speaking majority, thereby giving a counter-emphasis to the French cartography of Emmanuel de Martonne, which provided support for the Romanian point of view.[45]

---

[44] Anikó Kovács-Bertrand, *Der ungarische Revisionismus nach dem Ersten Weltkrieg: Der publizistische Kampf gegen den Friedensvertrag von Trianon, 1918–1931* (Munich: Oldenbourg, 1997), 45 ff.

[45] Seegel, *Map Men*, 66. Dániel Zoltán Segyevy offers a broad overview of Hungarian cartography in the period immediately after the war in *Térképművek Trianon árnyékában:*

In relation to the approaching peace negotiations at Paris, the memoranda and the statistics and maps necessitated professional preparation by experts with the required methodological experience and scientific credentials, who were also reliable and capable of discretion. Their elaborations combined established ideas of territoriality with wartime stereotypes in order to match as far as possible the interests of the victorious powers, and especially those of France. Under considerable time pressure, their main task was to anticipate opposing territorial claims and to put forward as coherent a set of arguments as they could. They were required to use scientific language for disseminating their national strategic interests among decision-making circles and fellow specialists working as scientific experts on the side of the Allies.[46] The official diplomatic channels designed to communicate territorial demands needed to be strategically flanked by informal contacts, press campaigns, and public information events. Actors such as Romer and Teleki can thus be characterized appropriately —following Neil Smith's description of Isaiah Bowman, one of the geographical experts responsible for the American position—as "policy entrepreneurs."[47]

At Versailles, Saint Germain, and Trianon, the conditions were quite different for Poland's case and Hungary's.[48] Once again, the maps[49] prepared by Romer and published in the form of an atlas in L'viv in 1921 made a great impression. Romer also organized 70 lectures in Paris and made great strategic use of his personal friendship with Isaiah Bowman, Emmanuel de Martonne, and the US specialist on Poland and Russia, Robert H. Lord.[50] Nevertheless, Polish policy by no means succeeded in imposing all its wishes on the proceedings.[51] Not only was there considerable irritation at Poland's lack of cooperation but, due to the political rivalry that existed between Roman Dmowski and Józef Piłsudski and their quite differing conceptions of what the Polish state should look like in the future, Poland's territorial program

---

Magyarország néprajzi térképe, 1918 [Mapwork in the Shadow of Trianon. The Ethnographic Cartography of Hungary, 1918] (Budapest: KSH Könyvtára, 2016), 89–112.

[46] On this topic, cf. Haslinger, *Nation und Territorium*, 237–267. Cf. also Frank Hadler and Tibor Frank, eds., *Disputed Territories and Shared Pasts: Overlapping National Histories in Modern Europe* (Basingstoke: Palgrave Macmillan, 2011).

[47] Painter and Jeffrey, *Political Geography*, 203. Thus, Romer became head of the geographical office of the Polish delegation to the peace negotiations in Paris and acted as their main specialist authority on the demarcation of borders in eastern Silesia and Upper Silesia. Mroczko, *Eugeniusz Romer*, 37 ff. Cf. also Misiniec, *Eugeniusz Romer*, 28.

[48] Górny, *Kreślarze ojczyzn*, 86–100.

[49] Schweiger, *Polens Zukunft*, 136.

[50] Kai Lundgreen-Nielsen, *The Polish Problem at the Paris Peace Conference: A Study of the Policies of the Great Powers and the Poles, 1918–1919* (Odense: Odense University Press, 1979), 166.

[51] On this, cf. Benjamin Conrad, *Umkämpfte Grenzen, umkämpfte Bevölkerung: Die Entstehung der Staatsgrenzen der Zweiten Polnischen Republik, 1918–1923* (Stuttgart: Steiner, 2014).

also appeared inconsistent to many. With no clear directives from Warsaw, the delegation gave contradictory statements on the territory they requested, and quite a few memoranda presented by the Polish delegation failed to express their arguments in academic language.[52]

The situation was even more difficult for the Hungarian delegation. It made its first appearance in Paris as late as January 1920, by which time all the important decisions had already been taken and many high-ranking Allied representatives had already left the peace conference. The territorial terms that had been arrived at were extremely harsh for Hungary, and the delegation led by Albert Apponyi aggravated this unfavourable position by adopting an entirely wrong negotiating strategy: in accordance with the national credo of a thousand years of continuity of Hungary's borders, the arguments put forward by the numerous Hungarian memoranda aimed at defending the country's full territorial integrity within its 1914 borders. This rather dogmatic attitude rendered Hungary unable to obtain rectifications in some segments of the border that had so far been delineated only provisionally.[53] In

---

[52] This can be seen even in these passages from the memorandum "Poland's Territorial Problems:" "The Polish problem is above all a problem of territory. Situated between Germany and Russia [...] Poland must herself become a solid State, completely independent, capable of self-protection and of development along lines of her own. [...] If Poland is to cope successfully with difficulties which are the direct result of her geographical situation, the following conditions are essential: 1) She must have an extensive territory and a large population. 2) The population must be sufficiently homogeneous to insure internal cohesion. 3) Her frontiers must correspond to geographical requirements, making her independent of her neighbours." On the eastern territories, the memorandum stated: "The Poles being the sole intellectual and economic force in that country. [...] The Ruthenians (Ukrainian), White Russian and Lithuanian majorities consist almost exclusively of small farmers and priests. [...] The Poles are at present not strong enough to rule with success the whole of the territory of the Eastern provinces. Considering there is no other cultured element of sufficient strength to replace them, the political future of these Eastern Polish provinces is a problem almost insoluble. The establishment of independent Lithuanian and Ukrainian states would lead to either anarchy or foreign, i.e., German, rule. The return of these provinces to Russia would no less surely lead to anarchy and to stagnation in intellectual as in economic spheres." "Poland's Territorial Problems," in *Ekspertyzy i materiały delegacji polskiej na konferencę wersalską 1919 roku*, ed. Marta Przyłuska-Brzostek (Warsaw: Polski Instytut Spraw Międzynarodowych, 2009), 85–111, here 85, 105, and 107.

[53] The "Memorial on the Frontiers of Hungary" for example, put it as follows: "It is no empty phrase, but a severely scientific truth that Hungary's frontiers, as of a thousand year's standing, enclose a geographical unit. New frontiers cannot be drawn within this boundary without disturbing the peace of the peoples inhabiting this basin. Within this most marked geographical unit, during a thousand years, each district and each people have been knitted with the others in so close a connection of traditions, history, culture, economics and traffic that the connection can only be sundered by ruthlessly destroying it and making for the unhappiness of the peoples living there. Not one of our nationalities has any right whatever to any regions of our country. [...] The new states formed according to the Conditions of Peace will be of a much more complicated structure than

the Hungarian case too, maps such as the *carte rouge* upon which the delegation based its negotiating position were rapidly reprinted in book form, to the extent that it went into its sixth edition by 1921. In his introduction to that publication, Pál Teleki retrospectively described the situation as follows:

From the moment we saw the way in which peace was settled with Germany, we had not the least hope of changing the minds and decisions of the conference [...]. Still we had to put our argument before them, even when our memoranda and maps remained closed and folded. It was our duty towards our nation, towards future generations and—towards our foes and judges. [...] they will see that by having mixed themselves into the affairs of the lands to the north of the Balkans they balkanized them too, instead of Europeanizing the Balkans. [...] We know the moment will come, when people will look around for remedies to repair the terrible confusion caused by the peace conference's lack of knowledge.[54]

In the years immediately following the war, the spatial sciences of the successor states of Austria-Hungary developed national territorial messages in two directions at once: an internal perspective was required to serve the needs of the integration of the state and an external perspective appealed for recognition and support for the national territorial program by the relevant protective powers (for Poland these were the Entente powers, while Hungary increasingly looked to Italy and Germany as its decisive points of orientation). At the same time, the post-imperial spatial sciences pursued two goals that were actually mutually contradictory in terms of the logic of their arguments: On the one hand, they strove towards international recognition as an independent scientific community, one which also provided the scientific underpinning for the foundation of the new nation state. On the other, geographers saw the need to visually eliminate all uncertainties about the new post-imperial political order and to authoritatively encourage people to identify with the new state territory. This was often accompanied with gestures designed to distance the new vision and territorial character of those states from spatial structures with dynastic connotations (such as historical lands or regions of partition).

While in Hungary, the Weimar Republic, and Lithuania revisionist cartographies emerged almost immediately after the peace treaties, geographers in Poland, Czechoslovakia, Romania, and Yugoslavia pushed ahead with cartographies that served the purpose of state integration. The main task of such cartographies was to map the new national territory in a uniform manner and

---

Hungary, from a nationalist point of view, and all the more unstable through the nationalities, filled with the spirit of irredentism, being in the case of a much higher degree of culture than their oppressors, who are on a Balkan level now." This is quoted from Jenő Cholnoky, *The Hungarian Peace Negotiations: An Account of the Work of the Hungarian Peace Delegation at Neuilly s/S, from January to March, 1920* (Budapest: Printing Office of Victor Hornyánszky, 1922), 75.

[54] Aladár Edvi Illés and Albert Halász, *Magyarország gazdasági térképekben: Gróf Teleki Pál béketárgyalást előkészítő iroda vezetője megbizásából szerkeztették Edvi Illés Aladár miniszteri tanácsos Halász Albert okl. vegyészmérnök, iparfelügyelő*, 6th ed. (Budapest: Pallas Nyomda, 1921).

to provide clarity and planning security for politicians and the general public.⁵⁵ Since all these states were composed of fragments of earlier state territories, thematic cartography pursued a post-imperial emancipatory mission, that is, by diverging sharply from imperial structures and practices and by securing the new territory as a new collective "we" space. The integration and re-organization of map making, however, also gave rise to logistical difficulties: there was a need to create new textbooks and educational materials as well as to redesign administrative and transport geographical maps to fit in with the new borderlines

In his contribution to a debate documented in the journal *Nauka polska*, anthropologist Jan Czekanowski noted that in relation to German and Russian science "our country [...] is a dependent province. [...] The creation of the elementary conditions to enable the development of an independent scientific life must become the subject of further efforts by active citizens."⁵⁶ In the same debate, Polish geographer Antoni Sujkowski therefore argued for a rigorous emancipation of the discipline from the military cartography of the former imperial powers. Sujkowski, a pre-war political activist who had participated in the work of the Polish delegation to the Paris Peace Conference and would become minister of religion and public education in 1926, referred to the very practical problem posed by the fact that all the maps available after the war were based on differing scales, cartographic symbols, and labelling conventions. His conclusion was that "We must make the effort to re-draw the kingdom cartographically."⁵⁷ Accordingly, Sujkowski not only called for funding to be provided to facilities already in existence and for the expansion of existing institutions and societies (such as the Towarzystwo etnograficzno-geograficzne), but also for the establishment of a new state cartographic institute. The newly resurrected Polish state held the responsibility in this regard, so his argument went, since the governments of every country in Europe, as well as that of the United States, provided support for their own equivalent structures via direct funding and nonmonetary assistance.⁵⁸

---

[55] Thus the preface of a German-language *Wirtschafts-Atlas des Tschecho-Slowakischen Staates* published in 1920 in Reichenberg appealed for a comprehensive new series of maps to be produced, using the following words, among other arguments: "[after the] collapse of the old empire on the Danube and the resulting distribution of its inheritance to its successor states [...] the regrouping and growth of what had now become 'domestic' production locations was total *terra incognita* [...]." Ernst Pfohl, *Wirtschafts-Atlas des Tschecho-Slowakischen Staates* (Reichenberg: Gebrüder Stiepel, 1920).
[56] Jan Czekanowski, "W sprawie portzeb nauk antropologicznych w Polsce," *Nauka polska* 1 (1918): 201–223, here 202 ff.
[57] Antoni Sujkowski, "Potrzeby nauki polskiej w zakresie geografii," *Nauka polska* 1 (1918): 155–164, here 159.
[58] As examples, Sujkowski cited the Smithsonian Institute and the Geographical Society in San Francisco, which had published works on the islands in the Pacific during the First World War. Ibid., 163.

During the Second Polish Republic, Eugeniusz Romer, who had now resettled at the University of Lwów (L'viv), proved to be *the* driving force behind the re-establishment of nation-state centered structures in cartography. Not only was he responsible for making the cartographic preparations for the peace negotiations held in Riga between Soviet Russia and Poland in March 1921, he also re-published his 1916 atlas in an expanded form in the same year.[59] Following the model provided by Justus Perthes in Gotha, Romer also established the *Książnica Atlas*, a series of popular maps dealing with the new Polish state, published under the auspices of the Cartographic Institute for Poland. Internationally, Romer was widely respected for his geographical expertise and was awarded prestigious posts and prizes (he replaced Penck as honorary member of the Royal Geographical Society of London, for example; honours in Germany and Austria, however, were to elude him).[60] But yet, above and beyond Romer's activities, topics relating to national geography were the subject of widespread attention in political discourse in Poland. Authors not only re-published earlier works after updating them to take account of a profoundly changed geopolitical situation.[61] They also published entirely new programmes for the development and structural and societal unification of Poland. What had been geopolitical visions and the imagining of a cultural mission before the war was now transformed into a discourse that had direct political implications, involving calls for state intervention and intensified interference into existing regional and local structures. Differences that had existed earlier between the academic milieux of the three regions of partitioned Poland-Lithuania lived on in a modified form in the post-imperial context, with one important difference with respect to the pre-war period: after the foundation of the state, references to historical borders as a point of orientation for the future development of Poland disappeared from almost all the studies of the early 1920s. At the same time, the thematic maps by which those works envisioned the reconstruction of the nation indirectly undermined the unifying rhetoric of the texts: the different levels of development from before the war became visually manifest.

One example among many is the case of Joachim Bartoszewicz, who had studied medicine in Warsaw, political science in Paris, and international law in L'viv, where his habilitation thesis dealt with inheritance tax. In his book

---

[59] Eugeniusz Romer, *Polski atlas kongresowy: Atlas des problèmes territoriaux de la Pologne* (L'viv: Nakł. Książnicy Polskiej Towarzystwa Nauczycieli Szkół Wyższych, 1921).

[60] Seegel, *Map Men*, 94 and 101.

[61] Among the numerous brochures and publications in book form, it is worth mentioning, merely as a sample, such diverse works as Jerzy Michalski, *Traktat pokojowy w Saint-Germain a obciążenie Polski* (Kraków: Nakł. Krakowskiej Spółki Wydawniczej, 1921); Antoni Sujkowski, *Geografia ziem dawnej Polski*, 2nd ed. (Warsaw: M. Arct, 1921); Adam Szelągowski, *Polska współczesna* (Warsaw: Biblioteka Dzieł Wyborowych, 1925); Jerzy Smoleński, *Przyrodzony obszar Polski i jego granice w świetle nowoczesnych poglądów* (Warsaw, 1926).

Walka o Polskę (The war for Poland), published at the height of the Polish-Russian war in 1920, he tells his readers:

> The achievement of suitable borders is [...] not the only task for the power and greatness of Poland. For this task a powerful internal organization and a planned economic policy is required. Without these [factors], the construction of a truly independent Poland is impossible. [...] Now, at a time when Poland finds itself in extremely difficult and dangerous circumstances, we must not forget that a political order, if it remains too liberal to trouble the comfort of particular individuals or social classes, could yet shake the house of state.[62]

Also, the anxiety that had developed during the war and in the context of the Paris Peace Conference due to the competing territorial programmes of the new neighbour states did not disappear. On the contrary. Given the level of mistrust and competition among some of the successor states of the Habsburg Monarchy, it becomes quite clear why the various systems of expertise of the interwar period functioned on the basis of intense mutual observation. Wherever geographers were engaged in state- or nation-building activities, they were very likely to rely on an "intentionally negative" perception of the cartographic production coming from those countries that were not regarded as allies or friendly neigbours and to pre-emptively undermine the arguments of the opposite side.[63]

With the *Polski przegląd kartograficzny*, Eugeniusz Romer initiated a journal that monitored the cartographic output of Poland and abroad, in Polish and French as well as English. In the foreword to the first issue, which appeared in October 1922, he expressed the view that in all the long history of Poland, cartography had never managed to achieve a European level of advancement. It was therefore high time, according to Romer, to establish a forum for the professional and impartial critique of cartographic production "in a language that is on an international level." It should, however, also act as an instrument to inform the international cartographic community of the errors of the "enemies of the Polish people and its civilization" and the official Polonophobic publications issued by governments hostile to Poland.[64] Consequently, Polish cartographers writing in the journal gave credit to German studies of a professional quality that lacked any political subtext and voiced harsh criticism of maps made by Polish geographers that did not satisfy basic academic requirements or that contained errors. The main subject of scrutiny, however, were German maps containing messages that questioned the international status quo. In order to discredit them, reviews contained a detailed list of errors and significant omissions and made ample reference to the authors' ignorance of the Polish literature and source material.

---

[62] Stanisław Kilian, *Myśl edukacyjna Narodowej demokracji w latach 1918-1939* (Kraków: Wydawnictwo Naukowe WSP, 1997), 49.
[63] On this topic, cf. the introduction in Haslinger, *Nation und Territorium*, 1-39.
[64] Eugeniusz Romer, "Avis de la rédaction," in *Polski przegląd kartograficzny* 1 (1923): 1-3, here 1.

Given the hugh gap between political beliefs about the extent of Hungarian national territory on the one hand and the terms of the Trianon peace treaty on the other, it is not at all surprising that what might be called cartographical anxiety about the post-war situation was at its greatest in Hungary among all the successor states of the Habsburg Monarchy. In the post-imperial context of the early 1920s, Hungarian geography unswervingly adhered to maintaining the country's pre-war territory, devoting much energy to demonstrating the apparent untenability of the new borders and to the collection of whatever scientific data that might be useful in supporting that case. A working meeting chaired by Prime Minister István Bethlen on 31 October 1922 came to the following conclusion:

> The task consists mainly in collecting materials that describe the internal weaknesses, problems, economic and political inability of the successor states to survive, [...] that highlight the impossible situation created by the peace treaty from a political, economic and cultural point of view, [and] that shine a light on the grievances of the minorities. [The materials] should illustrate that the new state entities, due to their internal structure, [and to] the lower /more Balkan/ level of culture, pose a constant danger to the peace of Central Europe and represent a huge slide backwards in the domain of economic and spiritual culture.[65]

In Hungary, the spatial knowledge accumulated up until 1914 now served as a backdrop for the documentation of the experience of loss and the consequences of the land's appropriation by the new successor states. Teleki, after his involvement as a cartographic expert at Trianon and his first term as prime minister between July 1920 and April 1921, remained the key figure in this effort. His fame meant that he was courted internationally as well.[66] Geographers in Hungary showed extreme creativity in developing a representational canon, one that may certainly be described as a cartographic narrative of trauma.[67] This isolationist, self-victimizing spatial vision corresponded to notions of an unnatural, inverted world.

---

[65] Anikó Kovács-Bertrand, *Der ungarische Revisionismus nach dem Ersten Weltkrieg: Der publizistische Kampf gegen den Friedensvertrag von Trianon, 1918–1931* (Munich: Oldenbourg, 1997), 131 ff.

[66] Teleki was a member of the same three-person commission tasked by the League of Nations with writing a report on a region disputed by Iraq and the Kingdom of Great Britain. Péter Kovács, "Paul Teleki ét le réglement de l'affaire de Mossoul dans la Société des Nations," *Miskolc Journal of International Law* 1, no. 2 (2004): 156–187. The work of the geographers de Martonne and Teleki at the Paris Peace Conference were compared to one another by Zoltán Krasznai, "Földrajztudósok az első világháború után: Emmanuel de Martonne es Teleki Pál," in *Léptékváltó társadalomtörténet: Tanulmányok a 60 éves Benda Gyula tiszteletére*, ed. Zsolt K. Horváth, András Lugosi, and Ferenc Sohajda (Budapest: Hermész Kör-Osiris, 2003), 345–366.

[67] On the Treaty of Trianon, cf. (among others) Balázs Ablonczy, "Trianon-legendák," in *Mitoszok, legendák, tévhitek a 20. századi magyar történelemről*, ed. Ignác Romsics (Budapest: Osiris K., 2002), 132–161; Miklós Zeidler, *Trianon* (Budapest: Osiris K., 2003); Miklós Zeidler, *Ideas on Territorial Revision in Hungary, 1920–1945* (Boulder, CO: Social Science Monographs, 2007).

One example of such creativity borne of frustration can be found in the publications of Ferenc Olay. He worked as a specialist at the Ministry of Education on the task of modifying textbooks. In this capacity he was also a Hungarian delegate to the International Committee on Intellectual Cooperation and the Committee for Moral Disarmament under the aegis of the League of Nations. His 1927 book *The Crisis Years of Hungarian Culture*, however, conveyed an entirely different set of messages: in his thorough documentation of the removal of Hungarian national monuments, the devastation or neglect of libraries and schools in the post-war turmoil, and the looting (by Romanian occupation forces, for example), a quasi-natural cultural hierarchy emerges, which had functioned before the war und was now turned upside down, with drastic consequences. Olay's work is therefore quite typical in its treatment of a theme generally expounded in Hungarian publications of the 1920s and 1930s, equating the transfer of sovereignty over territory with the destruction of cultural goods and educational infrastructures.[68]

Conclusion

As the above comparison of the two cases we have examined shows, even back in imperial times, Polish and Hungarian geographical writings and their visualizations of language regions and cultural boundaries were intertwined in a conflictive and negative way with the cartographies of neighbouring national communities. The maps that originated from that context became increasingly used as instruments in political debates. It was along this path that geo-narratives developed as spatial simulations for the different national communities within the Habsburg Monarchy. Since these discourses underpinned competing socio-economic as well as demographic security expectations, they soon began to undermine the notion of a common Habsburg space: Different crown lands or linguistic regions began to function as substitute spaces for national emancipation from which national statehoods would develop in the future.

After 1918, the imperial frame of reference either became obsolete and ceased to be part of state building discourses or it functioned as a dark herit-

---

[68] Ferenc Olay, *A maygar kultúra válságos évei 1918–1927* (Budapest: Magyar Nemzeti Szövetség, 1927). Cf. on map production Franz Sz. Horváth, "Karten als Fortsetzung der Politik mit anderen Mitteln: András Rónai und sein Mitteleuropa-Atlas," in *Osteuropa kartiert—Mapping Eastern Europe*, ed. Jörn Happel (Vienna: LIT, 2010), 187–199; Róbert Keményfi, "Grenzen, Karten, Ethnien: Kartenartige Konstituierungsmittel im Dienst des ungarischen nationalen Raums," in *Osteuropa kartiert—Mapping Eastern Europe*, ed. Happel, 201–214; Róbert Kemenyfi, "Karten machen—Macht der Farben: Zur Frage der Visualisierung des ungarischen nationalen Raums," in *Visualisierung des Raumes: Karten machen—die Macht der Karten*, ed. Sabine Tzsaschel, Holger Wild, and Sebastian Lentz (Leipzig: Leibniz-Institut für Länderkunde, 2007), 55–65.

age that had to be overcome. However, it is worth taking a look at the various dimensions of the "(post-)imperial:" depending on the context, authors tended to understand the expression "imperial" as referring to dynastic, federalist, and late feudal state structures. Due to the paradigm shift that had occurred in geopolitics around 1900, it had become easy to discredit such structures as outdated. References to imperial spaces, however, were also used to justify a civilizing mission of one's own nation vis-à-vis other groups. Just as nationally oriented geopolitics were not anti-imperial per se before 1914 (there existed spatial designs that were compatible with the spatial imaginaries of the Habsburg Monarchy or even developed a national mission on that basis), they were not post-imperial in nature after 1918. Rather, we can find various forms of geographical determinisms that were rooted in imperial patterns of thought and action, and thus developed nested cartographic mini-imperialisms under the umbrella of the nation state.

Translated from German by Jaime Hyland.

Guido Hausmann

BETWEEN COMPLICITY IN IMPERIAL POLITICS AND EXCLUSION
Politico-Geographical Sciences in the Tsarist Empire
and the Early Soviet Union

The rise in the social prominence of scientists is one of the great success stories of Russia between the eighteenth and the twentieth centuries. During this period, academics and scientists made themselves at home in Russia in several respects: institutionally, through their involvement in the Academy of Sciences, the universities, and in a growing number of scholarly bodies; socially as a category of persons employed in the civil service and promoted from there into the country's service nobility; and culturally, in terms of their intensive exchanges with other European scientists and countries (and, to a lesser extent, beyond Europe too). From the Russian point of view, such exchanges were really more about receiving new ideas than anything else during the eighteenth and early nineteenth centuries but were subsequently to develop for them into an increasingly active and participatory process. The tighter political regulation imposed on the sciences in Russia than in other states is one of the many special features of scientific activities in the country that set them apart from the rest of Europe. Such regulation was, however, imposed selectively, and research has not yet arrived at any empirically founded generalizations on the topic. On the whole, the rise of the sciences exemplifies the classic narrative that speaks of Russia as experiencing increasing Europeanization/Westernization/modernization and that makes it possible to explore the "emergence, expansion and internationalization of the institutional and communicative network of scientific productivity" in Russia during the period.[1] But such a perspective and such a valuation would in many respects provide no more than a superficial view, concealing as it would contradictions and counter-developments.

---

[1] This reflects the view of Mitchell G. Ash "Wissenschaftswandlungen und politische Umbrüche im 20. Jahrhundert: Was hatten sie miteinander zu tun?," in *Kontinuitäten und Diskontinuitäten in der Wissenschaftsgeschichte des 20. Jahrhunderts,* ed. Rüdiger vom Bruch, Uta Gerhardt, and Aleksandra Pawliczek (Stuttgart: Steiner, 2006), 19–37, here 20. I would like to thank Jörg Stadelbauer in Freiburg, and Zsuzsanna Török in Vienna, for their kind advice, and the Imre Kertész Kolleg in Jena for the opportunity they gave me to do research on the subject.

The essay that follows on the scientific work that came out of Russia both before 1917 and in the early days of the Soviet Union which explicitly defined itself as "politico-geographical" in nature will analyse and clarify some of the general characteristics of such work as compared with other European states.[2] It concentrates on providing an understanding of the basic notions and conceptual direction of political geography, of interpretations of Russia and the Soviet Union in politico-geographical terms, and, by way of an example, it also examines the significance and consequences of the political upheaval of the 1917–1922 period for the sciences. The guiding questions grappled with in this essay relate to three different levels of analysis.

As a first task, I examine the opportunities for, and restrictions imposed on, politico-geographical activities. I ask whether and to what extent academic politico-geographical activities were able to establish themselves in Russia and the early Soviet Union. This issue is examined in terms of expositions and currents within the discipline of geography that were described by their authors as "political geography" and were explicitly allocated to that field. Any such question of self-location within science should be separated from issues that involve the independent examination of the political content and political function of works on geography undertaken during the tsarist Empire and in the early days of the Soviet Union. Such content and function were certainly present in all the scientific work carried out in the discipline of geography within the Russian Empire. This essay, however, deals with an intellectual current within the discipline of geography that explicitly understood itself as "political geography." One can also speak of political geography in the narrower sense. Such a horizon is restricted but has the advantage of offering greater precision.[3]

---

[2] One might cite numerous recent publications that deal with the history of geography and of political geography, but they do not specifically include the case of Russia and the Soviet Union. In this regard, I refer only to Christian Holtorf, "Zur Wissenschaftsgeschichte von Geografie und Kartografie: Einleitung," *Berichte zur Wissenschaftsgeschichte* 40 (2017): 7–16 and Alan R. H. Baker, *Geography and History: Bridging the Divide* (Cambridge: Cambridge University Press, 2003), as well as to new works that take a critical look at the traditions of political geography and geopolitics: Anne Godlewska and Neil Smith, eds., *Geography and Empire* (Oxford: Blackwell, 1994); Barbara Christophe, ed., *Geopolitik: Zur Ideologiekritik politischer Raumkonzepte* (Vienna: Promedia, 2001); Rainer Sprengel, *Kritik der Geopolitik: Ein deutscher Diskurs, 1914–1944* (Berlin: Akademischer Verlag, 1996); Gearóid Ó Tuathail, Simon Dalby, and Paul Routledge, eds., *The Geopolitical Reader* (London: Routledge, 1998), especially the introduction by Ó Tuathail, "Thinking Critically about Geopolitics," 1–12; John Agnew et al., eds., *The Wiley Blackwell Companion to Political Geography* (Oxford: John Wiley & Sons, 2017).

[3] For a classification within the wider research context of spatial concepts and territoriality/territorialization, see Ulrike Jureit, *Das Ordnen von Räumen: Territorium und Lebensraum im 19. und 20. Jahrhundert* (Hamburg: Hamburger Ed., 2012). Jureit primarily examines the concept of space as "a formula for the self-description of societies" (eine Selbstbeschreibungsformel von Gesellschaften), 13, although her specific focus is on Germany.

Secondly, politico-geographical activities carried on within the tsarist Empire and in the early Soviet Union referred to a very wide variety of understandings of the term "political geography." One might argue that behind the expression are concealed two entirely different sub-disciplinary currents. On the one hand, the term relates to works in the eighteenth- and nineteenth-century European tradition of creating a description of a state. Such efforts differ markedly from work done at the end of the nineteenth and the beginning of the twentieth centuries that picked up a new trend in European geography which was labelled "political geography." It is therefore possible to tell two different stories: on the one hand a history of the transformation and decline of an old geographical sub-discipline and, on the other, a history of the rise of a brand-new geographical current. Questions need to be asked, however, about what links exist between the two currents, and especially about what linked them both with Russia as an empire.

Thirdly, works on political geography always make reference to the state or to statehood in its specific politico-geographical form. Yet the state is studied in a variety of different ways from a geographical point of view: in the eighteenth and early nineteenth centuries it was seen as a specific political unit or order that needed to be described in geographical terms, while later, in the late nineteenth and early twentieth centuries, it was seen as a processual order whose conduct was governed by the geographical conditions obtaining in a European and global context. Insofar as the Russian Empire understood itself as an imperial state, its imperial character also needed to be analysed as a function of political geography during the last years of the tsarist state. The Soviet Union, in contrast, saw itself as an anti-imperialist state and should be treated as an even more repressive political order than its tsarist predecessor. Accordingly, our examination will need to look at issues of continuity and change in the scientific output of the discipline of political geography.

The Older Tradition: Political Geography as a Way of Describing States

The first publications to refer to themselves as being politico-geographical in subject matter began to appear in Russia in the middle of the eighteenth century. At the time geography was already established as an academic discipline, with its own separate branch in the Academy of Sciences, which was founded in 1724. The discipline's enhanced institutional status and increased recognition reflected the benefits that geography had provided to the tsarist state over previous years by producing proprietorial descriptions of both its new and its old territories. Such work included, for example, two expeditions to Kamchatka to establish a state administration in Siberia and efforts to describe natural resources in the context of an economic policy whose logic included mercantilist elements. Significantly, political geography was put on an equal footing with physical geography and, most revealingly of all, was made a school subject at academic secondary colleges and cadet schools.

One prominent textbook author of the eighteenth century was Ivan Mikhailovich Grech (1732–circa 1772), who published a four-volume *Political Geography* for the pupils of the St. Petersburg Cadet Corps between 1758 and 1772.[4] He was the son of Johann-Ernst von Gretsch (1709–1760), who had immigrated to Russia from Prussia in the 1720s or 1730s before embarking on a career as a teacher in Petersburg. The work largely follows the pattern of a classical description of states. The chapter on France, for example, is purely descriptive, naming the various regions of the country along with the country's possessions outside its national borders, describing them all in geographical terms, and giving an introduction to France's political and administrative order.[5]

In this case, and indeed in Europe generally in the second half of the eighteenth century, political geography overlaps both conceptually and in terms of content with statistics, political science, and the discipline of describing states. The subject's task was descriptive and systematizing in nature and was essentially characterized in the contents of school textbooks: it was not a scientific discipline in the modern sense, whose business was to solve problems analytically.[6] This tradition was to prove durable and was to change only slowly, as can be seen from *Geography* by Aleksandr Dmitrievich Chertkov (1789–1858), an 1845 preparatory textbook for grammar school candidates. Chertkov retired from military service after 1828, subsequently turned to scholarly activities, and was later to build one of Moscow's finest private libraries on archaeology and Russian history. His presentation, dividing geography into the topics of mathematical, scientific, and political geography, contains an introductory first section, followed by a second section on the political geography of Europe, and a third on the political geography of the Russian Empire.[7] The political geography of Europe—the rest of the world is dealt with only summarily and marginally—is defined as an

external view of that part of the world, or of its surface, including details about special qualities of its air and about differing products of nature, the occupations of inhabitants and

---

[4] Ivan M. Grech, *Politicheskaia geografiia*, 4 vols. (St. Petersburg, 1758–1772).
[5] Grech, *Politicheskaia geografiia*, vol. 2 (St. Petersburg, 1761), chapter 5 (4–201 on the various regions of France, 204–260 on France's possessions outside its national boundaries in the narrowest sense, and 237–259 on the political order that ruled the country).
[6] See Jürgen Osterhammel, "Die Flughöhe der Adler: Räume und Sehepunkte zu Friedrich Hölderlins Zeit," in *Die Flughöhe der Adler: Historische Essays zur globalen Gegenwart*, ed. Jürgen Osterhammel (Munich: C. H. Beck, 2017), 223–244, here 237 and the relevant notes on school textbooks on "Lehre von der Politik," for example, in Hans Erich Bödeker, "'Wer ächte freie Politik hören will, muss nach Göttingen gehen': Die Lehre von der Politik in Göttingen um 1800," in *Die Wissenschaft vom Menschen in Göttingen um 1800: Wissenschaftliche Praktiken, institutionelle Geographie, europäische Netzwerke*, ed. Philippe Büttgen and Michel Espagne (Göttingen: Vandenhoeck & Ruprecht, 2008), 325–369, here 328. Both authors also provide significant additional bibliographical references.
[7] Aleksandr Chertkov, *Geografiia* (St. Petersburg, 1845), VI, 5 f.

the location of cities, administration, the faith, languages, and the degree of sophistication of the inhabitants of Europe.[8]

Chertkov emphasizes the relative importance of Europe as compared to other parts of the world and describes political geography as "a description of the parts of the world and of each country in detail."[9] His description of the Russian state starts out as being based on a given political order, but then peels off into a political statement: "the Russian empire takes first place not just due to its undeniable scale, but also through its strong influence on and participation in the affairs of the other European states."[10] The author refers to Russia's natural borders and names the seas around its northern and eastern edges as well as some of the mountain ranges and seas in the south, but in the following paragraphs relating to the west of Russia he restricts himself to naming the region's river sections. He does not get involved in any discussion of the relationship between natural and political boundaries. While the political space has natural features and is described in geographical terms, no particular natural features have any effect on Russia's political borders in the southern and western directions. The term "political" in this geography involves more than simply an emphasis on Russia's importance vis-à-vis the other states: it also includes a description of the internal administration of the state's territory. Yet Chertkov does not name the newly created Catherinian administrative units, instead dividing Russia into 14 major regions as entities partly defined by landscapes and partly in terms of administrative units.[11]

The older tradition of political geography as a description of the state can also be detected in textbooks from the middle and second half of the nineteenth century, including in a textbook by Nikolai Smirnov and another by Nikolai I. Zuev. In his *Geography*, whose tenth edition appeared in 1871, Smirnov begins by presenting the disciplines of mathematical and physical geography. After that, in the book's second section, which is on physical geography, he deals with various ethnic groups and then classifies the world's states in terms of their size, the number of their inhabitants, religion, and the type of statehood (distinguishing between monarchies and republics), here places special emphasis on the "form of supreme authority" in power in each state.[12] He emphasizes the greatness of Russia and classifies Russia as a kingdom headed by a monarch of unrestricted power at its very pinnacle: "that is why it is also referred to as an autocracy."[13] Finally, Smirnov once more ex-

---

[8] Ibid., 55.
[9] Ibid., 130.
[10] Ibid., § 2.
[11] Ibid., 133–135.
[12] Nikolai Smirnov, *Uchebnaia kniga geografii: Obshchie svedeniia iz geografii matematicheskoi, fizicheskoi i politicheskoi. Kurs nizshikh uchebnykh zavedenii, 10–izd. Vnov ispravlennoe* (St. Petersburg, 1871), 49 f.
[13] Ibid., 49.

plicitly points out the differences between physical and political geography, the former describing everything that the earth generates, while the latter deals with all transformations that can be traced back to man.[14] Nikolai I. Zuev came to a similar conclusion in 1888, though he presents his findings more neutrally, depicting political geography as the activity of describing the earth as inhabited by humans and characterizing Russia as an "absolute or unlimited monarchy."[15] This older tradition of political geography as a description of the state, which emphasizes the greatness of Russia, its monarchical form of government, and sometimes its political influence within Europe, was still in existence in the form of school textbooks at the end of the nineteenth century, a time when a new form of scientific discussion on political geography began, a discussion that carried a completely different understanding of the term.

The New Direction Taken by Political Geography at the End of the Nineteenth Century

Since the eighteenth century, the institutionalization of geography as a scientific discipline had been progressing slowly but steadily within the tsarist Empire. Important steps along this road were taken mainly during the first half of the nineteenth century: the Russian University Statute of 1835 established the subject at the universities of the empire and 1849 saw the foundation of the Imperial Russian Geographical Society in St. Petersburg, a scholarly association in which people from a variety of disciplines publicly discussed their expertise and presented their empirical findings and opinions on Russian geography to national and international readers in the society's own publications.[16] Geography, as an academic discipline in Russia in the second half of the nineteenth century, just as it was throughout Europe, was primarily oriented towards and influenced by the natural sciences, with geomorphology—the study of topographic forms—taking centre stage. The institutional division of the Russian Geographical Society into a number of branches or departments for the different parts of the country in the second half of the nineteenth century reveals a process of regionalization and of differentiation within the discipline, a phenomenon that can also be observed throughout Europe. However, these developments cannot be explained simply by intra-disciplinary factors; they were also affected by social and political processes. The leading Russian

---

[14] Ibid., 46–51.
[15] Nikolai I. Zuev, *Kratkoe obozrenie geografii matematicheskoi, fizicheskoi i politicheskoi: Kurs srednykh uchebnykh zavedenii* (St. Petersburg, 1888), 144 and 161.
[16] See Claudia Weiss, *Wie Sibirien "unser" wurde: Die Russische Geographische Gesellschaft und ihr Einfluss auf die Bilder und Vorstellungen von Sibirien im 19. Jahrhundert* (Göttingen: V&R unipress, 2007); Steven Seegel, *Mapping Europe's Borderlands: Russian Cartography in the Age of Empire* (Chicago: University of Chicago Press, 2012), 112–116.

geographers of the nineteenth century were strongly influenced by Germanophone geography: on the one hand by German and German-speaking academics and geographers working in Russia and, on the other, by young scientists from Russia who had spent time studying at German universities.[17]

From the 1880s on in particular, French and German geographers had been advocating innovations they referred to collectively as anthropo-geography, a field that might also be described as social geography. Some geographers required that the human being form part of geography. A lecture transcript from 1884-1885 by the Russian student A. T. Sokolov reveals that this trend was taken up directly by geography lecturers and students in Russia. Sokolov emphasizes the great importance of German geography, writing that the German geographer Friedrich Ratzel (1844-1904) in his "anthropo-geography shows that geography is primarily concerned with the relationship of man to the earth" and that political geography is primarily concerned with "statistics and ethnography."[18] The latter point seems to be conveying the traditional understanding of political geography.

Within the new field of social geography, however, a new political geography was to emerge soon enough (i.e., towards the end of the nineteenth century). It was to be closely associated with the name of Friedrich Ratzel, who published a work of that name in 1897 in which he explained the "dryness" of traditional political geography by asserting that "the facts of political geography still lie far too rigidly side by side and next to physical geography" and demanded that it be replaced by a reorientation in which "states at all stages of development are regarded as organisms in an inevitable connection with the soil and must therefore be regarded in geographical terms."[19] Just one year later, the widely consulted Russian-language Brokgauz-Efron encyclopaedia made a reference to Ratzel's publication in an entry under the heading "Political Geography."[20]

The expansion of the subject area occurred both in international discussions and in the context of political and imperial rivalries. It also became a reality as a result of a striving within the scientific community for the acceptance and dominance of geography, a discipline that was winning recognition almost everywhere in Europe during the last decades of the nineteenth century as a university subject and scientific discipline, a process of recogni-

---

[17] Thus, for example, Petr P. Semenov, vice president of the Russian Geographical Society, studied with Carl Ritter in Berlin. On Semenov, see W. Bruce Lincoln, *Petr Petrovich Semenov-Tian-Shanskii: The Life of a Russian Geographer* (Newtonville, MA: Oriental Research Partners, 1980).

[18] A. T. Sokolov, *Politicheskaia geografiia* (Nizhnii Novgorod, 1884-1885), 1-2, quotation on 2. It was not possible to ascertain whose lecture it was that Sokolov had noted down.

[19] See Friedrich Ratzel, *Politische Geographie: Geographie der Staaten, des Verkehres und des Krieges*, 2nd ed. (Munich: Oldenbourg, 1903). Foreword to first edition in 1897, IV-V.

[20] Anonymous, "Politicheskaia geografiia," in *Entsiklopedicheskii Slovar'*, vol. 24, ed. F. A. Brokgauz and I. A. Efron (St. Petersburg, 1898), 304.

tion that could be seen in its progressive institutionalization.²¹ At the same time, German geographers such as Ferdinand von Richthofen and Alfred Hettner also articulated glaring methodological deficits in the discipline, which they claimed had no clear subject matter, and at the beginning of the twentieth century called for a regionalization of the discipline through the creation of a comparative geography. Russian geographers imbibed this discussion and began to take an active part in it.²²

Controversial issues included questions, for example, on how much history the discipline of geography should effectively incorporate into itself, the weighting to be given to physical geography (the geography of the natural sciences) as against its social aspects, and where to make causal linkages between the various sub-fields of the discipline. This process of extending and demarcating the subject area was by no means limited to Germany; it was also clearly evident in French geographical discussions. French geographers working in political geography, such as Elisée Reclus (1830–1905), also exerted an influence on geography in Russia.²³

By "political geography" in this sense I am talking about a particular new direction or sub-discipline within the geographical sciences that surfaces at the end of the nineteenth century, initially in its German-speaking sphere, where the physical characteristics of space are related to territoriality and the formation of political space–up to and including ideas according to which physical space quasi-forces a particular territorial pattern. While the term

---

[21] See, for example, Horacio Capel, "Institutionalization of Geography and Strategies of Change," in *Geography, Ideology and Social Concern*, ed. David R. Stoddart (Oxford: Blackwell, 1981), 37–69, especially 59–65.

[22] Lev S. Berg, "Predmet i zadachi geografii," in *Izvestiia Imperatorskogo Geograficheskogo obshchestva* 51 (1915): 463–475, especially 465–472. On the International Geographical Congresses held from 1871 on (the first of them hosted in Antwerp) and their function as forums of national representation and international exchange, see Jörg Stadelbauer, *Berlin 1899: The Seventh International Geographical Congress. A Retrospective on Occasion of the 32nd IGC, Cologne August 2012* (Freiburg: Institut für Kulturgeographie, 2012). After Germany, Great Britain, and France, Russia had the fourth-largest representation at the event (among a total of 23 participants), see 20. The main defender of this comparative geography was Alfred Hettner, who published a book in 1905 after completing a journey through Russia. Cf. Alfred Hettner, *Das europäische Rußland: Eine Studie zur Geographie des Menschen* (Leipzig: Teubner, 1905). The third, expanded edition of Hettner's book appeared under the title *Rußland: Eine geographische Betrachtung von Volk, Staat und Kultur*, 3rd ed. (Leipzig: Teubner, 1916), and appeared also in Russian translation.

[23] For a brief introductory view of French geography, see Daniel Dory, "Géographie et colonisation en France durant la troisième république, 1870–1940," in *Science and Empires: Historical Studies about Scientific Development and European Expansion*, ed. Patrick, 323–329. Petitjean, Catherine Jami, and Anne Marie Moulin (Dordrecht: Kluwer, 1992)

"geopolitics," a term that first emerges at the beginning of the twentieth century, means something similar, it has its roots in the political sciences.

Kropotkin and Semenov-Tian-Shanskii—Faces of the New Current

Two well-known Russian geographers from the end of the nineteenth and beginning of the twentieth centuries who took up the subject of political geography provide examples of the opportunities and restrictions on the development of this new current during the last years of the Tsarist Empire.

Petr A. Kropotkin (1842–1921), a scion of old Russian nobility who had done his military service as a young man in Siberia after having completed his training in the Page Corps, undertook several expeditions to Eastern Siberia, publishing his findings in scientific journals. After studying mathematics and physics at St. Petersburg University from 1867 to 1872, he was quickly promoted to the position of secretary of the section for physical geography at the Russian Geographical Society and was to become highly regarded for his work on the glaciology and orography of northern and northeast Asia. With political views that tended towards anarchism and socialism, he managed to flee to Western Europe in 1876 after spending two years in prison. From then on, living mainly in France, he turned away from Darwinist and social-Darwinist thinking. Through lively exchanges with the French geographer and anarchist Elisée Reclus, he also turned his attention to political geography.[24] Kropotkin was not to return to Russia until after the fall of the autocracy in the February Revolution in 1917, and worked primarily in the preceding decades as a political philosopher and author of popular and programmatic scientific works, including writings on political geography.

In his 1885 work "What Geography Ought to Be," he characterized geography as a scientific discipline that as such shared in the revival of the natural sciences and even had the potential to help shape it:

Geography should be, first, a study of the laws to which the modifications of the Earth's surface are submitted: the laws—for there are such laws, however imperfect our present knowledge of them, which determine the growth and disappearance of continents; their present and past configurations; the directions of different upheavals—all submitted to

---

[24] For an introduction, see Rob Knowles, "Kropotkin and Reclus: Geographers, Evolution and 'Mutual Aid,'" in *Evolutionary Economics and Human Nature*, ed. John Laurent (Cheltenham: Elgar, 2003) 134–152 and Myrna M. Breitbart, "Peter Kropotkin: The Anarchist Geographer," in *Geography, Ideology and Social Concern*, ed. Stoddart, 134–153. For new biographical information, see Brian Morris, *Kropotkin: The Politics of Community* (Oakland, CA: PM Press, 2018); Philippe Pelletier, *Géographie et anarchie: Élisée Reclus, Pierre Kropotkine, Léon Metchnikoff et d'autres* (Paris: Éditions du Monde Libertaire, 2013).

some telluric laws, as the distribution of planets and solar systems is submitted to cosmical laws.²⁵

In and of itself, geography produces new knowledge, but it can also contribute to human progress in general if it is taught in schools in a new way. Nationality, for example, is a natural category and is offered as such, but Kropotkin rejects any tendency to assign differing evaluations to nationalities, along with any legitimizing function such evaluations might have for any particular political order through the discipline of geography. The anarchist summed up his view succinctly in these words: "Political frontiers are relics of a barbarous past."²⁶

As a scientific discipline and teaching subject in schools and universities, he divided geography into four main branches: orogeny (geomorphology), climatology, zoo- and phyto-geography, and various fields of ethnology. While the first three areas could be grouped under the heading of physical geography, the fourth area belongs to political geography ("what is partly taught now under the head of political geography").²⁷ And under that heading he mainly included topics within social and economic geography:

> the geographical distribution of races, beliefs, customs, and forms of property, and their close dependence upon geographical conditions; the accommodation of man to the nature that surrounds him, and the mutual influence of both the migrations of stems, in so far as they are dependent upon geological causes; [...] the raising of cities and the conditions of their development; the geographical subdivision of territories into natural manufacturing basins, notwithstanding the obstacles opposed by political frontiers [...].²⁸

For Kropotkin then, geography, including political geography, was anti-state and anti-imperialist in orientation as a scientific discipline. With such a radical attitude, he was unable to find a home, either personally or scientifically, in Russia. His works nevertheless reached an audience in Russia, and his pronouncements on political geography show that, from its very beginnings, politico-geographical (and later geopolitical) thought and writing formed part not just of an "imperialist discourse," but also of a critical counter-discourse.²⁹

In geographer Veniamin Semenov-Tian-Shanskii (1870–1942) one detects an entirely different life path as well as a completely different treatment of

---

²⁵ Peter Kropotkin, "What Geography Ought to Be," *The Nineteenth Century* 18 (1885): 940–956, quoted from https://onlinelibrary.wiley.com/doi/pdf/10.1111/j.1467-8330.1978.tb00111.x, here 10. For a short introduction, see Breitbart, "Peter Kropotkin," 144–146.
²⁶ Ibid., 3.
²⁷ Ibid., 7.
²⁸ Ibid., 9.
²⁹ See Ó Tuathail, "Thinking Critically about Geopolitics," 4. Far from being a phenomenon exclusive to the last few decades, the concept of "anti-geopolitics" has historical precursors and traditions.

the new trend in political geography.³⁰ Veniamin Semenov, the son of well-known geographer and long-time vice president of the Russian Geographical Society Petr P. Semenov, grew up in the midst of educated St. Petersburg society and, after completing his scientific studies at St. Petersburg University in 1893, quickly found a post in the Ministry of the Interior, working in the Central Committee for Statistics. It was in this post—and not as a university professor of geography—that he gained fame and prestige at the end of the nineteenth and the beginning of the twentieth century for his work in economic geography and, in general, popular geography. He was later to retrospectively characterize a series on the geography of various regions of Russia he had published as "a Russian Baedeker."³¹ Before the First World War, he published ground-breaking economic and socio-geographical publications on the *City and Village in European Russia* and *Trade and Industry in European Russia by Region* while holding various posts in the civil service and as a member of the Imperial Russian Geographical Society.³² His knowledge was thought politically relevant enough for deputies in the State Duma to ask him to produce a new edition of his study on *Trade and Industry by Region*, this time expanded to include the Asian regions of Russia, promising to provide the financing for the necessary research.³³ However, it cannot be said that in his work as a geographer he was explicitly politico-geographical before the First World War. Semenov was rather more an economic, social, and cultural geographer, though one whose works were also significant from a political point of view.

This trend was sustained during the First World War, with Semenov, as a member of a newly established Commission for the Exploration of Russia's Natural Resources (Kommissiia po izucheniiu estestvennykh proizvoditel'nykh sil strany, KEPS) at the Imperial Academy of Sciences, making his knowledge and energy available for activities of importance to the war economy. The KEPS—or rather its principal office under Vladimir I. Vernadskii—aimed to describe the geographical distribution (the location of deposits, for example), the extent, and the previous research into Russia's natural resources—that is, its minerals—in a series of publications. At the commission's first meeting in October 1915, Semenov, who represented the Russian Geographical Society on that body, called for urgent and wide-ranging geographical and statistical

---

30 About this in his biography, see Pavel M. Polian, *Veniamin Petrovich Semenov-Tian-Shanskii, 1870–1942* (Moscow: Nauka, 1989).
31 Veniamin P. Semenov-Tian-Shanskii, *To, chto proshlo*, vol. 1 (Moscow: Novyi Khronograf, 2009), 426.
32 Veniamin P. Semenov-Tian-Shanskii, *Gorod i derevnia v Evropeiskoi Rossii: Ocherk po ekonomicheskoi geografii* (St. Petersburg: Tip. V. F. Kirshbauma, 1910); Veniamin P. Semenov-Tian-Shanskii and N. M. Shtrupp, *Torgovlia i promyshlennost' Evropeiskoi Rossii po raionam: Obshchaia chast' i prilozhennaja*, 12 vols. (St. Petersburg: Tipografiia V. F. Kirshbauma, 1909–1911).
33 Semenov-Tian-Shanskii, *To, chto proshlo*, 554 and 571.

exploration to investigate the country's natural resources and their geographical distribution. He opined that the research could be carried out by the branches of the Academy of Sciences newly established in the various regions of Russia, through the use of topographical and civilian maps and via a new census. In his view, the necessary specialist knowledge, as well as the required mastery of different European languages, was more readily available in Russia than "among the highly educated and omniscient Germanic peoples," as he put it with some irony. Fellow geographers would have proposed such projects as far back as 15 years previously, but—so he implies—it seems it took the war to first produce the realization in Russia.[34]

It was during the First World War that Semenov directly took up the discipline of political geography for the first time, though he did so without explicitly responding to latest German-language politico-geographical works published during wartime. In two publications, in 1915 and 1916, he first pointed out that anthropological and political geography had as yet hardly met with any reception in Russia, partly due to censorship.[35] Political geography deals with natural rather than administrative borders, he tells his readers in an explicit reference to Friedrich Ratzel, objecting that in Russia borders have always been arbitrarily drawn from above. He also argues that there was still no satisfactory regional geography in Russia that could be regarded as the one methodical royal road in the eyes of the contemporary geography of the German-speaking world. In addition to the question of the relationship between physical-natural space and territorial sub-division, Semenov also ascribes decisive importance to ethno-geographical questions, which he examines from a global perspective. For him, the terms "Russian" and "East Slavic" name identical categories (thus subsuming Belarusians and Ukrainians under the former heading). A political geography of Russia should first of all explain the role of Russian migration in world history (though Semenov does not use the term "migration" [migratsiia], preferring to speak of "colonization" [kolonizatsiia]), and then secondly it should go on to describe the various territorial forms within Russia's realm and to identify what he called "useful territories," then thirdly should examine Russia's historical settlement nuclei or bases, fourthly it should research the transport routes used in Russia in con-

---

[34] See Veniamin P. Semenov-Tian-Shanskii, "O sovremennykh nuzhdakh osnovnykh istochnikov k poznaniiu Rossii," in *Otchety o deiatel'nosti komissii po izucheniiu estestvennykh proizvoditel'nykh sil Rossii sostoiashchei pri Imperatorskoi Akademii Nauk* (St. Petersburg: Tipografiia Imperatorskoi Akademii Nauk, 1915), 13–20, here 15.

[35] See, for example, Veniamin P. Semenov-Tian-Shanskii, "Vladimir Ivanovich Lamanskii kak antropogeograf i politikogeograf" (St. Petersburg, 1915). See Ute Wardenga's comparison with developments in German geography, to which Semenov also referred during this period, "'Nun ist Alles, Alles anders!': Erster Weltkrieg und Hochschulgeographie," in *Kontinuität und Diskontinuität der deutschen Geographie in Umbruchphasen: Studien zur Geschichte der Geographie*, ed. Ute Wardenga and Ingrid Hönsch (Münster: Institut für Geographie der Westfälischen Wilhelms-Universität Münster, 1995), 83–97.

nection with migrations to settle such bases, and fifthly should develop a cartography dealing with a broader range of topics.³⁶

So Semenov proposed a programme of research into the history of Russian migration as a function of the natural conditions under which that migration occurred. In his view, similarities existed between Russian migration (or colonization) routes and Anglo-American maritime trade routes, and also between Russia's historical settlement bases and Anglo-American maritime trade hubs. In the context of this discussion, Semenov also disapproved of any geographical or historical division of Russia into a European and an Asian part. By his lights, Russian Eurasia was not a periphery (okraina), but a historical chunk, or even an original constituent part, of Russia (korennaia russkaia zemlia). In these views, he was picking up on ideas that saw Russia as Eurasian which had become widespread in Russia during the end of the nineteenth and the beginning of the twentieth centuries.³⁷

Historian of geography Mark Bassin explores the political and ideological background to these ideas in his work. He concentrates on analysing Russia's changing geographical self-location. Numerous intellectuals and scholars understood the country as a physical-geographical and cultural-geographical entity in its own right—as Eurasia—and accordingly they turned against both its division into a European and an Asiatic Russia and its exclusive assignation to Europe, a designation that was still predominant in the eighteenth century.³⁸ Semenov subscribed to the latter view, arguing that European Russia had emerged from four historical core settlement locations which had combined together to form the state from an economic and cultural point of view. The four included the region of Galicia and Kiev-Chernigov, then the region of Novgorod-Petrograd, then Moscow and, fourthly the middle reaches of the Volga. According to his account, in addition to these four historical centres, four new settlement centres were appended in the Asian part of Russia, some of which are still developing dynamically at the time of Semonov's writing: the Urals and the Altai, mountainous Turkestan and eastern Siberia

36 Veniamin P. Semenov-Tian-Shanskii, "O mogushchestvennom territorial'nom vladenii primenitel'no k Rossii: Ocherk po politicheskoi geografii," *Izvestiia Russkogo Geograficheskogo Obshchestva* 51 (1915): 425–457. See also Veniamin P. Semenov-Tian-Shanskii, "Geograficheskaia soobrazheniia o razselenii chelovechestva v Evrazii i o prarodine slavian," *Zemlevedenie*, books I–II, (1916) 1–12.
37 See Stefan Wiederkehr, *Die eurasische Bewegung: Wissenschaft und Politik in der russischen Emigration der Zwischenkriegszeit und im postsowjetischen Russland* (Cologne: Böhlau, 2007), 36. According to Wiederkehr, Eduard Suess, the well-known Austrian geologist, coined the term "Eurasia" in the 1880s.
38 See, for example, his monograph, cf. Mark Bassin, *Imperial Visions: Nationalist Imaginations and Geographical Expansion in the Russian Far East, 1840–1865* (Cambridge: Cambridge University Press, 1999) or Mark Bassin's essay, "Russia between Europe and Asia: The Ideological Construction of Geographical Space," *Slavic Review* 50, no. 1 (1991): 1–17.

(centred around Lake Baikal). One aspect of Semenov's account—the extent to which he uses the idea of islands of settlement in the midst of "empty space" as a basis for that account, using a similar conceptual scheme as accounts of the seizure of lands by Western European colonial powers—needs further clarification.[39]

But Semenov certainly seized upon ideas about Russia's self-colonization that had been adopted by influential Moscow historians Sergei M. Solov'ev and Vasilii O. Kliuchevskii as a national topos during the second half of the nineteenth century. Following Susi K. Frank, this geo-deterministic and ethnocentric concept of migration can be understood as an imperial strategy that defines migration "as a process of the predestined appropriation of an 'interior space'" and legitimizes such expansion.[40] Semenov's programmatic remarks on Russia's political geography flow out into a commitment to politically legitimize the country's centralist political organization (as "uniform and indivisible") on the grounds that Russia would be destroyed by the creation of federal structures.

During the First World War, Semenov was also to publish a book on the regional subdivisions of European Russia according to natural and physical criteria, an effort considered a contribution to physical geography in which he combined natural landscape descriptions with regional studies in the tradition of Carl Ritter.[41] However, he leaves aside the question of Russia's western border, a topic that had immediate political relevance at the time in view of the constant shifts in fronts during the war. The combination of ethnic, physical, and political geography, as he first formulated it during the First World War, remained for Semenov, even in the throes of the political upheavals of 1917 and afterwards, a dominant theme in his (political) geographical thinking.[42]

Directions in Political Geographical Thought during the Early Soviet Union

The new political order in the Soviet Union had to have an impact on the capacities and limits of politico-geographical activity within the world of geo-

---

[39] Jureit, *Das Ordnen von Räumen*, 24.
[40] Susi K. Frank, "'Innere Kolonisation' und Frontier-Mythos: Räumliche Deutungskonzepte in Rußland und den USA," *OSTEUROPA* 53, no. 11 (2003): 1658–1675, here 1666.
[41] V. Semenov-Tian-Shanskii, *Tipy mestnostei Evropeiskoi Rossii i Kavkaza: Ocherki po fizicheskoi geografii v sviazi s antropogeografiei* (St. Petersburg, 1915). Petr P. Semenov had already arranged for parts of Ritter's geography—in particular the chapters concerning Asian Russia—to be translated into Russian: Karl Ritter, *Zemlevedenie Azii Karla Rittera: Geografiia stran vkhodiashchikh v sostav Aziatskoi Rossii ili pogranichnykh s neiu, pereveden i izdan [...] pod rukovodstvom P. Semenova*, vol. 1 (St. Petersburg, 1879).
[42] See, for example, Veniamin P. Semenov-Tian-Shanskii, "Edinstvo Rossii i velikoe tysiacheletnee vostochno-slavianskoe pereselencheskoe dvizhenie," *Puti soobshcheniia Severa* 6–8 (1919): 42–52.

graphical research as areas of scientific work that impacted politics became subject to stricter regulation under the new regime. It was not until the process of the Sovietization of science that began in the late 1920s and early 1930s, however, that radical and repressive sanctions were imposed by the state on such work. For the period that concerns us then, questions may be asked at an individual and institutional level on continuities and new approaches, and on the spaces for thinking and writing that were available at the time.[43] All this took place against a background of the further politicization of geography throughout Europe, with a focus on the legitimization of territorial claims and in particular of older or newer political borders.[44]

On Soviet Russian emigration during the interwar period, geographer Petr N. Savitskii (1895–1968) in particular minted a genuinely geopolitical or politico-geographical current within the Eurasian ideology, one that also connects with Semenov's wartime works. This current was situated in a post-imperial political context, but it continued, with some modifications, along the lines of existing ideologically scientific, yet imperial thinking. This way of thinking saw Russia and the Soviet Union as an "exceptional geographical world," proceeding from the "idea of historical convergence as a result of contact in space." The Eurasian thinking of the interwar period rejected the idea of the Urals as an artificial border, making reference to climate, botany, and soil science, and re-evaluating the concept of a Eurasian space in terms of culture, economics, and politics, but without explicitly making reference to such geopolitical thinkers as Halford Mackinder (1861–1947).[45]

For the early Soviet Union until 1934–1935, four trends can be identified in the field of geography, each with own protagonists, that involved taking up issues and topics relating to political geography in one way or another. However, one cannot speak of any actual debate over political geography, but rather of there being separate individual currents of thinking and writing, a fact that reveals the extremely restricted opportunities for scientific exchange that existed in the field during the 1920s and 1930s.

Firstly, Semenov's geographical work in the 1920s displayed both new approaches and continuities with his work during the last years of the tsarist Empire. On the one hand, he became involved in the field of regional studies

---

43 Ash, "Wissenschaftswandlungen und politische Umbrüche," 29.
44 See, for example, Guntram H. Herb, *Under the Map of Germany: Nationalism and Propaganda, 1918–1945* (London: Routledge, 1997); Herb, "Von der Grenzrevision zur Expansion: Territorialkonzepte in der Weimarer Republik," in *Welt-Räume: Geschichte, Geographie und Globalisierung seit 1900*, ed. Iris Schröder and Sabine Höhler (Frankfurt am Main: Campus, 2005), 175–203; Maciej Górny, *Kreślarze ojczyzn: Geografowie i granice międzywojennej Europy* (Warsaw: Instytut Historii PAN, 2017).
45 See Wiederkehr, *Die eurasische Bewegung*, 72–73 and 77–81. The work also provides further bibliographical references to the research conducted up until 2007. On Mackinder, see Brian W. Blouet, *Halford Mackinder: A Biography* (College Station: Texas A&M University Press, 1987).

that was flourishing at the time and founded a dedicated geographical museum for and in his hometown of St. Petersburg/Petrograd, of which he then became the director. The creation of the museum provides an example of institutional change that reached beyond the political upheaval of the 1917–1922 period. Yet until the end of the 1920s he also continued to think in terms of his ideas on political geography as he had articulated them during the First World War. Such continuity can already be clearly seen in a 1919 essay[46] of his as well as in his 1928 study *Raion i strana* (District and Land), in which he summarized his methodological commitments together with his substantive views. Accordingly, in one Russian review his work was juxtaposed with Alfred Hettner's synthesis of geography, which was published at the same time, since it encompassed the entire field of geography, all the way from mathematical to political geography, and, more importantly still, combined the themes of economic and political geography in a single "geographical synthesis."[47] Semenov's approach was both geo-deterministic and ethnocentric: in this respect he makes explicit mention of Friedrich Ratzel, whose influence over his work clearly eclipses that of other geographers.

For Semenov, geography was above all a science of natural boundaries and accordingly, and from a global perspective, he highlighted the regions with a Mediterranean coast (including the Black Sea), or a coast on the China Seas and the Sea of Japan, or on the Caribbean as centres of migration, of human civilizations, and of state-building, all of which, not coincidentally, were characterized by their clement climate. Semenov continues in the vein of his imperial politico-geographical thinking from before 1917 and, following Friedrich Ratzel's example, distinguishes between more and less highly developed cultures, and between the more powerful and weaker states that emerged as a function of their evolutionary, historical cultural development on the soil and in the space they occupied.[48] He then identifies the northern hemisphere as his starting point and the United States and Russia—countries with a proven ability to appropriate space for themselves through migration—as centres from where new political space was formed.[49] In this context he referred to the importance of Russian migration ("East Slavic colonization and domination") in Siberia and Central Asia ("the Muslim steppe") from a global perspective, appealing to Friedrich Ratzel in the following terms: "A characteris-

---

[46] Semenov-Tian-Shanskii, "Edinstvo Rossii."
[47] Veniamin P. Semenov-Tian-Shanskii, *Raion i strana* (Moscow: Gosudarstvennoe izdatel'stvo RSFSR, 1928); Alfred Hettner, *Die Geographie: Ihre Geschichte, ihr Wesen und ihre Methoden* (Breslau: Hirt, 1927). See the review by Nikolai N. Baranskii, "Prof. V. P. Semenov-Tianshanskii, 'Raion i strana,' GIZ, 1928 g. str. 311, tsena 5 rub.," *Planovoe khoziaistvo* 10 (1928): 283–284, quote from 283.
[48] See Jureit, *Das Ordnen von Räumen*, 147.
[49] Semenov, *Raion i strana*, 172.

tic geographical feature of all the territorial expansions of humanity on earth is the necessity to fight for space."⁵⁰

He classified the countries of Eastern Central Europe like Finland, Estonia, Latvia, Lithuania, Poland, Czechoslovakia, Hungary, and Romania as "weaker, second-tier" entities, calling them "buffer states," "without any particular autonomy." Such countries were, as he put it, "used by the Western Entente powers to isolate the USSR from Germany" and he compared their function to that of the territories lying between Turkestan and India in the second half of the nineteenth century.⁵¹ Such explicit statements, however, which also placed East Central Europe within the continuity of the imperial world politics of the nineteenth century, were exceptional in his work.

Secondly, Vladimir Eduardovich Den (1867–1933), an economic geographer and, like Semenov, a native of St. Petersburg who had studied at several German universities, graduated from Moscow University with a law degree and held the first professorship for economic geography in Russia at the Petersburg Polytechnic Institute from 1902 onwards. Den understood economic geography to be a science that worked in statistics, researching economic sectors or industries and studying the current situation of various branches of economic life all across their geographical reach. For him, economic geography formed part of economics rather than geography.⁵² He did not express any opposition to the Soviet system after 1917 and helped to develop the discipline of geography over the subsequent years but was neither ever a Bolshevik nor advocated any explicitly Soviet understanding of geography. His students were to occupy key positions at some universities and within the state's institutions of economic management in their efforts to shape the Soviet New Economic Policy (NEP).⁵³

Den is interesting in this context, because in the years between 1925 and 1927 he was permitted to make two research trips abroad, one to Germany and another to Sweden.⁵⁴ He was the first Russian geographer to deal in depth with issues of contemporary geopolitics, particularly with the writings of the Swedish political scientist and politician Rudolf Kjellén (1864–1922) and German geographer Friedrich Ratzel. Den wrote two large-scale critical, though not quite polemical, essays on geopolitics for the Soviet magazine *So-*

---

50   Ibid., 190.
51   Ibid., 162.
52   On the conference speech given by V. V. Cheparukhin, see "V. E. Den i sovremennaia Rossiia," in *Izvestiia Russkogo Geograficheskogo Obshchestva* 2 (1994): 89–91; V. V. Cheparuchin, "Vladimir Eduardovich Den: Izvestnyi i neizvestnyi," in *Deiateli russkoi nauki XIX–XX vekov*, vol. 2 (St. Petersburg: Bulanin, 2000), 53–75; A. I. Chistoabev, *Geografy-Ekonomisty: Sudby i nasledie uchenykh* (Tiumen': Izd. Vektor Buk, 2003), 8–25, especially 8–18.
53   Ibid., 9–10.
54   Cheparuchin, "Vladimir Eduardovich Den," 59.

tsialisticheskoe ustroistvo.⁵⁵ In these essays, he pointed out the various attitudes towards this new field of knowledge in Germany and throughout Europe which some would likely consider under the heading of political geography. He gives a biography of Kjellén, outlining the development of his thinking on a state science as understood from a legal to a biologistic perspective. He stresses Ratzel's influence during the First World War and deals with various aspects of Kjellén's ideas before arriving on balance at a negative critical assessment in which he contrasts the differences between Kjellén and Lenin in their understanding of the state. Because, as he put it, for Kjellén "any idea of a class struggle or of the class character of the contemporary state was entirely alien." In Den's view, Kjellén's "realism" was to be celebrated, but the restrictions he put on the content of the various fields of the new science were unsatisfactory.⁵⁶ His broader criticism was directed primarily against Kjellén's understanding of geography as a systematic rather than a chorological science, and as a discipline exclusively directed towards the natural sciences.⁵⁷

But the journal refused to publish either of the two essays and they both disappeared into obscurity, destined to be published only in post-Soviet Russia. From 1926 on, the accusation was ever more frequently levelled against Den, along with his fellow campaigners and students, that what they were engaged in was bourgeois science. Den resigned from his office in 1931 partly of his own accord and partly under duress, and several people close to him were repressed or committed suicide, though some of his pupils did manage to hold onto their professorships over the following decades.⁵⁸ Den's geography was directed neither towards geopolitics nor ethnological geography. Indeed, his 1922 book *New Europe* was rather more typical of his work: it described and statistically presented the states of Europe either in their modified form or as brand new entities.⁵⁹ However, a look at Den's scientific work will demonstrate the narrow limits within which scientific production and discussion were confined in the early days of the Soviet Union. While Semenov was unhindered and uncritical in his references to Friedrich Ratzel, no critical discussion of Kjellén's geopolitics (which, it should be noted, took in a comparative treatment of Lenin and thus implicitly also of the Soviet Union) was open to Den.

The third direction is represented by Moscow economic geographer Nikolai N. Baranskii (1882–1961), who from 1926 on openly attacked Den and the direction taken by his research, along with other geographers, claiming in 1931 that he himself was developing and defending a genuinely Soviet geog-

---

55 See V. E. Den, "Uchenie Rudol'fa Chellena o predmete i zadachakh geopolitiki," *Izvestiia Russkogo Geograficheskogo Obshchestva* 129, no. 1 (1997): 26–38 and no. 2, 28–41.
56 Ibid., no. 1, 36.
57 Ibid., no. 2.
58 Cheparuchin, "Vladimir Eduardovich Den," 61 f.
59 See Vladimir Eduardovich Den, *Novaia Evropa* (St. Petersburg: Pravo, 1922).

raphy, an approach that was to finally displace Den's approach in 1934–1935. At the same time, he established geography at the universities (even managing to set the subject up with a separate faculty at Moscow State University in 1938), and made it a separate school subject. Unlike Semenov and Den, the younger Baranskii had been a Bolshevik—indeed he had been a professional revolutionary since 1898—and had not worked in geography in any university or scholarly association before 1917 and was not a member of intellectual circles in either St. Petersburg or Moscow. A figure vaguely known to Lenin and Stalin, he decided in the early 1920s against pursuing a career in the party in favour of science, and quickly made a name for himself through a number of publications (which included a textbook).[60]

At the centre of his research programme Baranskii placed the study of different economic regions and their relationships with one another (raionirovanie, raioniki) using Marxist terminology. He also created explanatory publications from an economic-geographical perspective on the first five-year plan.[61] Awarded the first professorship in economic geography at Moscow State University in 1929, he went on to determine Soviet geography from the early 1930s right up into the 1950s.[62]

But the outcome of the power struggle within the discipline of geography remained undecided at the end of the 1920s. For example, when Baranskii published the Russian translation of the main work on methodology by leading German geographer Alfred Hettner in 1930 and provided it with an introduction, he too was accused of "Hettnerism" (getnerianstvo).[63]

---

[60] An initial overview of his professional life is provided by Julian G. Saushkin, *Nikolai Nikolaevich Baranskii, 1881–1963* (Moscow: Nauka, 1971); Zakhar G. Freikin, *Nikolai Nikolaevich Baranskii* (Moscow: Mysl', 1990).

[61] See, for example, Nikolai Baranskii, *Ekonomicheskaia geografiia Sovetskogo Soiuza: Obzor po oblastiam gosplana* (Moscow: Gosudarstvennoe izdatel'stvo, 1926); Nikolai Baranskii, *Ekonomicheskaia geografiia SSSR: Obzor po oblastiam gosplana* (Moscow: Gosudarstvennoe izdatel'stvo, 1927); Nikolai Baranskii, *Sotsialisticheskaia rekonstruktsiia oblastei, kraev i respublik SSSR: V postanovleniiakh partiinikh i sovetskikh organov*, ch. II (Moscow: Gosudarstvennoe sotsial'no-ekonomicheskoe izdatel'stvo, 1932); Nikolai Baranskii, *Ekonomicheskaia geografiia SSSR: Uchebnik dlia 8–go klassa* (Moscow: Gosudarstvennoe uchebno-pedagogicheskoe izdatel'stvo, 1935).

[62] On this, see Chistobaev, *Geografy-Ekonomisty*, 10 f. His memoirs, written in 1948 and published in 2001, offer only an embellished, highly polished view of his professional life. See Nikolai N. Baranskii, *Moia zhizn' v ekonmgeografii* (Moscow: Geograficheskii fakultet MGU, 2001). For a general orientation to the topic, see also Adolf Karger, "Die Diskussion über das Wesen der geographischen Wissenschaft in der Sowjetunion," in *Bilanz der Ära Chruschtschow*, ed. Erik Böttcher et al. (Stuttgart: Kohlhammer, 1965), 51–74.

[63] See Freikin, *Baranskii*, 36–37; Al'fred Gettner, *Geografiia: Ee istoriia, sushchnost' i metody* (Moscow: Gosudarstvennoe izdatel'stvo, 1930), "Introduction," 3–6. The German original is Hettner, *Die Geographie*.

Yet in his introduction, Baranskii had commented critically on Hettner. Although Hettner had spoken out in favour of a "geography of man" and, in this context, had presented "disciplines such as the geography of races as well as economic, political and cultural geography" on an equal footing, his ideas lacked any proper social geography (sociography). Hettner answered neither "the question of the relationship between nature and human society" nor the question of a geography of the division of labour.[64]

The reasons why Baranskii, unlike other Old Bolsheviks (i.e., people who joined the Bolshevik faction before 1917), was not persecuted in the 1930s are not entirely clear. It was widely rumoured that Stalin had ordered his people to "Leave Baranskii in peace" (Baranskogo ne trogat').[65] One thing he did not do was develop his own political geography, at least not for the Soviet Union. In fact, for him the category of politics remained subordinate to that of economics. On the other hand, he abstained from making any criticism of Semenov in a review of the latter's late work *Raion i strana*, published in 1928.[66]

The fourth and final direction is represented by the geographer Stepan Rudnyc'kyi (1877–1937), a native of western Ukraine. Until the end of the First World War, he had presented works in Austria-Hungary and Germany on the physical geography of Ukraine as well as politico-geographical works that made positive reference to Friedrich Ratzel and legitimized the political independence of Ukraine from a geographical point of view.[67] He belonged to the circle of the school of the Viennese and (from 1905 on) Berlin geographer Albrecht Penck (1858–1945).[68] Rudnyc'kyi survived in the first half of the

---

[64] Gettner, *Geografiia*, 4 f.
[65] See Chistobaev, *Geografy-Ekonomisty*, 11. See also Heinrich Täubert's uncritical appreciation of Baranskii as an outstanding economic geographer, "Zum 80. Geburtstag von N. N. Baranskii," *Petermanns Geographische Mitteilungen* 106 (1962): 75–77.
[66] See note 47.
[67] In this context, see especially Stephan Rudnyzkyj, *Zur politischen Geographie der Ukraine* (Vienna: Verlag des Ukrainischen Korrespondenzblattes, 1916). Rudnyc'kyi here takes up Ratzel's political geography as an argument against a traditional description of the state, "that not the state itself, but its connection with the soil should be the object of politico-geographical activity. [...] Space and its natural conditions provide the initial basis of the state, the second being that part of humanity that forms the state in the sociological sense. How these two elements—soil and man—grow together into a more or less homogenous, idiosyncratic organism—the state—that is the main question for all political geography." Ibid., 3.
[68] On Rudnyc'kyi, in recent Ukrainian research see the (sometimes rather uncritical) contributions by Oleh I. Shablyi, *Akademik Stepan Rudnyc'kyi: Fundator ukraiins'koji heohrafiji* (Lviv: Lvivskij derzhuniversytet, 1993) and Ihor Ditchuk, ed., *Heohraf-akademik Stepan Rudnyc'ky: Vydatni postati Ternopillia* (Ternopil: Navchalna knyha – Bohdan, 2007). For an introduction, see also Guido Hausmann, "Das Territorium der Ukraine: Stepan Rudnyc'kyjs Beitrag zur Geschichte räumlich-territorialen Denkens über die Ukraine," in *Die Ukraine: Prozesse der Nationsbildung*, ed. Andreas Kappeler (Cologne: Böhlau, 2011), 145–157; Steven Seegel, "Remapping the Geo-Body: Transnational Dimensions of Stepan Rudnyts'kyi and His Contemporaries," in *The Future of the Past: New Perspectives on*

1920s, socially, economically, and scientifically marginalized in exile, in institutions in Vienna and Prague. His German scientific network from the period up until 1918 had become largely meaningless by that time. Then, in 1926, he accepted an offer to take on a leading role in establishing and shaping geography as an academic discipline in Soviet Ukraine. Thus, he applied for and subsequently headed a new Ukrainian Geographical Institute in Kharkiv and took an active and leading part in the work of scientific associations and journals. He was elected to the All-Ukrainian Academy of Sciences in 1929 and quickly took on a leadership role there as well. From his private letters during this period, one can clearly discern how strongly he understood Soviet Ukraine to be a Ukrainian state, though at the same time he also suffered from some social exclusion as a western Ukrainian living in Kharkiv. It can be seen from his proposal for the foundation of a new Ukrainian Geographical Institute in Kharkiv how seriously he took his research interests and the geography of Ukraine and how insistently he persevered with that work in the context of the way in which geography was developing in Europe (and in German, French, and English-language geography in particular).[69] His institutional and scientific commitments are further reflected in a memorandum in which he argues for the creation of a comprehensive description of the regions of Soviet Ukraine, a cause he had taken up from similar efforts being made simultaneously by historian Mychailo Hrushevs'kyi. Despite all this activity, he was not to write any further politico-geographical works during this period.

A few years later—in 1933—Rudnyc'kyi was arrested, sentenced to five years in a prison camp, and sent to northern Russia (to Solovki, among other camps), where he was to perish in 1937. The main allegation against him was espionage, and especially of having "contacts with German right-wing nationalist and fascist circles."[70] Also counting against him were his nationalist published writings from the period of the First World War and the years that followed, though that work had not seemed to matter in his appointment to Kharkiv only a few years earlier.[71] The case of Rudnyc'kyi is more than simply an example of how an entire generation of Ukrainian intellectuals and scien-

---

Ukrainian History, ed. Serhii Plokhy (Cambridge, MA: Harvard University Press, 2016), 205–229. On Albrecht Penck, see Hans-Dietrich Schultz, "'Ein wachsendes Volk braucht Raum': Albrecht Penck als politischer Geograph," in *1810–2010: 200 Jahre Geographie in Berlin*, ed. Bernhard Nitz, Hans-Dietrich Schultz, and Marlies Schulz (Berlin: Geographisches Institut, 2011), 99–153.

69 The project proposed by Rudnyc'kyi is described in Pavlo Shtojko, ed., *Lystovannja Stepana Rudnyc'koho* (Lviv: Naukove tovarystvo im Shevchenka, 2006), 211–220.

70 The relevant documents have been assembled in Oksana I. Babak, Vasyl' M. Danylenko, and Jurij V. Plekan, eds., *Praga-Charkiv-Solovki: Archivno-slidcha sprava Akademika Stepana Rudnyc'koho* (Kiev: Instytut ukrajinskoji archeohrafiji ta dzereloznavstva NAN Ukraiiny, 2007); the quote from Rudnyc'kyi's charge sheet appears on 186.

71 Ibid., 41–48.

tists were annihilated during the Soviet effort to Stalinize science: it provides the clearest possible case of the violent suppression of all politically independent thought in the field of science in the Soviet Union in the 1930s.

A 1935 secret police document tells us that the process of the Sovietization of geography was largely complete by that time. Under the Stalinist regime, the expression "political geography" could only mean the active provision of support for Stalin's politics, and especially for his economic policy, through the discipline of geographical science. So the act of allocating Rudnyc'kyi's work to geopolitics facilitated his exclusion and oppression:

Geopolitics is a purely fascist science. [...] Geopolitics as a science is officially recognized in Germany, Poland, Finland and Japan, and that is mainly where it flourishes. Prof. Rudnyc'kyi was an open advocate of geopolitics within the USSR [...]. Aside from him there were no other open geopolitical scientists in the USSR, though the Hettnerist tendency, which was essentially disguised geopolitical science, was very widespread, and continues to this day to have representatives in leading geographical institutions of the USSR.

The NKVD document goes on to explicitly name V. E. Den and V. P. Semenov-Tian-Shanskii—and even Baranskii—as the most important protectors of the "Hettnerians" in the Moscow Geographical Society.[72]

Conclusions

The foregoing discussion of examples of politico-geographical activities within the Russian Empire and in the early days of the Soviet Union demonstrates firstly and most importantly the narrow confines within which the study of this particular current within geography had to work. No real debate on this trend in thinking happened either before or after 1917 as there was no political freedom and thus no freedom of research in the country at the time. The older form of political geography from the eighteenth and nineteenth centuries, which saw the subject as a description of the state, remained a marginal phenomenon that was even neglected in the country's traditional picture of itself before 1917 in the programmatic formulation of a new political geography. Individual voices show how any political geography with an anti-imperial orientation could only develop in exile (think of Kropotkin and of the Eurasians), or at least outside the borders of Russia (as in the case of Rudnyc'kyi), but also that even political geography oriented to suit imperialism managed to establish itself only to a limited extent in Russia itself (think of

---

[72] The document is reproduced in the annex to Cheparuchin's article, "V. E. Den i sovremennaia Rossiia," 90 f. Karl Schlögel refers to the influence of Karl Haushofer (and, through Haushofer, of Halford Mackinder) on the "imperial geopolitics" of the Soviet Union during the interwar period. See Karl Schlögel, *Berlin—Ostbahnhof Europas: Russen und Deutsche in ihrem Jahrhundert* (Berlin: Siedler, 1998), 255–272, especially 268, and Milan Hauer, *What is Asia to Us? Russia's Asian Heartland Yesterday and Today* (London: Routledge, 1992).

Semenov). It is thus only with certain reservations that the politico-geographical sciences can be seen as a symbolic resource for the Russian Empire.[73]

Secondly, this essay also outlines the lines of thought of the geographers included in the text. It thus reveals a constant interest in new developments in European geography and especially in the geography of the Germanophone countries, including in the emergence of a new discipline of political geography (as personified, for example, by Friedrich Ratzel, Alfred Hettner, and Rudolf Kjellén). Also detectable is creativity in the appropriation of outside developments by Russian (Semenov, Den) and Ukrainian (Rudnyc'kyi) geographers, all of whom have thus far remained largely unknown to research on and critical analysis of the politico-geographical and geopolitical heritage of the period.

Thirdly, it shows that the fall of the tsarist Empire and the emergence of the new Soviet order should not be seen as a total rupture. Fractures, continuities, and new beginnings were all discernible in the 1920s. What did mark a clear break, though, was the period of the Stalinization of the sciences at the end of the 1920s and the beginning of the 1930s.

In any case, the politico-geographical works presented here merit being examined critically in both the pan-European and the global scientific context of the time, as well as being included in a critical reappraisal of national and imperial traditions in Russia and Ukraine. This value judgement applies both to the interwar period and to post-Soviet times.[74]

Translated from German by Jaime Hyland.

---

[73] On this topic, see Ralph Jessen and Jakob Vogel, "Einleitung: Die Naturwissenschaften und die Nation: Perspektiven einer Wechselbeziehung in der europäischen Geschichte," in *Wissenschaft und Nation in der europäischen Geschichte*, ed. Ralph Jessen (Frankfurt am Main: Campus, 2002), 7–37, here 13.

[74] For an introduction and overview, see Olga Pavlenko, "Geopolitical Visions in Russia: The Post-Soviet Interpretations," in *Russlands imperiale Macht: Integrationsstrategien und ihre Reichweite in transnationaler Perspektive*, ed. Bianka Pietrow-Ennker (Vienna: Böhlau, 2012), 103–120; Mark Bassin and Gonzalo Pozo, eds., *The Politics of Eurasianism: Identity, Popular Culture and Russia's Foreign Policy* (London: Rowman & Littlefield International, 2017). In general, see Jonathan Crush, "Post-colonialism, De-colonization, and Geography," in *Geography and Empire*, ed. Godlewska and Smith, 333–350, here 336–337.

Volker Zimmermann

CRIMINOLOGY AND IMPERIAL DIVERSITY
The German Empire, the Habsburg Monarchy, and Tsarist Russia in Comparison

Over the course of the nineteenth century and in various European countries, criminology developed as a science dealing with the genesis, specificity, and prevention of crime as well as with criminals themselves, the reasons for their actions, and their punishment and societal reintegration.[1] It was initially not a discipline with a uniform self-conception, dedicated research institutions, or university chairs, but instead comprised a broad spectrum of criminal law experts, jurists, psychiatrists, statisticians, police experts, and other persons engaging with the topics of crime and criminals in general terms.[2] The heterogeneity of this group and the varied and often controversial communication by its members thus constitutes an essential characteristic of the science of criminology originating at the time.[3]

Like the representatives of other disciplines, early criminologists were influenced by their societal and national frames of reference as well as by their specific professional backgrounds. This means we can safely assume that the

---

[1] I wish to thank Professor Marina Mogilner (University of Illinois at Chicago) for valuable suggestions and comments on the sections of this essay concerning Russia.

[2] The term "criminology" only became established slowly over the course of the twentieth century as a self-description of a scientific discipline. While the periodical *Archiv für Kriminologie* had existed in the German-speaking area since 1916 and the *Institut für Kriminologie* was established at the University of Vienna in 1923, criminological research remained a subdomain of the science of criminal law in Germany until the middle of the twentieth century. The first chair of criminology was established in Heidelberg in 1959, the first corresponding research institute in Tübingen in 1962. In the United States, on the other hand, the American Institute of Criminal Law and Criminology was founded as early as 1909 in Chicago. The Criminological Institute established in St. Petersburg in 1908 was a department of the local Psycho-Neurological Institute, but it can also be viewed as one of the first scientific institutions of its kind. Although the Kriminalistisches Institut founded in Graz in 1913 during the time of the Habsburg Monarchy and engaging with the auxiliary sciences supporting criminal law did not have the term "criminology" in its name, it was likewise highly significant for the development of the discipline.

[3] An excellent overview of the history of criminology in several countries is provided by the volume by Peter Becker and Richard F. Wetzell, *Criminals and Their Scientists: The History of Criminology in International Perspective* (Cambridge: Cambridge University Press, 2006).

respective form of government they lived and worked under likewise shaped their perspectives and research approaches. For example, an imperial context with its heterogeneous circumstances in regard to space and population would likely have had an impact on the formulation of questions about the genesis of crime or the development of prevention strategies. This assumption will be tested in the following by way of analysing criminology in the three continental European empires: the German Empire, the Austrian part of the Habsburg Monarchy, and the Russian Empire.[4]

One key feature of empires is their multinational population and its geographic distribution. This statement by no means refers only to autochthonous inhabitants of often far-flung colonies who—as in the case of the British Empire—provided the majority of the population with ethnicities differing from that of the people in the motherland. Rather, the three continental empires to be examined in this essay were characterized by highly diverse populations and the cohabitation of various ethnic groups within (with the exception of the relatively small German overseas colonies) largely contiguous territories. The theory to be tested is that this very circumstance may likewise have influenced perspectives and approaches in criminological research.

Existing studies on the developmental history of criminology generally assume a shift in perspective occurring towards the end of the nineteenth century: while the so-called "classical school" based on the equality of all individuals and their free will (including that to commit crimes), which placed social control along with criminal law and its reform at the centre of its deliberations, had maintained an authoritative position until this time, an increasing number of authors now began to focus on perpetrators as persons with specific individual and environmentally triggered deviations. Criminal anthropologists in particular considered offenders to be a danger for the moral state of society due to their mental and sometimes also physical deviance.[5]

Against this background, the multiethnicity of empires posed a particular challenge for criminological research, for who or what could be considered a yardstick for a "normal," law-abiding population in light of this ethnic and cultural diversity—especially since the national frame of reference had been continuously gaining importance in considerations on the definition of a state's population over the course of the nineteenth century. Historical re-

---

[4] While the characteristic feature of being a continental empire is undisputed for the Habsburg Monarchy and the Russian Empire, application of the same description to the German Empire has only been discussed in recent years. There are, however, many reasons to view it not simply as a nation state but likewise as a continental empire. Cf. Philipp Ther, "Deutsche Geschichte als imperiale Geschichte: Polen, slawophone Minderheiten und das Kaiserreich als kontinentales Empire," in *Das Kaiserreich transnational: Deutschland in der Welt, 1871–1914*, ed. Sebastian Conrad and Jürgen Osterhammel (Göttingen: Vandenhoeck & Ruprecht, 2004), 129–148.

[5] Peter Becker, *Verderbnis und Entartung: Eine Geschichte der Kriminologie des 19. Jahrhunderts als Diskurs und Praxis* (Göttingen: Vandenhoeck & Ruprecht, 2002).

search has hitherto dealt with this question only sporadically, with attention being focussed instead on the treatment of populations in the colonies and people of colonial origin migrating to the motherland by legal systems and the police. What is more, the increasingly (post-)colonial gaze has been directed primarily at the British Empire.[6]

Analysis of the criminological approach to diversity in the three continental empires examined in this essay should likewise offer instructive insights, however. The topics of crime and punishment were actively engaged with in each of them, with some authors postulating supposed specific criminality by different ethnic groups, among other ideas. It should be noted that a joint scientific discourse developed in publications and learned journals in the German Empire and the Austrian part of the Habsburg Monarchy as a result of the common use of the German language among their politically dominant nations. This discourse was shaped by developments abroad and in turn influenced criminology in other countries. Transnational influences also affected criminological research in the Russian Empire, which developed with a slight temporal delay.

In the following, fundamental developments in criminology will be briefly sketched. One focal point of this overview will be criminal anthropology, since it was this branch that tabled the topic so important for the question of imperial specificities of criminological research: that of the "race," the *Volkstum* (roughly, ethnicity), or the "nationality" (likewise with the meaning of ethnicity) of delinquents. Subsequently, selected criminological approaches that discussed the putative link between race/ethnicity and crime will be presented. The analysis of German-language criminology will be based on primary sources, some of which will be cited at length to illustrate the authors' patterns of argumentation. The presentation of the Russian criminological positions will primarily use the research literature on the subject, of which there is significantly more available than for the German-language case studies.[7] The conclusion will discuss the possibilities and limitations offered by an

---

[6] This topic had hitherto been represented primarily in criminal history research; in very recent times, historians performing research on Russia have also addressed it. An overview of colonial perspectives in the historiography of crime can be found in Paul Knepper, *Writing the History of Crime* (London: Bloomsbury Academic, 2016), 203–228 (the chapter on "Empire and Colonialism") and Shaun L. Gabbidon, *Criminological Perspectives on Race and Crime*, 2nd ed. (New York: Routledge, 2010), 181–197 (the chapter on "Colonial Perspectives on Race and Crime"). On research concerning the British Empire, cf. as recent paradigmatic studies Preeti Nijhar, *Imperial Reflections: Criminality and the Constitution of "Dangerous" Groups in Colonial India and Victorian England* (PhD dissertation, University of London, 2006) and Preeti Nijhar, "Imperial Violence: The 'Ethnic' as a Component of the 'Criminal' Class in Victorian England," *Liverpool Law Review* 27 (2006): 337–360.

[7] A reference to the two most recent studies on the topic shall suffice: Riccardo Nicolosi and Anne Hartmann, eds., *Born to be Criminal: The Discourse on Criminality and the Practice of Punishment in Late Imperial Russia and Early Soviet Union. Interdisciplinary*

"imperial perspective" on the criminological research of the late nineteenth and early twentieth centuries.

Criminology and Criminal Anthropology in Three European Continental Empires

The topic of crime, criminals, and punishment steadily gained significance in the (mass) press, publishing, literature, and science in most European countries as well as in North America over the course of the nineteenth century.[8] The ongoing industrialization and its concomitant phenomena like growing cities and the rapid modernization of transport and communication—along with the accompanying massive changes to social structures—caused significant disconcertment in many people. Industrial cities overpopulated as a result of immigration were considered hotbeds of depravity, and newspapers were frequently full of reports of violent crimes. It is certainly correct to view the fear of crime as one of the key elements of the debate about the state of society.

This situation was inevitably also reflected in scientific approaches to the phenomenon of crime. Efforts to investigate the conditions of its genesis as well as measures for its prevention and the question of how to handle delinquents had been undertaken beginning in the eighteenth century at the latest, but an intensive and systematic discourse only began to develop during the following century. The abovementioned heterogeneity of persons participating in this discourse facilitated a variety of different research approaches—largely owing to the specific perspectives offered by the very diverse professions and academic disciplines the contributors worked in. As a consequence, representatives of sociological, psychological, anthropological, and other patterns of analysis competed for interpretive dominance within the criminological discourse, which eventually led to a veritable flood of corresponding publications in the course of the establishment and professionalization of the discipline during the final third of the nineteenth and the beginning of the twentieth century.

Approaches (Bielefeld: Transcript, 2017) and Ilya Gerasimov, *Plebeian Modernity: Social Practices, Illegality, and the Urban Poor in Russia, 1906–1916* (Melton: Boydell & Brewer, 2018), 143.

[8] Cf. as a comparative presentation on this topic Jörg Schönert and Ulrich Broich, eds., *Literatur und Kriminalität: Deutschland, England und Frankreich, 1850–1880* (Tübingen: Niemeyer, 1983) and Jörg Schönert, ed., *Erzählte Kriminalität: Zur Typologie und Funktion von narrativen Darstellungen in Strafrechtspflege, Publizistik und Literatur zwischen 1770 und 1920* (Tübingen: Niemeyer, 1991) and Isabella Claßen, ed., *Darstellung von Kriminalität in der deutschen Literatur, Presse und Wissenschaft 1900 bis 1930* (Frankfurt am Main: Peter Lang, 1988).

Numerous professional journals served as platforms for scientific exchange. Alongside joint conferences, they offered a framework of sorts, at least for representatives of individual schools of thinking, who often formed heterogeneous groups themselves. The criminological discourses were by no means restricted by national boundaries—neither in terms of the reception of literature in foreign languages, nor in terms of the publication of findings made by authors from other countries or of meetings during international conferences. As was the case in other scientific disciplines, this led to intensive transnational exchange on positions and topics, especially from the last third of the nineteenth century onwards.

The consequences were far-reaching, for the abovementioned dominance of the "classical school of thought" began to be challenged—starting in the 1880s at the latest—not just by criminal sociologists, but also by a school of criminal anthropology based on (at least demanded and alleged) empirical research. A key initial figure in this context was the Italian psychiatrist Cesare Lombroso, who advanced the view of criminality caused by anlage and presented a typology of criminals in his main work *L'uomo delinquente* (The criminal human) published in 1876.[9] He claimed that certain in part inherited physical and mental aberrations predisposed some individuals to a career as criminals. According to Lombroso, this placed some perpetrators at a lower level of evolution, their behaviour being "atavistic," while others had become degenerate as a result of specific circumstances (e.g., alcoholism or malnourishment). Lombroso's book established his reputation as the most influential criminal anthropologist of the nineteenth century.

Although many criminologists shared the opinion that mental or physical deviance were at least a partial cause for criminal actions, Lombroso's theories of atavism and born delinquents were met with much scepticism.[10] His colleague Enrico Ferri eventually succeeded in building a bridge between the two camps: as a criminal sociologist, he connected his preceptor's approach with milieu theories, which were particularly common among French thinkers and concentrated on social and societal factors thought to promote criminal actions. This combination of criminal anthropology and criminal sociology resonated with many criminologists especially in the German Empire, since many German authors considered the focus on a single theory too limiting.

Ferri provided a solution for linking different criminological notions revolving around "pathological" deviance or the social and economic causes emphasized by many French and German-speaking authors. His influence

---

[9] Cf. Mary Gibson, *Born to Crime: Cesare Lombroso and the Origins of Biological Criminology* (Westport, CT: Praeger, 2002).

[10] On the reception of Lombroso in Germany, cf. Richard F. Wetzell, *Inventing the Criminal: A History of German Criminology, 1880-1945* (Chapel Hill: University of North Carolina Press, 2000), 39-72.

furthered the development of the German "Vereinigungstheorie" (unification theory), according to which only a multilayered perspective of criminality could deliver meaningful results and promising concepts for prevention. The Austrian-born criminologist and penologist Franz von Liszt, who worked in Germany and will be discussed in more detail below, was a prominent and influential proponent of this school of thought.[11]

Irrespective of certain differences in detail, this description applies to German-language criminology in the German Empire as well as in the Habsburg Monarchy. Its best-known representative in the latter was Hans Gross, whose *Handbuch für Untersuchungsrichter, Polizeibeamte, Gendarmen u.s.w.* was translated into several languages (e.g. into English as *Criminal Investigation: A Handbook for Magistrates, Police Officers and Lawyers*) and saw multiple editions far into the twentieth century.[12] Gross can be considered a key figure in Austrian criminology and German-language criminology in general. And although he criticized Lombroso for the latter's often abstruse, pseudoscientific conclusions, he was nevertheless a proponent of the emerging field of criminal anthropology—as evidenced in the name and content of the influential *Archiv für Kriminal-Anthropologie und Kriminalistik* (Archive for Criminal Anthropology and Criminalistics) he published beginning in 1898.[13]

In the Russian Empire, the fascination with crime and criminalistics was likewise ubiquitous from the nineteenth century onward. Although the process of industrialization was not as advanced as in the German Empire and the Austrian part of the Habsburg Monarchy, Russian society was undergoing rapid and dramatic change as well—including the expansion of cities and their importance for the discourse on crime. With a small temporal delay, the criminal anthropology debate was thus also finding its way into the Russian scientific landscape.[14] And as in the German-speaking countries, it offered a significant advantage over traditional approaches by allowing the deviance of delinquents to be interpreted as a symptom of an unhealthy society. The Rus-

---

[11] Silviana Galassi, *Kriminologie im Deutschen Kaiserreich: Geschichte einer gebrochenen Verwissenschaftlichung* (Stuttgart: Steiner, 2004), 190–225, especially 191.

[12] Hans Gross, *Handbuch für Untersuchungsrichter, Polizeibeamte, Gendarmen u.s.w.*, 2nd ed. (Graz, 1894).

[13] On Gross, cf. for example Christian Bachhiesl, "Die Grazer Schule der Kriminologie: Eine wissenschaftsgeschichtliche Skizze," *Monatsschrift für Kriminologie und Strafrechtsreform* 91 (2008): 87–111.

[14] For a concise overview of the development of Russian criminology (with special consideration of gender history), see Sharon A. Kowalsky, "Continuity and Change: Russian and Early Soviet Criminology and the Criminal Woman," in *The Oxford Handbook of the History of Crime and Criminal Justice*, ed. Paul Knepper and Anja Johansen (New York: Oxford University Press, 2016), 416–432, here 416–422. For more detail, see Sharon A. Kowalsky, *Deviant Women: Female Crime and Criminology in Revolutionary Russia, 1880–1930* (DeKalb: Northern Illinois University Press, 2009), 21–48.

sian criminologist Ignatii Platonovich Zakrevskii, for example, wrote in 1891 that the theory of criminal anthropology was

first and foremost sociological, seeking to determine, with the help of anatomy, psychiatry, history, and statistics, the origins and meaning of crime in a series of manifestations of social life. According to this school, since crime is to the social body what disease is to the human body, its teachings represent a kind of course of social pathology and therapy.[15]

Many Russian scientists naturally also viewed Lombroso's ideas with skepticism,[16] but criminal anthropology ultimately took hold in all three continental empires as a variant of the modern criminological discourse, albeit with differing distinctive features. And as was the case in the German Empire and the Austrian part of the Habsburg Monarchy, tsarist Russia experienced a transition from the "classical school of thought" and its concept of criminals as "fallen" persons to the notion of "impeded" persons advanced by criminal anthropology.[17] Some authors like, Dmitrii Dril', applied approaches similar to those of the German unification theory, combining criminal anthropology and sociology in their work.[18] What is more, numerous authors in all three empires subscribed to the idea of society as a "body" threatened by "pathological" and social aberrations in the same way the human body was threatened by diseases—and that it was the task of science to diagnose these harmful influences and contribute to their eradication.

Ethnic Diversity of Empires as a Criminological Challenge

Ethnic diversity had a significant impact on criminology in empires for several reasons. Firstly, these bodies politic were strongly characterized by the diversity and close proximity of ethnicities within them due to their colonies and the immigration from them to the motherland as well as—in the case of the continental empires—due to their areal expansiveness. In addition and independently of this circumstance, "race" or ethnicity had become a subject of criminological considerations. One of the most prominent authors to deal with this topic in the nineteenth century was once again Cesare Lombroso. Having previously only mentioned the aspect of "race" unsystematically, he eventually addressed it in more detail in the fifth edition of his book *L'uomo delinquente*. According to Lombroso, members of certain races had a stronger predisposition for committing crimes due to physical and mental underdevelopment, degeneracy, and lack of morals. Other Italian criminologists shared

---

[15] Quoted from Daniel Beer, *Renovating Russia: The Human Sciences and the Fate of Liberal Modernity, 1880–1930* (Ithaca, NY: Cornell University Press, 2008), 98.
[16] Ibid., 102 f.
[17] Becker, *Verderbnis und Entartung*.
[18] Leftist Russian criminologists in particular were inspired by Franz von Liszt, among others. Cf. Kowalsky, *Deviant Women*, 212, note 59.

similar views, and soon an increasing number of authors in various other countries were discussing these ideas as well.[19]

Among them were faithful followers like the German psychiatrist Hans Kurella, who translated Lombroso's works and propagated his theories. As a senior physician at the provincial asylum in Brieg, he published his own *Naturgeschichte des Verbrechers* (Natural history of the criminal) in 1893.[20] According to the book, "race and nationality [are] important biological factors of crime in which the sum of a heredity covering many generations is expressed."[21] Although most of Kurella's German colleagues subscribed to a different view in this regard,[22] race and ethnicity were increasingly being taken into consideration in criminological studies. This can be traced in part to common stereotypes concerning groups of persons who were present in several European states and differed from the respective societal majorities in ethnicity, culture and/or religious denomination, or confession. Jews and gypsies, for example, were traditionally suspected of being harmful to the majority society—the former as artificers of fraud, the latter as vagabonds and thieves.

This circle was now being expanded, however. For example, discussions about alleged links between race and criminality in the United States inspired some European authors.[23] Lombroso thought the "primitive, wild instincts" of "negroes"[24] responsible for the high murder rates in the United States, and the German Paul Näcke likewise pointed to biological influences putatively causing a criminal predisposition among "negroes."[25] These positions were doubtless fuelled by the biologism-based racism that was booming at the

[19] Cf. Gibson, *Born to Crime*, 97–126 (the chapter on "Race and Crime"). Lombroso's work referred not only to groups to which a harmful influence on society had already been ascribed for a long time, like the so-called "gypsies." In fact, he also provided examples of population groups in the south of Italy that allegedly produced increased numbers of criminals. In them, Lombroso saw African and Oriental influences at work that were damaging the moral substance of the southern Italian population.

[20] Hans Kurella, *Naturgeschichte des Verbrechers: Gründzüge der criminellen Anthropologie und Criminalitätspsychologie für Gerichtsärzte, Psychiater, Juristen und Verwaltungsbeamte* (Stuttgart, 1893).

[21] Ibid., 153.

[22] Cf. my own contribution, Volker Zimmermann, "'Der Einfluss des slavischen Elements': Zeitgenössische Erklärungen für die Kriminalität im Osten des Deutschen Kaiserreichs," in *Die Deutschen und das östliche Europa: Aspekte einer vielfältigen Beziehungsgeschichte*, ed. Dietmar Neutatz and Volker Zimmermann (Essen: Klartext, 2006), 131–147, as well as more generally on criticism of Lombroso's ideas about the connection between racial degeneration and crime, Galassi, *Kriminologie im Deutschen Kaiserreich*, 201–203.

[23] Daniel Schmidt, "Zahl und Verbrechen: Kriminalstatistiken im internationalen Dialog," in *Die Internationalisierung von Strafrechtswissenschaft und Kriminalpolitik, 1870–1930: Deutschland im Vergleich*, ed. Sylvia Kespers-Biermann and Petra Overath (Berlin: Berliner Wissenschafts-Verlag, 2007), 126–139, here 136–138.

[24] Ibid., 136.

[25] Paul Näcke, "Rasse und Verbrechen," in *Archiv für Kriminal-Anthropologie und Kriminalistik* 25 (1906): 64–75, here 66.

time. As a result, Näcke saw "a root of crime [in the] race-biological structure itself."[26]

The terms *Rasse* (race), *Nationalität* (nationality, in the meaning of ethnicity), *Volkstum* (ethnicity, national characteristics) or *Volksstamm* (people; literally, tribe, again with the connotation of ethnicity) were often used synonymously, however, and terminological clarity was therefore lacking.[27] For example, unification theoretician von Liszt likewise stated:

> Race is to be considered first and foremost as the most important social group. It is race that has determined the education and the initial development of social life. [...] It seems to me entirely beyond doubt that the formation of criminality is also determined by racial influences.[28]

However, von Liszt used the word "race" primarily as a synonym for *Volkstum*, which was a common term at the time; he also described the Bavarians as a separate "race," for instance. In short, the discourse around the turn of the century was by no means unequivocal and well-defined, and misunderstandings were therefore inevitable.

Since studies and deliberations on crime and legal developments in other countries were also published regularly in German and Austrian professional journals, the authors participating in these discussions were able to receive a multitude of different thoughts and positions from across Europe. Richard Weinberg, for example, a physician and private lecturer working at the University of Dorpat in the Russian Empire, wrote the following:

> For if every race of humans represents a more or less discrete and defined mental-physical type developed through heredity and adaptation, and if this type ultimately appears as the source and root of the social and moral instincts, it stands to reason to assume differences in regard to moral qualities between larger anthropologically delimited groups of humans.[29]

None of the protagonists of the criminological discourses in the three empires argued that a connection between race/ethnicity and crime would automatically lead to criminality irrespective of other circumstances. But over time, the discussion developed a dynamic of its own, placing the topic on the agenda for good. What is more, it possessed a clearly transnational character from the very beginning, which highlights its relevance within the criminological scientific discourse being established in the nineteenth century. The positions

---

[26] Ibid. Emphasis is Näcke's.
[27] For Austria, cf. Brigitte Fuchs, *"Rasse," "Volk," Geschlecht: Anthropologische Diskurse in Österreich 1850–1960* (Frankfurt am Main: Campus, 2003), 131.
[28] Franz von Liszt, "Die gesellschaftlichen Faktoren der Kriminalität: Vortrag am 21.9.1902 auf der Petersburger Tagung der I. K. V.," *Zeitschrift für die gesamte Strafrechtswissenschaft* 23 (1903): 203–216, here 211. The acronym I. K. V. stands for "Internationale Kriminalistische Vereinigung."
[29] Richard Weinberg, "Physische Degeneration, Kriminalität und Rasse," *Monatsschrift für Kriminalpsychologie und Strafrechtsreform* 2 (1905/06): 720–730, here 720.

held by participants within the three empires in this regard nevertheless exhibited specific peculiarities.

The German-Speaking Area

A few years after the foundation of the German Empire, the government in Berlin decreed the introduction of empire-wide statistics on crime so as to obtain an overview of the new body politic in terms of order and security. Until then, criminal statistics had been collected separately by the individual federated entities. The publication of the first overall statistics of persons convicted of crimes in the German Empire in 1882 revealed a phenomenon upon which the Statistisches Reichsamt in Berlin commented as follows:

> It is not yet appropriate in this first year of processing [...] to create maps of crime [...]. [...] It can however be stated already that [such maps] would in all probability reveal a relatively continuous decrease in crime from east to west; [...].[30]

According to the statistics, there was apparently an east-west gradient in the German Empire in terms of crime. That several million Polish-speaking citizens lived in the eastern provinces was soon accepted as a logical explanation for the unfavourable statistics in the east. The statistics did not report the native or colloquial language of the convicted persons, however, and this allowed traditional negative stereotypes and supposedly objective data to be linked—with serious consequences. For in order to prove—soundly, from the contemporary point of view—the alleged specifically Polish propensity for criminality, the scientists interested in the topic overlaid the crime statistics with language statistics for the affected areas. In this manner, statisticians, criminologists, and nationalistic German publicists used a criminal geography approach to construe problem areas presumed to pose a danger for the morality of the body politic.

The actual reasons for the high numbers of convicted persons in the eastern regions are difficult to determine from the available sources. Besides repressive practices by the police and judiciary, which led to conflicts between the Polish population and the powers of the state, recurring violent altercations between Polish and German citizens in particular may have been partially responsible for a greater incidence of battery convictions. In addition, regions marked by social distress and hardship—as large parts of the eastern Prussian provinces were at the time—frequently experience an increased occurrence of property crime.

Despite these correlations, which were likewise discussed at the time, some authors perceived the high conviction rates to be the result of putative

---

[30] *Statistik des Deutschen Reichs, Neue Folge*, Band 8, *Kriminalstatistik für das Jahr 1882, Neudruck der Ausgabe 1884*, ed. Reichs-Justizamt and Kaiserliches Statistisches Amt (reprint Osnabrück: Zeller, 1973), 60 ("Erläuterungen zu den Uebersichten II bis IV").

specificities of the "national character," respectively the "nationality" or "race" of the Polish: as a supporter of Lombroso's theories, for example, Hans Kurella considered the Polish population of the empire to be degenerate due to its historical development. In light of low wages, insufficient and unhygienic living space and malnutrition,

> congenial feelings, modesty, honesty, self-restraint cannot develop and be passed on, the population must degenerate, as is so often the case in the eastern provinces of Prussia with their distressingly high, distressingly growing crime.[31]

Kurella claimed that large numbers of the "Poles sucked dry by the landed gentry" over the past centuries had intellectually and morally imbruted through multiple generations,[32] and that this "degeneration" was the real reason for the high crime rates in the eastern provinces, for "normally dispositioned" humans would not commit crimes due to social hardship alone. In this case, not even the civilizing efforts of the Germans, who in Kurella's opinion were more highly developed, could remedy the problem—for while the latter continued to evolve, the Poles remained at the same low level of development.[33]

Although such views were met with criticism in the German Empire and the notion of criminally degenerate "races" was mostly rejected,[34] the topic was nevertheless on the table. Over the following years, several regional studies were conducted that not only included the usual analyses of the structural conditions of convicted persons like profession, age, gender, or confession, but also addressed the question whether the factor of "nationality" or *Volkstum* could at least in part explain the higher numbers of convictions in the eastern provinces. The discussion made use of ever more detailed statistical data that allowed systematic comparisons down to the level of individual *Landkreise* (administrative districts). Ultimately, however, most authors agreed that the poor social and economic situation of the Polish population and the often violent clashes resulting from Polish-German antagonism were more relevant to the high crime rates than nationality.[35]

The latter in particular seems noteworthy, as it meant that ethnic conflict situations were being explicitly perceived as factors for explaining criminality.

---

[31] Kurella, *Naturgeschichte des Verbrechers*, 172 f.
[32] Ibid., 179.
[33] Ibid., 139.
[34] Particularly vehement disagreement came from Viktor Aschaffenburg, *Das Verbrechen und seine Bekämpfung: Criminalpsychologie für Mediciner, Juristen und Sociologen. Ein Beitrag zur Reform der Strafgesetzgebung* (Heidelberg: Carl Winter's Universitätsbuchhandlung, 1903), 26–45.
[35] On this, see in detail Zimmermann, "'Der Einfluss des slavischen Elements'" and Volker Zimmermann, "Race and Ethnicity in German Criminology: On Crime Rates and the Polish Population in the Kaiserreich, 1871–1914," in *The Persistence of Race: Continuity and Change in Germany from the Wilhelmine Empire to National Socialism*, ed. Lara Day and Oliver Haag (New York: Berghahn, 2017), 129–153.

A treatise on crimes committed in the districts Marienwerder and Thorn within the Western Prussian *Regierungsbezirk* (a larger administrative unit) stated that

> the Poles should not in complexu be portrayed as bad people. It is however quite natural that, where two nationalities face each other in significant strength, many frictions develop between them that then lead to crimes, especially when, unfortunately on both sides, the hatred against nationalities is ignited. And there is certainly more delinquency by the Poles than by the Germans, since the former feel oppressed by the latter.[36]

Even some of the authors who argued in a more balanced fashion, however, could not completely detach themselves from the notion that a certain "national character" could have an influence on the development of criminality. The idea that Poles (as well as other Slavic peoples) were particularly prone to violence was widespread: In his 1910 study on crime in the province of Posen, Walther Stöwesand asserted the following: "The Pole is described as vain and conniving by all who know the country. With a strong physique and great nimbleness, he is animated and irritable, dissolute and hot-tempered."[37] Like most others, Stöwesand rejected Kurella's theory of degeneration, but he nevertheless sought an explanation for the high incidence of battery convictions in the province of Posen during the final third of the nineteenth century.

The discussion about crime in the east of the German Empire thus integrated seamlessly with the commonly held notion of putatively uncivilized "indigens" in the east[38]—as corroborated by unspecified persons who "knew the country." This view held by criminologists was shared by some judicial officers as well. The head prosecutor of Gleiwitz in Upper Silesia, for instance, reported the following to Berlin in the year 1890 concerning the Polish part of the population in his jurisdiction: "It is to be hoped that the German culture will eventually be able to reduce the inclination towards criminality existing in the national character [of the Poles]."[39] This statement can be considered representative of an attitude towards the Polish as well as the Lithuanian inhabitants of the empire held primarily by German nationalist judicial officers, politicians, and publicists.[40]

---

[36] Bruno Blau, *Kriminalstatistische Untersuchung der Kreise Marienwerder und Thorn: Zugleich ein Beitrag zur Methodik kriminalstatistischer Untersuchungen* (Berlin: J. Guttentag Verlagsbuchhandlung, 1903), 126–127.

[37] Walther Stöwesand, *Die Kriminalität in der Provinz Posen und ihre Ursachen* (Stuttgart: F. Enke, 1910), 86.

[38] On this, see e.g., Kristin Kopp, *Germany's Wild East: Constructing Poland as Colonial Space* (Ann Arbor: University of Michigan Press, 2012).

[39] Quoted from Becker, *Verderbnis und Entartung*, 328.

[40] Cf. Nachtrag zum Generalbericht des Oberstaatsanwalts in Königsberg vom 7. Juli 1894 inklusive Bericht "Zur Kriminalität der Litauer" und Stellungnahmen der Staatsanwaltschaften bei den Landgerichten Tilsit und Memel, GStA PK (Geheimes Staatsarchiv Preußischer Kulturbesitz) Berlin, I. HA (Hauptabteilung), Rep. (Repertorium) 84a, Nr. 2641, Bl. 1021–1071.

Comparisons with the colonially influenced patterns of interpretation in criminal policy as encountered in Victorian England suggest themselves in this context.[41] All of these cases are ultimately about the establishment of societal hierarchies, about the relationship between the foreign and the dominating section of the population, about the question of primacy within the state, about the equality of all citizens, and about the neutralization of supposedly harmful influences on the "own" nation or *Volkskörper* (national body). An alleged moral superiority of the German population was the starting point for all considerations, as evidenced in the Gleiwitz head prosecutor's statement that the Poles were to be civilized—or Kurella's that this was no longer even possible.

A compounding aspect was the fact that some authors thought the migration of workers from the eastern reaches of the empire into the Rhenish-Westphalian industrial regions was "exporting" the crime problem. Violent crimes and theft, which according to the former *Amtsgerichtsrat* (district court councillor) Paul Frauenstädt from Breslau bore "a typical character trait of Slavic criminality," were suddenly more noticeable, especially in districts with immigrated Polish populations. Frauenstädt considered this sufficient evidence for the assumption that "the main factor" lay "without a doubt in the influence of the *Slavic element* on the local figuration of criminality."[42] He ended his deliberations with a statement that clearly exposed his nationalist political motives:

> While it was already a bad political mistake that will come back to roost to drive a German-averse wedge into these long-standing core German lands by heavily drawing on Polish elements for the Rhenish-Westphalian mining and ironworks industry, the coming statistical years—regrettably—will most likely also have to report a further deterioration of the previously so favourable crime conditions in the Rhenish-Westphalian district connected to the intrusion of Polishness.[43]

Some German-speaking criminologists in the Austrian part of the Habsburg Monarchy likewise ascribed a greater inclination for committing crimes to certain population groups. As was the case in the German Empire, almost all authors considered the notion an indisputable fact in regard to so-called "gypsies," and Jews were similarly targeted by anti-Semitic circles.[44] Voices construing connections between ethnicity and conviction rates regarding the Slavic population could be heard in Austria as well—even though, as was the case in the German Empire, nationality was not reported in the Austrian

---

41   See e.g. Nijhar, "Imperial Violence," 337–360.
42   Paul Frauenstädt, "Die preußischen Ostprovinzen in kriminalgeographischer Beleuchtung," *Zeitschrift für Sozialwissenschaft* 9 (1906): 570–583, here 574.
43   Ibid, 583.
44   Daniel Vyleta, *Crime, Jews, and News: Vienna 1895–1914* (New York: Berghahn, 2007); Leo Lucassen, *"Zigeuner:" Die Geschichte eines polizeilichen Ordnungsbegriffes in Deutschland, 1700–1945* (Cologne: Böhlau, 1996).

crime statistics. One example for this argumentation is an essay on the Austrian statistics for 1888 by Carl Beurle, an Austrian jurist and politician who supported the so-called Greater German solution. On the basis of his own analysis of statistics on language use and convictions, Beurle ascribed a higher proclivity for violent crime to Slovenians and a higher proclivity for theft to Czechs. He concluded that his results were indicative of a "greater natural predisposition of a nationality towards a certain offence."[45]

As was the case in the German Empire, the reference value in such studies was always the German-speaking population, whose lower conviction rates seemed to prove it to be less criminal beyond dispute, at least statistically. The Austrian prosecutor general Hugo Högel voiced the following opinion concerning the crime statistics for 1914: "Decisive is the membership in a *Volksstamm*." According to Högel, the statistics served as an "expression of the traits [of the ethnicities]," like a tendency towards "brawls" or, in more general terms, a certain "impulsiveness."[46] It was not a ubiquitous position, however: another prominent representative of criminalistics in the Habsburg Monarchy, Hans Gross, mentioned ethnicity as a criminological category only in regard to the so-called "gypsies"—albeit at considerable length.[47] He apparently did not consider the concept worthy of discussion concerning other sections of the population.

A further influential Austrian criminologist at the time was Hugo Herz, who is considered a pioneer of German-language criminal sociology. To assume a biological, physical, or racial "predisposition" towards crime seemed nonsensical to him. And though he too saw higher conviction rates among non-German population groups, Herz attributed this circumstance first and foremost to social and economic factors. He investigated the economic development of the affected regions along with education levels, professions, age, and alcohol consumption. And while he did speak of a "Czech invasion" into German cities or of the upswing of a backward Czech rural population thanks to German influence, he did not corroborate these circumstances with biologistic ideas, instead arguing that they were the result of social developments. He also considered some German-speaking Austrian regions to be underdeveloped for the same reasons.

In any case, Herz found ethnically mixed regions a rewarding object of investigation. Of particular interest to him were the Bohemian lands, in which he declared a "world-historical battle between Germans and Slavs" was rag-

---

[45] Carl Beurle, "Einige Ergebnisse der österreichischen Kriminalstatistik," *Zeitschrift für die gesamte Strafrechtswissenschaft* 8 (1888): 325–341, here 341.
[46] Hugo Högel, "Kritik und Reform der Kriminalstatistik," *Archiv für Kriminal-Anthropologie und Kriminalistik* 58 (1914): 1–69, here 30–31.
[47] Gross, *Handbuch für Untersuchungsrichter, Polizeibeamte, Gendarmen*, 328–349 (the chapter on "Die Zigeuner, ihr Wesen, ihre Eigenschaften").

ing. The region thus offered a potential area of conflict and exhibited a specific development in terms of crime. The differences between the

> Slavic peasant population continually dissociating from the rural districts and the German population rooted in the cities of the old cultural centres keeps the tempers in uninterrupted tension, incessantly changes the ethnographic and social composition of the populace, fragments national conflicts into class struggles and vice versa. From these foundations a different criminality than in all other areas of the empire develops as a matter of course.[48]

It was obvious to Herz that "the foreign immigrants eventually succumb to the economic and cultural influences of their environment," raising them to a higher level and thus resulting in lower crime rates. However, the most important aspects of the topic to him were the socioeconomic consequences of migration and the social adaptation issues associated with it on the one hand and the existing national conflicts on the other. The latter intensified especially during the final third of the nineteenth and the early twentieth century—just as criminological research was booming. It is precisely during this period that possible links between ethnicity and criminality became a subject of research in the Austrian part of the Habsburg Monarchy as well, albeit to a significantly lesser degree than in the German Empire.

At around this time we therefore also find a common denominator in many Austrian and German studies dealing with criminality: multiple authors pointed to the consequences of conflicts between nationalities, including a greater incidence of assault and battery in the affected regions (which likely applied to the eastern Prussian provinces in the German Empire, but not to the Bohemian lands that were—at least statistically—unremarkable in terms of violent crime). In the eyes of other authors, however, the national conflicts seemed to facilitate or reinforce the attribution of criminality based on ethnicity. German nationalist circles in particular used crime-related strategies to agitate against their political adversaries. During a city council meeting in the northern Bohemian city of Reichenberg in 1890, for example, the German nationalist politician and jurist Engelbert Jennel declared in regard to Jews and Czechs that these "foreign elements of the populace" were detrimental to the city because "as the crime statistics prove, [they] also [negatively] affect the average sense of morality."[49]

A further commonality can be found as well: as mentioned above, the combination of statistics on convictions and spoken languages was the primary instrument for constructing a correlation between ethnicity and criminality in both empires. This was not enough for some authors, however: with a view to the now more intensive discussion on ethnicity as a presumptive criminogenic factor, they wanted the question answered more authoritatively.

---

[48] Hugo Herz, "Die Kriminalität in den österreichischen Sudetenländern," *Monatsschrift für Kriminalpsychologie und Strafrechtsreform* 6 (1909/10): 205–245, here 205 f.

[49] Isa Engelmann, *Reichenberg und seine jüdischen Bürger: Zur Geschichte einer einst deutschen Stadt in Böhmen* (Berlin: LIT, 2012), 32.

Numerous German as well as Austrian authors therefore called for future crime statistics to include data collected on the "nationality" (i.e., ethnicity, not citizenship) of persons convicted of crimes.[50] Their wish would not be granted, however; the native language of convicted delinquents was not included in the conviction statistics until the end of the German Empire and the Habsburg Monarchy.

Summarily, one may take away the impression that the debate was somewhat less intensive in Austria than it was in the German Empire. And indeed, the question of a correlation between cultural underdevelopment and criminal behaviour certainly played a more important role in Germany. The conflict of nationalities in the eastern Prussian provinces—especially since it was promoted by the state—apparently had an amplifying effect, in part due to inter-ethnic disputes that were often settled violently and in part due to political and police repression. Taken together, these two factors explain the higher conviction rates in the affected regions at least to some extent. In the Austrian part of the Habsburg Monarchy, on the other hand, efforts toward national reconciliation were increasingly being made despite the contention between the nationally conscious Czechs and the German Moravians and German Bohemians, and this likely resulted in fewer convictions as well as a more moderate discussion on crime and criminal policy. In both cases, no significant evidence can be found of members of non-German sections of the population being systematically discriminated against by the judiciary system (aside from the national-political repression practiced especially in the German Empire).[51]

Russia

During the nineteenth and early twentieth centuries, various authors in the Russian Empire likewise supported the idea of underdeveloped and/or societally damaging ethnic groups, with Caucasians, Tatars, and Jews being the prime targets. For sociologist and historian Maksim Maksimovich Kovalevskii, for instance, the behaviour of Caucasians was incompatible—due to what he considered premodern attitudes toward law and justice—with legal con-

---

[50] Among them were Hugo Högel, "Kritik und Reform der Kriminalstatistik," *Archiv für Kriminal-Anthropologie und Kriminalistik* 58 (1914): 30–36; Johannes Galle, "Zur Methode der Kriminalstatistik," *Monatsschrift für Kriminalpsychologie und Strafrechtsreform* 6 (1909): 586–591.

[51] This is corroborated by the hitherto existing results of my project "The Criminality of Others: Ethnically Based Attribution of Criminality in the Eastern Provinces of Prussia and the Bohemian Lands, 1871 to 1914" funded by the Deutsche Forschungsgemeinschaft (DFG), which I have been working on since July 2016 at the Collegium Carolinum, Research Institute for the History of the Czech Lands and Slovakia, in Munich.

cepts that should apply to modern societies.[52] The head statistician of the Ministry of Justice in Moscow, Evgenii N. Tarnovskii, argued similarly, portraying the "bloodthirsty tribes of Turkmen, Kurds, Malays, and Ashanti" as so underdeveloped that murder was "almost legal" to them.[53] Ethnic and cultural diversity in the imperial context was thus an object of particular attention and interpretation by criminology and criminal anthropology in tsarist Russia as well.

This may come as no surprise considering the Russian Empire's tremendous ethnic and cultural diversity. It was for this reason that the abovementioned Austrian criminologist von Liszt considered this body politic to be a suitable laboratory for the investigation of possible correlations between "race" and criminality.[54] And indeed, indications of a higher crime rate among certain ethnic groups were readily found through superficial examination—for example in the shape of criminal geography comparisons showing that the incidence of murder and robbery in the Caucasus was nearly three times as high as in the European part of Russia.[55] Data on the nationalities of convicted persons were available for extended periods of time in tsarist Russia, however, so that more precise statements could be made in this regard than in the German Empire or the Habsburg Monarchy.[56]

In light of these data, statisticians like Tarnovskii discussed the question of the nationality of convicted persons along the same lines as their German and Austrian colleagues.[57] The problems with interpreting higher rates of crime were likewise similar to those in the other two countries. Starting in the late nineteenth century, for example, there was a rising number of Polish persons convicted of crimes,[58] and in the years from 1909 to 1913 at least, Poles represented the most frequently convicted group in relation to their numbers.[59] Religious denomination also seemed to point towards cultural differences in terms of delinquency, with Muslims being sentenced significantly more often than Christians by percentage from 1909 to 1913 as well. But it is also precise-

---

[52] Cf. Marina Mogilner, "The Empire-Born Criminal: Atavisms, Survivals, Irrational Instincts, and the Fate of Russian Imperial Modernity," in *Born to be Criminal*, ed. Nicolosi and Hartmann, 31–62, here 35; Marina Mogilner, *Homo Imperii: A History of Physical Anthropology in Russia* (Lincoln: University of Nebraska Press, 2013), 337.
[53] Quoted from Stephen P. Frank, *Crime, Cultural Conflict, and Justice in Rural Russia, 1856–1914* (Berkeley: University of California Press, 1999), 60.
[54] Liszt, "Die gesellschaftlichen Faktoren der Kriminalität," 212.
[55] Jonathan W. Daly, "Criminal Punishment and Europeanization in Late Imperial Russia," *Jahrbücher für Geschichte Osteuropas* NF 48 (2000): 341–362, here 348.
[56] Frank, *Crime*, 54.
[57] Ibid., 76.
[58] Ibid., 76 f. This also applies to gender-specifically recorded data, since Polish women were overrepresented in the Russian crime statistics. See Stephen P. Frank, "Narratives within Numbers: Women, Crime, and Judicial Statistics in Imperial Russia, 1834–1913," *Russian Review* 55 (1996): 541–566, here 561–563.
[59] Gerasimov, *Plebeian Modernity*, 143.

ly these two examples that reveal an important reason for higher crime rates, namely the repressive policies in regard to various minorities applied at certain times, which affected the official crime statistics but were naturally not evidence of a putative criminal inclination of specific population groups.[60]

The potential in terms of reservations and prejudices towards groups with other nationalities was huge, however—and it was not limited to Russian-speaking authors, as evidenced by the abovementioned example of psychiatrist and anthropologist Richard Weinberg. Weinberg dealt with the connection between "race" and criminality in several treatises, among them the essay "Physische Degeneration, Kriminalität und Rasse" (Physical degeneration, criminality and race) for the *Monatsschrift für Kriminalpsychologie und Strafrechtsreform* (Monthly Journal for Criminal Psychology and Penal Law Reform) published in the German Empire.[61] Although well aware of the issues attached to a generalized application of the term "race" to the very diverse population groups in tsarist Russia, he still believed it useful for illustrating distinctions between the criminal behaviours of different sections of the population. He too used statistical data as a basis for ascribing a higher predisposition to criminality to Poles, among others.

Furthermore, Weinberg advanced similar views regarding the alleged tendencies of certain groups toward violence to those expressed by Stöwesand concerning the Poles in the province of Posen, claiming himself to be an "expert" on the example of Estonians and Latvians as a result of his own experiences: "I know both peoples, since I am born in the country, from long observation and can say that the Estonian is generally inclined to strike out quickly, a peculiarity that is not part of the Latvian's temperament."[62] He considered such "temper anlagen" to have "biological roots"[63] and thought that one of the most dangerous forms of degeneration leading to criminality was "racial mixture." As was the case in the German Empire, there were apparently different varieties of the debate on degeneracy and criminality. In most cases, however, the predominant opinion was that nationalities suspected of being particularly criminal could be "civilized."

In general, the discussion (and the criminal policy practice) focused on those sections of the population that were considered the most dangerous to the political and cultural integrity of the empire, like Poles, Jews, Muslims, or Tatars. But the group of potentially suspicious persons could easily be expanded as well: in a spectacular trial against members of the small Finno-Ugric minority of the Votiaks from the Volga region during the 1890s, for example,

---

[60] Frank, *Crime*, 76 f.
[61] Weinberg, "Physische Degeneration." Among other aspects, Weinberg also dealt with the putative physical and mental specificities of (especially the brains of) Jews, Latvians, Estonians and Poles.
[62] Weinberg, "Physische Degeneration," 724.
[63] Ibid.

Robert Geraci showed how a cultural threat was constructed through allegations of a ritual murder of an ethnically Russian victim—and the role that criminal anthropology played in them.[64] The Votiaks were considered uncivilized, and the case became a major political and media event in part because it was ultimately about the question of the "Russification" of the empire's minorities and the self-image of the dominant Russian nation.[65]

Persons of Russian nationality also ranked highly in the conviction statistics, however. It is against this background that a specific form of the criminal anthropology debate in the Russian Empire must be viewed, for the problem of backwardness or underdevelopment was mentioned by many of its authors as one that affected nearly all nationalities.[66] Many experts referred to the degeneration theory of French psychiatrist Bénédict A. Morel.[67] While a certain fascination with such theories could be observed throughout Europe, it was linked in the Russian case to liberal notions concerning the reform and modernization of the empire in all areas of political, economic, and social life.[68] Among other things, the existing extreme social inequality was thought to promote crime.

Lombroso's work was regularly received in Russia as well, but his theories on atavism were criticized by most authors. While they also believed in the physical and mental deviance of criminals, they considered it to be the result of degeneration caused by social circumstances. This variant of criminal anthropology established itself in Russia, especially among psychiatrists who were interested in the sociological origins of crime and achieved a great deal of influence and prestige as authorized court experts. According to them, the danger lay in an inherited degeneration affecting almost all parts of the Russian population and spreading quickly due to economic and societal changes, impeding the country's modernization.[69] The entire population was thus threatened from the inside by deviant groups of all types (independent of

---

[64] Robert Geraci, "Ethnic Minorities, Anthropology, and Russian National Identity on Trial: The Multan Case, 1892–1896," *Russian Review* 59 (2000): 530–554.

[65] The actual perpetrator was an ethnic Russian. He had committed the murder under the (initially warrantable) assumption that it would be attributed to the Votiaks and he would be able to benefit economically from their removal from a particular village. Ethnicized criminalization was thus the calculated goal of a crime in this case.

[66] This corresponds to the findings of Alexander Etkind, according to which the Russian Empire was simultaneously engaged in external (in the newly acquired territories) and internal (in the Russian-speaking core territories) colonization as a civilizing mission. For more detail, see Alexander Etkind, *Internal Colonization: Russia's Imperial Experience* (Cambridge: Polity, 2011).

[67] Cf. on this, as well as on Russian criminology in general, Beer, *Renovating Russia*, and Louise McReynolds, *Murder Most Russian: True Crime and Punishment in Late Imperial Russia* (Ithaca, NY: Cornell University Press, 2012).

[68] Cf. in detail Beer, *Renovating Russia*.

[69] Ibid., 48–50.

ethnic backgrounds), and crime was to be interpreted as an attack on the Russian *Volkskörper* that could also come from within.

Remarkable in this context is the approach of psychiatrist Praskov'ia Nikolaevna Tarnovskaia, a supporter of Lombroso's theories who examined female murderers and prostitutes, taking great care to ensure that the scrutinized group of women were—in Tarnovskaia's opinion—ethnically unambiguously Russian. In this way she believed she would be able to determine a genuinely "Russian" norm and deviance; the most significant differences she found were between a less developed rural population and a more advanced urban population. According to Tarnovskaia, this also implied that other ethnic groups could exhibit their "own" norms and deviations, however, and that there thus existed no exclusive criminological reference point of a Russian culture from which other nationalities diverged in a negative sense.[70]

In the years after the revolution of 1905, and in the context of the excesses of violence associated with it, the discussion about a specific Russian "wildness" was particularly intense.[71] Such notions of a putative degeneration of the Russian population as a whole could also be combined with racist hierarchizations. Some Russian military psychiatrists saw a special criminal-political danger originating from Jews—whom they had pegged as the lowest category on their imperial scale of societal value—and various other minorities, for their "degeneration" was considered biologically rooted, while that of ethnic Russians was a cultural degeneracy and could thus be overcome. As a result, Russians were thought to rank at the top of the value scale.[72]

That the Russian majority nation was by no means threatened primarily by allegedly criminal ethnic minorities in the empire was proven by investigations into the oft-discussed problem of hooliganism. Originating in Great Britain, this term has been used since the nineteenth century to refer to offences like assault or vandalism committed by groups. Hooliganism in Russia had—at least in the perception of the government and the public—reached such an intensity by the beginning of the twentieth century that the Ministry of the Interior in Moscow decided to appoint an investigative committee. The committee's final report submitted in 1913 showed that Poles, Germans, and Tatars could barely be linked to these types of offences at all—instead, they were most common in areas with almost exclusively Russian populations.[73]

Ultimately, in Russia—as was the case in the German Empire and the Habsburg Monarchy—it was not the specificities of individual nations or ethnicities that were responsible for higher or lower crime rates, but a combination of political, social and economic factors. And in Russia, too, the dis-

---

[70] Cf. Mogilner, "The Empire-Born Criminal," 36 f. on Tarnovskaia's research. On the question of the interpretation of rural crime, cf. in particular Frank, *Crime*.
[71] Cf. Mogilner, "The Empire-Born Criminal," 39–43.
[72] Ibid., 47–53.
[73] Neil B. Weissman, "Rural Crime," *Slavic Review* 37 (1978): 228–240, here 230.

cussion about the atavism or degeneration of certain peoples was nothing more than a variant of the debate about the root causes of crime. The image propagated by nationalist authors of a Russian nation taking a leading role over population groups with other nationalities is comparable to the attitude of German national conservative circles. The rigorous battle against national movements fought in both the Russian and German Empires corroborates this theory further—not least because such repressive measures targeted the Polish-speaking parts of the population in both empires first and foremost, causing them to be accordingly overrepresented in the criminal statistics.

While these political and social premises resulted in an obvious inequality among different sections of the population, Marina Mogilner has pointed out that some Russian scientists identified norms and corresponding deviations separately for different ethnicities, thereby effectively accepting the diversity of the Russian Empire.[74] In the same vein, it should be mentioned that despite the judiciary reform of 1864, the Russian government left existing customary law untouched in parts of the Caucasian and Central Asian provinces, which resulted in a degree of legal heterogeneity accompanying the ethnic diversity in these peripheral regions of the empire.[75]

Finally, recent studies on judiciary practice indicate that, in regard to nationality, non-Russian subjects in tsarist Russia were not discriminated against per se by the courts.[76] Imperial hierarchization apparently did not necessarily imply abandonment of the rule of law. One important reason for this may have been that in the continental dominions of the three empires analysed for this essay—with certain restrictions in the Russian case—most inhabitants held an imperial citizenship and no special criminal law for "autochthonous" population groups had been established, as was the case in the overseas colonies, for example, by Great Britain, but also by the German Empire.[77]

---

[74] Mogilner, "The Empire-Born Criminal."
[75] This had presumably occurred involuntarily, since a modern rule-of-law pretension hardly seemed enforceable with the available resources. Cf. Jörg Baberowski, *Autokratie und Justiz: Zum Verhältnis von Rechtsstaatlichkeit und Rückständigkeit im ausgehenden Zarenreich, 1864–1914* (Frankfurt am Main: Klostermann, 1996), 385–427, 786 f.
[76] Cf. Stefan B. Kirmse, "Dealing with Crime in Late Tsarist Russia: Muslim Tatars and the Imperial Legal System," in *One Law for All? Western Models and Local Practices in (Post-) Imperial Contexts*, ed. Stefan B. Kirmse (Frankfurt am Main: Campus, 2012), 209–241, and Jane Burbank, "The Ties That Bind: Sovereignty and Law in the Late Russian Empire," in *One Law for All?*, ed. Kirmse, 153–179.
[77] On the German case, cf. Ralf Schlottau, *Deutsche Kolonialrechtspflege: Strafrecht und Strafmacht in den deutschen Schutzgebieten, 1884 bis 1914* (Frankfurt am Main: Peter Lang, 2007), which also includes a comparative look at the Netherlands, France and Great Britain (361–375, the chapter on "Das europäische Kolonial(straf)recht"). On the peculiarities of the Russian case (with subjects of "native" and of "foreign" origin), cf. Dieter Gosewinkel, *Schutz und Freiheit? Staatsbürgerschaft in Europa im 20. und 21. Jahrhundert* (Suhrkamp: Berlin 2016), 67–83 (the chapter on "Ein fragmentierter Status: Untertan-

Conclusion

Ethnic and cultural diversity was a political and societal challenge for empires—but it also posed a scientific challenge, especially in regard to criminology. This was due in part to the differing social and cultural contexts discussed in connection with crime and criminal policy. We have seen that this background increasingly spurred the interest of authors of criminological texts in the German Empire, the Habsburg Monarchy, and the Russian Empire alike during the late nineteenth century. Nationalist publicists and scientists in particular considered the respective dominant majority population to be the society worthy of protection from criminals.

In the German Empire, this meant that discourses on crime were often linked to a self-image of a civilizationally more developed German nationality and backwards or "uncivilized" sections of the population (e.g., Poles or Lithuanians in East Prussia). This view was supported by, among others, criminal anthropologists who were proponents of Lombroso's theories of innate criminality. And while these ideas were unable to attain interpretive dominance in the discourse on crime, the meticulously collected and analysed data of the conviction statistics offered substance for a seemingly more objective discussion of the topic. Some authors who argued in a more differentiated fashion based on a criminal geography approach likewise subscribed to the notion of deviance traceable to ethnic peculiarities—for example because they thought that "fervour" determined by ethnicity or culture generated a higher incidence of violent crimes. Such views were thus also shared by some "experts" in *different* criminological schools of thought, albeit in differing intensity.

The discussion in tsarist Russia was similarly varied: as was the case in German-language criminology, we see a change in perspective from "fallen" delinquents to "prevented" criminals, with psychiatrists in particular adopting ideas from criminal anthropology.[78] In contrast to the German Empire and the Habsburg Monarchy, however, the Russian gaze additionally focused explicitly on the "own" (Russian) population, which was likewise thought to be threatened by underdevelopment and degeneration. Nevertheless, the context of a multinational empire led to ethnic hierarchization by Russian nationalists as well, and the discourse on ethnic reasons for criminality in the Russian Empire was thus similarly traceable to the self-perception of a dominant majority nation, which clearly shaped the discussion on the causes (underdevelopment) and possible prevention ("civilization") of crime.

Certain differences in detail notwithstanding, these results correspond to those of American and British studies on the topic of "race and crime." Among other aspects, the latter focus on the ways in which criminal policy dealt with ethnic groups in colonial and postcolonial societies, where the con-

---

schaft im russischen Zarenreich"). On at least the pretension of creating a uniform judiciary, cf. Baberowski, *Autokratie und Justiz*.

[78] Cf. Mogilner, "The Empire-Born Criminal," 43–46.

struction of "immoral" and "uncivilized" population groups and their strict control by the judiciary, police and military played an important role.[79] This form of "othering" following postcolonial approaches and practiced—among others—by "experts" from various scientific fields served to order societies hierarchically and legitimize rule over territories inhabited by populations with other nationalities. Corresponding tendencies are detectable more frequently and in greater intensity among some imperial German and Russian authors than among German-speaking authors from the Habsburg Monarchy.

The differences within German-language criminology may be traceable to opposing concepts of state in the German Empire and the Habsburg Monarchy. The political and social elites in the German Empire had imperial, but unarguably also nation-state pretensions. In contrast to this form of nation-building, the Habsburg Monarchy pursued a policy of state-building, as Pieter M. Judson emphasizes.[80] The Czech population, for example, was politically, legally, and economically in a significantly better position than the Polish population in the eastern Prussian provinces. Circumstances like these may have had a radicalizing effect in Germany and a moderating effect in Austria in terms of the shifting of discourses on crime towards ethnicity. The more repressive practice regarding minorities in imperial Russia, particularly beginning in the final third of the nineteenth century, may likewise partially explain the more radical views on criminal "others" expressed by some authors there.

In none of the three empires examined, however, did a position imputing a *general* predisposition towards crime to certain ethnic groups truly permeate—with the important exception of the so-called "gypsies." Ultimately it was only a few, albeit often influential authors who thought it was possible to discern some form of cultural backwardness or even biologically founded degeneration of certain parts of the population as a cause for crime. It should also be noted that in principle, all discourses followed the contexts and logic (especially also the transnational ones) of scientific discourse mentioned in the introduction. This meant that like the imperial practice in the three cases studied, the criminological views and positions inevitably also differed, resulting in a relatively broad spectrum of proposed answers to the question of the influence of imperial diversity on crime and criminality. The majority of authors focused on problems relating to social, national and economic policy.

In many studies, the discussion about connections between ethnicity and crime is traceable to general prejudices and/or migration processes that have

---

[79] Gabbidon, *Criminological Perspectives on Race and Crime*, 181–197.
[80] Judson referred to the Habsburg Monarchy as an "interesting example of state-building that took place separately from the kind of cultural and political practices of nation-building that characterized many other European states during the same period." Pieter M. Judson, "Marking National Space on the Habsburg Austrian Borderlands, 1880–1918," in *Shatterzone of Empire: Coexistence and Violence in the German, Habsburg, Russian, and Ottoman Borderlands*, ed. Omer Bartov and Eric D. Weitz (Bloomington: Indiana University Press, 2013), 122–135, here 131.

led, and continue to lead, to ethnicity-related criminalization tendencies in other forms of government. Similar ascription and hierarchization processes were discernible in the United States—including a racially justified leading role of the "white" population.[81] What is more, ethnicity-related criminalization exists regardless of epoch, as the repeatedly arising discussions about crime by "foreigners" or "aliens" in the early modern period as well as in modern-day democratic (even non-post-colonial) societies prove. Finally, it was criminal anthropologists from Italy (i.e., not from an empire) in particular who shaped the discussion on whether certain sections of a population were more likely to commit crimes than others due to their ethnic imprinting. The imperial context can therefore naturally not be viewed as an exclusive nor even as a sufficient explanation for the ethnicization of discourses on crime.

What *could* make multinational empires susceptible to such discussion, however—besides their greater ethnic and cultural diversity as compared to other bodies politic (with the exception of the United States)—were the associated national movements emerging since the nineteenth century and their demands for political and cultural self-determination. Such conflicts were becoming increasingly characteristic of all three empires examined in this essay, challenging the claims to power previously viewed as self-evident and thereby representing a threat to imperial security—especially when the regions in which the population groups in question resided were located at the edges of the respective empire.

In Germany and Russia, national conflicts thus resulted in increased conviction rates due to repression and violent inter-ethnic disputes as well as in intensified discussions about possible links between ethnicity and crime (particularly in the case of the Poles). In Austria (e.g., in the case of the Czechs), these discourses were at least influenced. There were noticeable differences in the criminological assessment of national groups in this context, however: owing to the low crime rates in the respective settlement areas, Danes in the German Empire and Germans in the Russian Empire were considered unproblematic groups in terms of criminal policy. National-political conflicts apparently only affected criminological discourses if they could be linked to conspicuous trends in the crime statistics and/or traditional stereotypes of underdevelopment and otherness. With a view to this connection of power-political and civilizing/missionary agendas in particular, imperial rule in European continental empires can be interpreted as a context-setting category for ethnicity-related discourses on crime.

Translated from German by Stephan Stockinger.

---

[81] An enlightening and (owing to persisting discrimination and criminalization) pessimistic view onto the developments in the United States is offered by Paul Butler, "One Hundred Years of Race and Crime," *Journal of Criminal Law & Criminology* 100 (2010): 1043–1060.

Marianne Klemun

INTERWOVEN FUNCTIONALITIES BETWEEN EMPIRE AND SCIENCE
IN THE HABSBURG MONARCHY
A Comparison of Nineteenth-Century Geology and Botany

Emphasizing the functionality of science in a political setting, this essay provides an overview of the central topic of this volume in which we are debating the imperial meanings of science and the spaces of their negotiation in Central and Eastern Europe. In order to avoid any unilateral explanation, I will bring together two contrasting considerations. First, if we highlight the importance of sciences in relation to imperial frameworks, this implies that science serves imperial ambitions and is central to the construction or reshaping of an empire. In this sense, empires depend on different kinds of knowledge and different kinds of science: spatial, legal, social, political, cultural, and natural.[1] Second, we might argue that scientists and, more particularly, scientific organizations profit by establishing their status or reputation through fulfilling imperial tasks.

I see and analyse both aspects as being interwoven. Science and imperial frameworks are resources for each other.[2] Mitchell Ash speaks in this case of "mobilizing resources:"[3] These are concepts which help to see science not as being instrumentalized by politics alone and as opening up complex perspectives. The concept of empire and its implementation involves many aspects. Referring to empire as a geographical, physical, political, and cultural unit, I intend to outline the importance of different types of spatial dimensions as multilayered or interwoven entities in shaping the subject of analysis.[4]

---

[1] See, in general, for early times László Kontler et al., "Introduction," in *Negotiating Knowledge in Early Modern Empires: A Decentered View*, ed. László Kontler et al. (New York: Palgrave Macmillan, 2014), 1–22.

[2] Mitchell Ash, "Wissenschaft und Politik als Ressourcen für einander," in *Wissenschaften und Wissenschaftspolitik: Bestandsaufnahmen zu Formationen, Brüchen und Kontinuitäten im 20. Jahrhundert*, ed. Rüdiger vom Bruch and Brigitte Kaderas (Stuttgart: Franz Steiner Verlag, 2002), 32–51.

[3] Ibid., 32–36.

[4] With regard to the spatial dimension, I prefer the concepts based on sociology. See Martina Löw, *Raumsoziologie* (Frankfurt am Main: Suhrkamp Taschenbuch Wissenschaft, 2001); Susanna Rau, *Räume: Konzepte, Wahrnehmungen, Nutzungen* (Frankfurt am Main: Campus, 2013).

This essay focuses on two different fields of science—botany and geology—and on two organizations in the Habsburg Monarchy: the Geological Survey (k. k. Geologische Reichsanstalt), founded in 1849, and the Zoological-Botanical Society (k. k. Zoologisch-botanische Gesellschaft) in the years between 1848 and 1867. During this time of political recovery after the revolution of 1848, the state attempted to construct a cultural and political unity headed by the dynasty, the bureaucracy, and religion. It is well known that historians, philologists,[5] cartographers, and geographers contributed to establishing this notion of unity. Apart from these fields, natural history also displayed such a political functionality in the service of the empire, although this may not seem obvious, because nature was seen by naturalists more or less as being nationalized. Nationalizing nature was very attractive for many of the scientists in the field of natural history. This history-of-science analysis pursues questions in two opposing directions: in what way did natural sciences contribute to create the unity of the Empire, and in what way did they themselves benefit from this?[6]

The political goal of unity of different nationalities was propagated by Emperor Franz Joseph and his slogan *viribus unitis*.[7] Thus, a *Kulturnation*[8] (cultural nation) under German-speaking leadership was constructed, encompassing many different language-based national identities. I argue that in the case of geology this phenomenon of unity was also manifested within the epistemology and practice of the Geological Survey which conceptualized the whole state in a geologically defined and abstractly unified space which sup-

---

[5] See, in general, William M. Johnston, *Österreichische Kultur- und Geistesgeschichte: Gesellschaft und Ideen im Donauraum 1848 bis 1938*, 4th ed. (Vienna: Böhlau, 2006); William M. Johnston, *The Austrian Mind: An Intellectual and Social History, 1848–1938* (Berkeley: University of California Press, 1972); William M. Johnston, *Der österreichische Mensch: Kulturgeschichte der Eigenart Österreichs* (Vienna: Böhlau, 2010).

[6] Ralph Jessen and Jacob Vogel, "Die Naturwissenschaften und die Nation: Perspektiven einer Wechselbeziehung in der europäischen Geschichte," in *Wissenschaft und Nation in der europäischen Geschichte*, ed. Ralph Jessen and Jacob Vogel (Frankfurt: Campus, 2002), 7–37.

[7] A number of visual and lyrical works of the period employ this motto. The basic idea can already be found in Franz Grillparzer's well-known poem in homage to Radetzky. *Treue und Eintracht der österreichischen Völker, Viribus Unitis* is the title of a lithograph by Franz Kollarz from 1849 (after a drawing by Josef A. Hellich). This is an allegorical depiction of the eighteen peoples living within the Monarchy. Men in national costumes are assembled around Emperor Franz Joseph, above them in the ornamental band the seven ministries are symbolized, and below, the coats of arms of all crownlands. See Siegfried Nasko, ed., *Österreich unter Kaiser Franz Joseph I.: Historische Sonderausstellung im Schloß Pottenbrunn* (St. Pölten: Verlag des Magistrats der Stadt St. Pölten, 1979), particularly 102.

[8] See particularly Johannes Feichtinger, *Wissenschaft als reflexives Projekt* (Bielefeld: Transcript, 2010), 39.

planted traditional territorial and spatial frameworks. And even in botany there was a reference to unity, as will be shown below.

Both research institutions analysed here conformed, in the way science was practiced, to international standards and experienced ever-greater recognition by the political elite of the Habsburg Monarchy as well as by the court. Both entities and their scientific activities served the idea of unity in their rhetoric and their research goals, but they were not equally successful in doing so. The two institutions claimed to base their research on all the lands of the monarchy as their single objective, and both were prepared to integrate naturalists in principle from all countries of the monarchy. This essay shows the differences between geology and botany, focusing on the tension between ideal and reality. Whilst the state-founded survey achieved to fulfilment its quintessential task—the geological investigation of all territories of the Habsburg Crown through stratigraphy, the language of geology,[9] in the time up to 1867 and coordinated all knowledge acquired in an overall map—the bourgeois-dominated botanical society (k. k. Zoologisch-botanische Gesellschaft) did not manage to trace the *flora* of the empire as a whole. The difference between these organizations was that, apart from the fact that the Geological Survey was state-funded, the Geological-Botanical Society—in the hands of a bourgeois public movement—was based on a different culture of practices in fieldwork and different space-related activities. In this way, the story presented here reflects more profound epistemological as well as organizational differences between geology and botany.

The task of the Geological Survey was directly connected to concrete physical space. While, on a general level, stratigraphy depended on international exchange,[10] fieldwork in the regions was defined, in general, by the heads of the Geological Survey, who determined how layers of rock were to be recorded in maps. The support of laypeople in the provinces was appreciated by the centre in Vienna, who depended on such help. There was a conscious effort from the headquarters in Vienna to intensively involve local experts in the fieldwork. But this cooperation in the field, in spite of the goal of standardization, was sometimes more and sometimes less intense in actual practice.

Botanical collecting, apart from being shaped by an internal concept of species and taxonomy, was in the sense of epistemology a space-decontextualizing activity. Collecting, however, was an eminently local practice and thus

---

[9] See Marianne Klemun, "National 'Consensus' as Culture and Practice: The Geological Survey in Vienna and the Habsburg Empire, 1849–1867," in *The Nationalization of Scientific Knowledge in the Habsburg Empire, 1848–1918*, ed. Mitchell Ash and Jan Surman (London: MacMillan, 2012), 83–101, here 97–101.

[10] See Mott T. Greene, *Geology in the Nineteenth Century: Changing Views of a Changing World* (Ithaca, NY: Cornell University Press, 1982), 120; David R. Oldroyd, *Thinking about the Earth: A History of Ideas in Geology* (Cambridge, MA: Harvard University Press, 1996).

connected to regional and local identities and concerns. Plants found in the field were transformed into herbarium specimens, determined as species according to universal botanical knowledge. And the list of plants reached readers and collectors from a variety of social levels within and beyond learned societies.[11] The k. k. Zoologisch-botanische Gesellschaft encouraged botanists from all territories of the monarchy to participate in the objective of collecting and describing all the plants from every region,[12] but it did not have the means to bring these activities together in one single publication dedicated to the empire as a whole. Collecting plants was an open process in which botanists themselves chose where to work, whereas the geological survey developed a systematic strategy in the context of which space was surveyed according to a plan shared by all surveyors (the topographical maps).

Flora Writing Practices

Focusing on botany, *flora* writing should be the scientific practice to explore first. This genre depends on a special concept of species and also on collecting activity and a defined space of fieldwork. This space might be a city with its surroundings, a natural landscape, a continent, an island, or a politically or symbolically defined territory such as a monarchy or any other kind of state.[13]

For example, the Jesuit Franz Xaver Wulfen (1738–1805) did his fieldwork in Carniola, Istria, Goricia, and Carinthia, and gave his *flora* the title *Flora Norica*. It was published in 1858—long after his death—and referred to Roman history. This *flora* was edited under the auspices of the k. k. Zoologisch-botanische Gesellschaft. Being aware that the space he referred to did not correspond to any political territory of in his time, Wulfen used the Roman geographical radius, which provided an umbrella for his places of fieldwork, and this was still accepted by the editors in 1858.[14]

In 1848, when the existence of the Habsburg Empire was threatened by the tensions between dynasty and empire, state and countries, people and nationalities, Joseph Karl Maly's (1797–1866) *Enumeratio plantarum phanero-*

---

[11] See, in case of parts in Germany, Denise Phillips, *Acolytes of Nature: Defining Natural Science in Germany, 1770–1850* (Chicago: The University of Chicago Press, 2012), 74.

[12] Marianne Klemun, "Die Gründung des 'Zoologisch-botanischen Vereins' 1851: Eine 'Kathedrale' der Naturgeschichte und Biologie in der wissenschaftsorganisatorischen Landschaft der Habsburgermonarchie," *Verhandlungen der Zoologisch-Botanischen Gesellschaft in Österreich* 138 (2001): 255–270.

[13] Marianne Klemun and Manfred A. Fischer, "Von der 'Seltenheit' zur gefährdeten Biodiversität: Aspekte zur Geschichte der Erforschung der Flora Österreichs," *Neilreichia* 1 (2001): 85–131.

[14] The Roman province of "Noricum," which reached the Danube, covered Wulfen's space of experience. However, we must point out that the territory of Krain (today's Slovenia) belonged to Friuli and the lands of the coast belonged to "Italia" as well as "Pannonia" in ancient Roman times.

gamicarum imperii Austriaci universi[15] (Enumeration of the phanerogamic plants of the entire Austrian Empire) was published. This work attempted to produce a *flora* for and of the whole empire. Maly spoke explicitly of the *Herrschaftsgebiet Österreichs* (Austrian Dominion):

> The Austrian Dominion extends in longitude from 42°21' to 51° north and in latitude from 26° to 44°15' east. In the west its frontiers reach Bavaria, Switzerland and Piedmont, in the north it joins Bavaria, Saxony, Prussia and Poland, in the east it is surrounded by Russia and Turkey and in the south it joins Turkey, the Adriatic and Italy. The Austrian Dominion includes the following provinces: the Archduchy of Austria (including Salzburg), Bohemia, Carinthia, Slovenia, Croatia, Dalmatia, Galicia with Bukovina, Hungary with the Banat, Istria, the coastal area, Lombardy, Moravia, Austrian Silesia, Slavonia, Styria, Transylvania, Tyrol and Venice.[16]

It was the first time that an Austrian *flora* had been defined through clear external boundaries according to the shape of the whole state as an Empire. Later the Austrian botanist August Neilreich (1803–1871), who himself had worked on a *flora* of Upper Austria, criticized Maly's endeavour because of the fact that he copied lists of plants from Koch's *flora* on Silesia, although these plants were not found in the Austrian part of Silesia.[17] This argument can be explained with reference to Neilreich's serious interest in the regional *floras* instead of superficially representing the Empire as a whole. Although Maly's *flora* was incomplete and not precise regarding the borders of the Empire, it greatly enhanced his reputation.

Floras and Visions of the Imperial Territory

Let us look briefly back to the past, in order to stress the change of *floras* according to their political functions and spatial contexts. The following analysis of the historical development of floristry in Austria proceeds from two perspectives: in the first place, I am interested in determining what spatial concept underlay the floral inventory of that time; and secondly, I ask which different spatial-territorial-natural conceptual worlds in those changing times led to the scientific study of plants.

The making of inventories of plants in the region around Vienna began with Carolus Clusius in the seventeenth century, and his approach shaped the practice in the region in the following period.[18] Up to the eighteenth century,

---

15 Karl Maly, *Enumeratio plantarum phanerogamicarum imperii Austriaci universi* (Vienna, 1848).
16 Ibid., 9–10. Translation from Latin by Klemun.
17 August Neilreich, *Nachträge zu Maly's Enumeratio plantarum imperii austriaci* (Vienna, 1861).
18 Johann Heniger, "Clusius und die ungarischen und österreichischen Pflanzen im Leidener Universitätsgarten," in *Festschrift anlässlich der 400jährigen Wiederkehr der wissenschaftlichen Tätigkeit von Carolus Clusius (Charles de l'Escluse) im pannonischen Raum*,

fieldwork done by individuals was held in great esteem by botanists. For example, Thaddäus Haenke, a Bohemian botanist who travelled widely in the eighteenth century, prepared a *flora* in 1788[19] which listed the lands where he had collected plants: Bohemia, Austria,[20] Styria, Carinthia, Tyrol, Hungary. These lands were connected by Habsburg rule, and the *flora* thus reflected the fragmented political structure of the monarchy. It was during the time after 1848 that this consciousness of putting the parts together to form an imperial whole was stressed, even by botanists. But there was an intermediate step: in Joseph August Schultes' *Austrian Flora*,[21] published in 1814, the term Austrian Empire (Österreichisches Kaiserthum) was used for the first time as a title in a *flora* dedicated to Austria as a space.

The Patent of 15 August 1804, in which Emperor Franz II (reg. 1792–1806), in fear of the impending loss of the supreme title as a consequence of the activities of Napoleon, designated himself as "Regent of the House and the Monarchy of Austria" and similarly as "Sovereign of Austria" with the title of "Hereditary Emperor of Austria," did not refer explicitly to the empire. This was because the connecting link between the various "independent kingdoms and states" and the "unified Austrian body politic"[22] continued to be seen solely as the House of Austria and the person of the monarch. Although in constitutional terms the "Empire" was first referred to only in the draft constitutions of 1848 and 1849, this term—in the sense of an act of governance—entered public usage, as is also documented in Lichtenstern's *Atlas of the Austrian Empire* (1805). Intellectuals in particular frequently spoke of the "Austrian Hereditary States"[23] before the turn of the century. Both terms may be viewed as an indication that the traditionally heterogeneous, side-by-side existence of the different territories under the guardianship of the hereditary ruler would continue to apply but that at the same time the self-image of a union was beginning to manifest itself.

It is in the context of this upcoming rhetoric of unity that we should understand Schultes' plan to record, in his own express terms, "the whole area of

---

ed. Burgenländisches Landesarchiv (Eisenstadt: Burgenländisches Landesarchiv, 1973), 149–167.

[19] Thaddäus Haenke, "Observationes botanicae in Bohemia, Austria, Styria, Carinthia, Tyroli, Hungaria factae," in *Collectanea*, vol. 1, ed. Nicolaus Jacquin (Vienna, 1786), 1–96.

[20] Austria meant Lower Austria (today Niederösterreich).

[21] Joseph August Schultes, *Österreichs Flora: Ein Handbuch auf botanischen Excursionen, enthaltend eine kurze Beschreibung der in den Erbstaaten des österreichischen Kaiserthumes wildwachsenden Pflanzen*, vol. 2 (Vienna, 1814).

[22] Grete Walter-Klingenstein, "Was bedeuten 'Österreich' und 'österreichisch' im 18. Jahrhundert: Eine begriffsgeschichtliche Studie," in *Was heißt Österreich? Inhalt und Umfang des Österreichbegriffs vom 10. Jahrhundert bis heute*, ed. Richard Georg Plaschka (Vienna: Verlag der Österreichischen Akademie der Wissenschaften, 1995), 149–220.

[23] Ibid.

the flora of the Empire."[24] Indeed, Schultes turned his attention to a number of territories that he had previously not taken into account. This was facilitated by his workplace in Cracow, where he taught for a short time at the Jagiellonian University as a professor of botany and chemistry and from where he was able to travel to territories such as the Kingdom of Galicia and Lodomeria (which had belonged to the Habsburgs since 1772) and the Kingdom of Croatia and Slavonia (a Habsburg territory since 1699). In the interim, moreover, the first special *floras* for these areas had been produced.[25] Schultes reacted to the loss of territory in his *flora* by the following words:

> In view of the fact that since the first publication of this work a number of political changes have caused a small reduction in the southern and south-western regions of the Austrian *flora* [here he is referring to the "Illyrian Provinces" and the "Kingdom of Italy" created by Napoleon], I have replaced these by including plants that grow wild in Hungary and [...] in Eastern Galicia and in Bohemia.[26]

Besides the dynastic concepts of imperial unity mentioned above, there existed other, quite different concepts of organizing imperial space in botanical terms. An example is provided by Wilhelm Daniel Koch's (1771–1849) *Synopsis Florae Germanicae et Helveticae* (Survey of the flora of Germany and Switzerland),[27] which influenced subsequent generations of botanists, as it was republished a number of times until 1892–1897.[28] Koch's concept of space ignored state boundaries but implicated a symbolic cultural space, the German language. In the connection between Switzerland and territories of the *Deutsche Bund* (including the countries of the Österreichisches Kaisertum), Koch saw no naturally existing boundaries except that with Hungary, for which he used the description *allmählige Übergänge* (gradual transitions). For Koch, his space covered a "flora of Central Europe," an "Alpine Flora," and a "southern flora."

Only a few years later, the Austrian botanist August Neilreich, a lawyer employed by the Vienna municipality, criticized the idea that botany should be divided into two parts, according to political issues, although this made no sense in plant-geographical terms. Before 1848 the territories under the Habsburg dynasty were divided into two parts: one belonging to the "German Confederation," the other to the lands outside it in the east. In terms of plant geography, this politically inspired division did not make sense to Neilreich, because the *floras* of the south and east were different from those at the heart

---

[24] Schultes, *Österreichs Flora*, IX.
[25] This was Willibald Besser, *Primitiae Florae Galiciae austriacae utriusque: Encheiridion ad excursiones botanicas* (Vienna, 1809). With regard to Hungary, the works by Paul Kitaibel and Graf Waldstein (1803–1812), which were published in Vienna, constitute an important reference, although there the locations of the plants are only indicated in very general terms.
[26] Schultes, *Österreichs Flora*, IX. Schultes used the old German form "Ungern" instead of "Ungarn."
[27] Wilhelm Daniel Koch, *Synopsis Florae Germanicae et Helveticae* (Frankfurt, 1837).
[28] In its third edition (Leipzig, 1892–1897), it was revised by E. Hallier and R. Wohlfahrth.

of the Habsburg territories. It is clear that Neilreich tried to establish a broad agreement between political and plant-geographical conditions in what from today's perspective seems an inaccurate way:

> The fact that Koch took German federal territory as the basis for his survey is the accidental reason why Austria, in both political and botanical respects, is divided into two dissimilar halves; namely the western territories, i.e. those belonging to the German Federation, and the southern and eastern territories, i.e. those that are excluded from the same territories, and this is a division that is much suggesting in terms of plant-geography.[29]

The Zoologisch-Botanische Gesellschaft

In the Crown Lands (Kronländer), the museums[30] and societies newly founded in the nineteenth century promoted the search for plants in local places to underline the identities of the Crown Lands. Meanwhile, one of the leading naturalists, who was also a professor at the University of Vienna, Franz Unger (1800–1870),[31] made a proposal to the newly founded Academy of Sciences in Vienna that a *flora* should be produced for the whole Empire.[32] To this end it was necessary to establish a central herbarium, and this task was finally taken over by the k. k. Zoologisch-botanische Gesellschaft, founded in 1851. The importance of the society lies in the fact that it was the first specialized natural history institution in Vienna and that it fulfilled political functions in terms of propagating the unity of the Habsburg Monarchy.

At its first meeting, one of the founders of the society claimed that it was a "cathedral"[33] among other learned societies of the territories. Its rules read that research on botany and zoology should be spread over the whole territo-

---

[29] "Der Umstand, dass Koch seiner Synopsis das deutsche Bundesgebiet zu Grunde legte, ist die zufällige Ursache, dass Oesterreich wie in politischer so auch in botanischer Beziehung in zwei sehr unähnliche Hälften zerfällt, nemlich [!] in die westlichen, d. i. die zum deutschen Bundesstaate gehörigen, und die in die südlichen und östlichen, d. i. die von demselben ausgeschlossenen Länder, eine Eintheilung welche auch in pflanzengeographischer Beziehung Vieles für sich hat." Neilreich, *Nachträge zu Maly's Enumeratio*, 2.

[30] Marlies Raffler provides an overview of this topic. Sie her *Museum – Spiegel der Nation? Zugänge zur Historischen Museologie am Beispiel der Genese von Landes- und Nationalmuseen der Habsburgermonarchie* (Vienna: Böhlau, 2007).

[31] See in particular Marianne Klemun, ed., *Einheit und Vielfalt: Franz Ungers, 1800–1870. Naturforschung im internationalen Kontext* (Göttingen: V&R Vienna University Press, 2016).

[32] See *Almanach der kaiserlichen Akademie der Wissenschaften für das Jahr 1851* (Vienna, 1851, published annually) and [Anonymus], "Commissionsbericht über die botanische Erforschung des Königreichs Baiern und Vorschläge für eine ähnliche Einführung in der österreichischen Monarchie," *Sitzungsbericht der Akademie der Wissenschaften* 1 (1850): 316.

[33] Georg von Frauenfeld, "Gründende Versammlung am 9. April 1852," *Verhandlungen des k. k. Zoologisch-botanischen Vereins* 1 (1852): 1.

ry of the monarchy, as a part of a serious programme of investigation. In fact the society functioned only as a platform where data collected in the regions were put together. Information went only from the periphery to the centre, and this continually puzzled botanists in the regions. I may emphasize that this provoked the wish among some to "strike back" from the periphery or challenge the hegemony of the centre. Perhaps to counter such tendencies, even poems were published to mobilize the cooperative unity of all botanists and zoologists:

If the noblest in a land can unite
to support one another with wise council.
When all these appear united to us,
Who have shone individually in word and deed,
If for this they share knowledge, hope and opinion,
Then hail to Thee, o happy land,
Hail to Thee, o city,
Where they are bound in noble dispute,
For blessings flow from Thee for all times![34]

I would like to mention one example of how the regions made themselves heard. Franz Freiherr von Hausmann zu Stetten (1810–1878) published a *Flora von Tirol*,[35] written after the first museum of the province, the National Museum Ferdinandeum, was established in 1823. In this flora, the author mentioned thirty-five local collectors[36] who supported the new museum and in the course of a few years had collected 17,000 specimens. A comparison of this herbarium with that of the Joanneum in Graz, which collected about 14,000 specimens in about thirty years of existence (this herbarium also included plants from southern Europe, however),[37] demonstrates the productivity of its collectors. The high number of specimens in the Tyrol Herba-

---

[34] "Wenn eines Landes Würdigste sich einen,/zu unterstützen sich mit weisen Rath,/ Wenn all die verbunden uns erscheinen,/Die einzeln schon geglänzt in Wort und That,/ wenn sie dazu tauschen Wissen, Hoffen, Meinen,/Dann Heil du, glücklich Land,/Heil dir, o Stadt!/Wo sie verbunden sind zu edlem Streiten,/Denn Segen strömt aus dir für alle Zeiten!" J. F. Castelli, "Lebewohl: An die Versammlung deutscher Naturforscher und Ärzte," in *Bericht über die Versammlung deutscher Naturforscher und Ärzte in Wien 1832*, ed. Joseph Johann von Littrow (Vienna, 1832), 81.

[35] Franz Hausmann, *Flora von Tirol* (Innsbruck, 1851).

[36] The same phenomenon is found in other regions of Europe. See Charles W. J. Withers and Diarmid A. Finnegan, "Natural History Societies, Fieldwork and Local Knowledge in Nineteenth-Century Scotland: Towards a Historical Geography of Civic Science," *Cultural geographies* 10 (2003): 334–353.

[36] Bettina Schlorhaufer, "Museumsraum Provinz: Die Gründung des Tiroler Landesmuseums Ferdinandeums und sein gesellschaftliches Umfeld," in *Museumsraum—Museumszeit*, ed. Gottfried Fliedl (Vienna: Picus Verlag, 1992), 31–48.

[37] Detlef Ernet, "Zur Geschichte der Botanik am Joanneum in Graz im 19. Jahrhundert 1997," *Mitteilungen der Abteilung für Geologie, Paläontologie und Bergbau am Landesmuseum Joanneum* 55 (1997): 103–122, here 108.

rium can be explained by reference to the initiative of Ludwig Freiherr Ritter von Heufler (1817–1885), who advertised in local papers for collections to be made for the sake of the museum. He was a young lawyer in the service of the Municipality of Innsbruck. Public lectures helped laymen to become familiar with the aims of the museum. At that time Heufler identified with the museum's aim of being a centre of botany for Tyrol. However when he later became an official at the Ministerium für Kultus und Unterricht in Vienna, he became involved with the life of societies in Vienna and took up an imperial perspective. The example of Heufler thus suggests that the particular geographic centre of a protagonist's life played an important role in determining his orientation, including whether his focus was local or imperial.

Overall, when it comes to botanical collecting and *flora* writing, the focus on smaller regions and local conditions remained the main motivation of Habsburg botanists in the nineteenth century, and the ideal of combining botanical knowledge to create an imperial whole was, despite attempts to do so, never realized.

Geology

A completely different example in terms of the spatial orientation of knowledge production is provided by the Geological Survey in Vienna (founded in 1849). This institution centralized knowledge production by sending geologists to the periphery and by bringing together all the knowledge into one representative map and large, synthesizing publications.

By bringing together the geological record of different countries assembled according to standardized principles of fieldwork[38] and documentation, a culturally and nationally diverse mosaic of territories was supposed to be united in geological representation. The geologists themselves were aware of this interweaving of power, politics, and science, and oriented their activities to the requirements of the *imperium*. The goal was to transform all the different countries within the Habsburg Monarchy into a systematic and geologically coherent space, modelled as a unified, scientifically defined entity. The participating geologists saw this effort as a great opportunity to attract attention to their work in the public sphere.

---

[38] See, in general, Kristian H. Nielsen et al., "Studying Scientists and Scholars in the Field: An Introduction," in *Scientist and Scholars in the Field: Studies in the History of Fieldwork and Expeditions*, ed. Kristian H. Nielsen et al. (Aarhus: Aarhus University Press, 2012), 9–28, here 11; Robert Kohler, "Place and Practice in Field Biology," *History of Science* 40 (2002): 189–210; Henrika Kuklick, "Personal Equations: Reflections on the History of Fieldwork, with Special Reference to Sociocultural Anthropology," *Isis* 102 (2011): 1–33, here 1.

Stratigraphy provides a map with an abstract temporal dimension[39] that is both naturally determined and at the same time profoundly political. The common efforts of mapping by way of using local manpower would "allow the Royal and Imperial Geological Survey to be seen as an embodiment of a large, single Austria,"[40] as Haidinger, the head of the survey, put it in a speech. Activities in the different lands were commented on in Vienna as follows: "When the Royal and Imperial Geological Survey was founded, it had to accept, as one of its first concerns, the need to transplant to the individual Crown Lands the impetus that existed in the imperial capital towards an increase in the knowledge of the country, so as to awaken general interest in the great tasks that have been set and to gain strong cooperation everywhere in seeking a solution to these tasks."[41]

The geologists of the Imperial Survey understood geological mapping as a paradigm of "harmony," "agreement,"[42] and "consensus," that is, as cooperation between the Crown Lands of the monarchy.[43] Without this cooperation with the regional authorities, it would have been difficult for the imperial institution to fulfil the extensive and very labour-intensive task of mapping. In the provinces, however, there was considerable skepticism towards this pathos of unity and political centralism. All in all, one can observe different degrees of cooperation and dedication in different territories of the monarchy.

Before turning to the question of the spatial concepts guiding geological research, let us first look briefly at the practices and epistemic attitudes of the state-employed geologists in the survey. The geologists, who were brought together in sections and stationed in different regions, always worked together in teams consisting of three field geologists. Within this framework, one leading person always brought together the results for a larger area. During the collection process, field diaries were kept and stratigraphic findings were

---

[39] On this phenomenon in cartography, see David Gugerli and Daniel Speich, *Topografien der Nation: Politik, kartografische Ordnung und Landschaft im 19. Jahrhundert* (Zurich: Chronos, 2002), esp. 114 and 132.

[40] GBA_AZ_1850_216, Archiv der Geologischen Bundesanstalt (Archive of the Geological Survey, hereafter Archiv der GBA) Vienna. For more on this topic, see Klemun, "National 'Consensus' as Culture and Practice," 83; Marianne Klemun and Thomas Hofmann, eds., *Die k. k. Geologische Reichsanstalt in den ersten Jahrzehnten ihres Wirkens: Neue Zugänge und Forschungsfragen* (Vienna: Verlag der Geologischen Bundesanstalt, 2012).

[41] GBA_AZ_1851_425_1, Archiv der GBA Vienna.

[42] This term is found in many letters and notebooks written by the geologists of the survey, especially by Peters. See Brief von Carl Peters an Franz Hauer (Letter from Peters to Hauer), Klagenfurt, 10.7.1854, Inv.-Nr. A00209-B.112, Bibliothek der Geologischen Bundesanstalt, Wissenschaftliches Archiv (Library of the Geological Survey) Vienna.

[43] Klemun, "National 'Consensus' as Culture and Practice," 90.

immediately noted into the copied maps of the general staff, by using standardized colour coding.[44]

After completion of this stage of work, these materials were sent from the fieldwork site to Vienna. The senior geologist at the Geologische Reichsanstalt in Vienna, Franz Hauer, checked the diaries and maps and, in case of irregularities, attempted to clarify issues by postal communication. If controversies arose in the field, these were first discussed by the two cooperating field geologists, and an attempt was made to reconcile the differences. Then the team leader would establish conformity with the maps of neighbouring territories. Finally, everything was discussed and coordinated during winter meetings in Vienna.

The adaptation of the data produced in the field—one might also speak of the making of "collations"[45] of data—took place, therefore, at different levels and different stages of the work: in the team, among the assistant geologists and the section leader, and at the headquarters. This procedure guaranteed the quality control for which the institutions were responsible. It was here that a culture of "consensus" came into practical and also epistemological existence. What emerges is an institutional-collective process by means of which the various discoveries and data—at first uncertain and subjectively obtained by way of observation—were compared with each other and brought together to produce a result. This way data, in the form of collective knowledge based on observation, gained the status of being accepted and valid.

Difficulties in the harmonization of findings were manifold. They arose from having to bring together observations based on different theories. Such differences in theoretical outlook existed not only among the inspection geologists appointed by the survey in Vienna, but also between them on the one hand and local interested parties and specialists working on the periphery quite independently of the survey on the other hand.

In accordance with the political attitude at the headquarters of the survey, considerable value was placed on cooperation with the scientific associations that had developed in the provinces at a civic level, with the local administration, and the local experts. Cooperation with the latter experts was simply indispensable: without their previous studies and local knowledge the survey could hardly function. There was, therefore, a conscious effort by the headquarters in Vienna to intensively involve local experts.

---

[44] See the research on these notebooks in Marianne Klemun, "Administering Science: The Paper Form of Scientific Practice and Geological Fieldwork," *Earth Sciences History* 33, no. 2 (2014): 279-293.

[45] I borrowed this term from Staffan Müller-Wille, who coined it for Linné's botany. See Staffan Müller-Wille, *Botanik und weltweiter Handel: Zur Begründung eines Natürlichen Systems der Pflanzen durch Carl von Linné, 1707-1778* (Berlin: Verlag für Wissenschaft und Politik, 1999).

But this cooperation, in spite of the requirement for unity, produced very mixed success. This ranged from a very close and constructive collaboration including the exchange of personnel, where local experts were included in the inspections almost as equal partners (as in the case of Ehrlich in Upper Austria), to a broad use of local knowledge from the Viennese side (as in Bohemia) and an absolute refusal to cooperate on the part of specialists in the province (as in the case of Carinthia). The local bearers of knowledge, for their part, either looked at the work of the Vienna-based institute positively as an enrichment of their own research activity and as an opportunity to improve their own status through keen participation, or they approached it with scepticism and an unwillingness to subordinate themselves to the orders of the headquarters.

Nonetheless, the survey managed to cover an enormous imperial space with its great geological diversity and, within a few years, to produce an overall image in form of a map (1867) incorporating the landscapes from Lombardy to Bukovina and from Dalmatia to the estuary of the Elbe. [46]

Conclusion

The result of the geological survey, in the form of a map (Geological Survey Map of the Austrian-Hungarian Monarchy, 1867),[47] depended on an elaborate set of negotiated relationships in the fieldwork in which the geologists took part, and this involved a practical as well as an epistemic culture of "agreeing" on questions of procedure.[48] Geological work, moreover, was produced in the context of a political culture of agreement that was represented by emperor Franz Joseph's slogan *viribus unitis*. It was more than a merely symbolical act that the Emperor's emblem with this slogan was part of every title page of the survey's publications.

The notion of Empire was also important in the context of botanical work, and especially in the attempts to establish a *flora* of the empire as a whole. If compared with the geological survey, however, it brought together local identities and knowledge in different ways. In botany, local forces became a dynamic factor that attempted to strike back against the Empire.

The two fields of science analysed here were characterized by different conceptions of space in "doing science" and in "constructing the empire." This implicated distinctive patterns of scientific practice in relation to changing political relations between Imperial province and center. The notion of empire thus had different functions. In the case of geology, it mostly operated

---

[46] GBA_AZ_1850_216, Archiv der GBA, Vienna. And many sources from letters of the geologist.
[47] Franz von Hauer, *Geologische Übersichtskarte der Österreichisch-Ungarischen Monarchie: Nach den Aufnahmen der k. k. geologischen Reichsanstalt* (Vienna, 1867–1871).
[48] Klemun, "National 'Consensus' as Culture and Practice," 90.

as a resource of cooperation and had a legitimizing function for both imperial politics and science. In the case of botany, on the other hand, it was also used as a resource of resistance to centralizing imperial politics. In both cases, however, it proved to be effective.

Jan Arend

FROM TSARISM TO NAZISM
Evolution and Transfer of Agro-Colonial Knowledge

The goal of the contribution to follow is to show by means of an example how scientific knowledge has been transferred between differing imperial contexts in Eastern Europe over wide temporal and cultural distances. To be specific, the essay discusses a transfer of a body of knowledge acquired by the Russian Empire of the late nineteenth century over to National Socialist Germany during the Second World War, a process that resulted in that knowledge being of service in the German imperialist policy of conquest in Eastern Europe. The knowledge whose "journey" through space and time is tracked in this text affected a core area of imperial policy: the colonization of arable land. Since the last decades of the nineteenth century in the Russian Empire, responsibility for this knowledge had been entrusted to a collective of experts in the fields of natural history and agricultural science referred to as *pochvovedy* (soil scientists).

An inquiry such as this into the transfer of knowledge between two different imperial contexts may serve to illuminate one particular dimension of the ambiguity present in modern science communication: in numerous branches of knowledge, scientists found themselves tangled up in a field of tension between the competing objectives of contributing to international scientific cooperation on the one hand and of serving national interests on the other. Their careers were thus marked out by "dual loyalties."[1] At international congresses, scientists of various nationalities exchanged knowledge which, under certain circumstances, had the potential to prove strategically important for the interests of their own state. The case presented here seizes upon the example of German agricultural specialist Hermann Stremme (1879–1961) to illustrate how the themes of international cooperation and competition between nations, between academically motivated scientific communication and the exploitation of knowledge as a strategic asset for purposes of war, could coexist within the biography of one individual scientist.

This contribution can be considered as a follow-through of recent research on the dissemination of knowledge between empires that has already illustrated how certain aspects of imperial politics were capable of being gen-

---

[1] Nikolai L. Krementsov, *International Science between the World Wars: The Case of Genetics* (London: Routledge, 2005), 125–126.

eralized and made transferable from one imperial project to another. Such was the case, for example, in relation to techniques used to reign over culturally and ethnically diverse populations, and also to techniques applied in appropriating new land in enterprises of imperial expansion or conquest, that is, the topic upon which this contribution concentrates.[2]

The aforementioned existing research on the inter-imperial transfer of knowledge has pointed to the key role of individual actors who, as boundary crossers, constituted the agencies responsible for conveying and receiving such knowledge. It has been argued that knowledge transfers between empires have relied less on institutionalized channels of exchange than they have on personal initiative.[3] The case recounted here supports this interpretation: in essence, the story reveals itself as a series of encounters—some direct and some mediated—between three individual scientists. The first of these was Vasilii Dokuchaev (1846–1903), charismatic founder of the Russian school of soil science at the end of the nineteenth century; the second is represented by his pupil Konstantin Glinka (1867–1927), who figures in the story as a highly communicative and internationally well-networked representative of the school to the outside world. The third main character is provided by the aforementioned German agricultural scientist Hermann Stremme, who, in close communication with Glinka, began to acquire the knowledge accumulated by Russian soil scientists from the beginning of the 1910s on, and who later placed that knowledge at the disposal of National Socialists in their efforts to impose their policy of conquest in Eastern Europe.

Vasilii Dokuchaev and Soil Science in the Imperial Context of the Late Tsarist Empire

During the final years of the nineteenth century, under the leadership of Vasilii Dokuchaev, professor of mineralogy at the Imperial University of St. Petersburg, an innovative new direction began to emerge in the Russian Empire in the exploration of soils under the umbrella of natural and agricultural science. Among the most important innovations achieved by this new Russian discipline of soil science (pochvovedenie) were the methods it applied to the classification and mapping of soils. The fact that Russia was an empire was of

---

[2] Daniel Schönpflug and Martin Aust, eds., *Vom Gegner lernen: Feindschaften und Kulturtransfers im Europa des 19. und 20. Jahrhunderts* (Frankfurt am Main: Campus, 2007); Rikarda Vul'pius, Aleksei Miller, and Martin Aust, eds., *Imperium inter pares: Rol' transferov v istorii Rossiiskoi Imperii, 1700–1917* (Moscow: Novoe Literaturnoe Obozrenie, 2010); Susan Gross Solomon, "Circulation of Knowledge and the Russian Locale," *Kritika: Explorations in Russian and Eurasian History* 9, no. 1 (2008): 9–26.

[3] Vul'pius, Miller, and Aust, eds., *Imperium inter pares*; Volker Barth and Roland Cvetkovski, eds., *Imperial Co-operation and Transfer, 1870–1930: Empires and Encounters* (London: Bloomsbury Academic, 2015).

great importance for the development of these methods: while questions of the cognitive integration and colonization of newly conquered land did not provide the starting point for the activities of Dokuchaev and his pupils, they quickly realized that their knowledge of soil cartography might be of great benefit to the Russian state as an imperial power with a growing agricultural economy. The imperially tinted fields of application presented by the opening up of new agricultural land and the planning of new settlements, among other tasks,[4] provided important contexts in which the ideas of Russian soil science were to develop.

At the initial stage of its development, there was a genuine interest in soil as a "natural body" within Russian soil science. Dokuchaev and his students looked to create a systematic classification reflecting natural soil types. In this respect, they followed the example of other natural sciences that employed classification methodologies, like botany and zoology. Thus, the act of allocating a soil to a type was not based on agricultural criteria, but on the morphology of the relevant soil. The prestige of field science—the study of natural phenomena in their natural environment—formed a central part of the scientific self-image of Russian soil scientists.[5] The new Russian specialists attached comparatively little value to the examination of soils in the laboratory. The legendary fertile black earth soils found in the south and southeast of the Tsarist Empire took on particular prominence as the object of investigation for Russian soil science.[6]

---

[4] For the Russian soil scientists there were of course other applications for their research—using soil evaluations as a basis for calculating land taxes, for example. On this topic, see Catherine Evtuhov, "The Roots of Dokuchaev's Scientific Contributions: Cadastral Soil Mapping and Agro-Environmental Issues," in *Footprints in the Soil: People and Ideas in Soil History*, ed. Benno P. Warkentin (Amsterdam: Elsevier, 2006), 125–148.

[5] On science in the field, cf. Kristian H. Nielsen, Michael Harbsmeier, and Christopher Ries, eds., *Scientists and Scholars in the Field: Studies in the History of Fieldwork and Expeditions* (Aarhus: Aarhus University Press, 2012); Keith R. Benson, "Field Stations and Surveys," in *The Cambridge History of Science*, vol. 6, *Modern Life and Earth Sciences*, ed. Peter J. Bowler and John V. Pickstone (Cambridge: Cambridge University Press, 2009), 76–89.

[6] On the history of Russian soil science, cf. A. A. Fedotova, "Geobotanicheskie issledovaniia o chernozeme F. I. Ruprekhta," *Voprosy istorii estestvoznaniia i tekhniki*, no. 1 (2008): 22–34; David Moon, *The Plough That Broke the Steppes: Agriculture and Environment on Russia's Grasslands, 1700–1914* (Oxford: Oxford University Press, 2013), 52–54, 61–62, 75–83; David Moon, "The Environmental History of the Russian Steppes: Vasilii Dokuchaev and the Harvest Failure of 1891," *Transactions of the Royal Historical Society* 15 (2005): 149–174; Catherine Evtuhov, *Portrait of a Russian Province: Economy, Society, and Civilization in Nineteenth-Century Nizhnii Novgorod* (Pittsburgh: University of Pittsburgh Press, 2011), 19–20, 170–173, 176–177; Jonathan D. Oldfield and Denis J. B. Shaw, *The Development of Russian Environmental Thought: Scientific and Geographical Perspectives on the Natural Environment* (London: Routledge, 2016), 48–77; Jan Arend, *Russlands Bodenkunde in der Welt: Eine ost-westliche Transfergeschichte, 1870–1945* (Göttingen: Vandenhoeck & Ruprecht, 2017).

During Dokuchaev's time, Russia was an expanding imperial state, reaching out from its southern and eastern borders into the Caucasus, Central Asia, and China. This territorial expansion went hand in hand with a growth in the amount of land in agricultural use. Even in areas that had been in Russian hands for some time, including southern Siberia, for example, much agricultural land had yet to be opened up. Especially in regions east of the Urals there was much potential agricultural land, some of it very fertile.

And in the Tsarist Empire, the creation of new productive land was accompanied by massive migration of peasants. The hope of making a living in the new lands enticed peasants from the more densely populated centers of the empire. Peasants increasingly migrated from the overpopulated Black Earth region on the European side of the Empire to its Asiatic lands. The tsarist government had been promoting this migration since the middle of the nineteenth century. It saw this population movement as an effective means of countering the dangers of overpopulation and of preventing famine crises on the one hand, and of consolidating Russian rule in Asia on the other.[7] In the press and other fora of the time, a large portion of the Russian educated class discussed the "question of colonization," that is, the question of how best to control the migration flows of peasants in regard to the continued survival, prosperity, and expansion of the empire.[8]

The imperial context in which Russian soil science developed was of central importance to how it did so. Russian soil science became a trailblazing science of exploration closely associated with state colonialism. Dokuchaev and his students developed techniques for recording soil conditions over extensive areas. By taking soil samples and analyzing them morphologically, they arrived at assessments of soils in relation to their suitability for agricultural use, thus always equipping them as well to arrive at an assessment of their "suitability for colonization" (kolonizatsionnaia prigodnost').[9] The data so collected were summarized in soil overview maps. Such techniques and methods proved of great value in colonial projects to open up new lands in the Tsarist Empire.[10]

Dokuchaev kept such fields of application for his knowledge of soils always in mind. In his journalistic texts he criticized, for example, the fact that those regions of the south and east of the Tsarist Empire that had been conquered or settled since the sixteenth century had been insufficiently re-

---

[7] Cf. Willard Sunderland, "The 'Colonization Question': Visions of Colonization in Late Imperial Russia," *Jahrbücher für Geschichte Osteuropas* 48, no. 2 (2000): 210–232, here 213–217.

[8] Ibid.

[9] Leonid Prasolov, "K izucheniiu vertikal'nykh pochvennykh zon v Tian-Shane," *Pochvovedenie* 11, no. 1 (1909): 90–92, here 90.

[10] I explore this issue more thoroughly in Arend, *Russlands Bodenkunde*, 87–90.

searched.¹¹ In Dokuchaev's view, this criticism applied generally to research into the natural resources possessed by the new regions, and thus included in his criticism the state of research on local coal and iron deposits as well as on their soils. In an 1880 text Dokuchaev described the lack of knowledge of Russia's "natural capacities" as "hardly comprehensible and hard to forgive."¹² His view was a majority opinion among the elites of the fading Tsarist Empire. In the final decades before the revolution, the government intensified the efforts it had been making since the eighteenth century to produce an inventory and to develop the natural resources of the empire more systematically.¹³

At the beginning of the twentieth century, the government of the Tsarist Empire had a separate administrative body to deal with settlement issues, the *pereselencheskoe upravlenie* (Resettlement Administration). This institution became an important initiator of soil research projects in the empire. Between 1907 and 1917 it organized more than one hundred soil-science expeditions covering the entire Asian side of the empire. The soil scientists employed in this research, led by one of Dokuchaev's students, Konstantin Glinka, concentrated in particular on regions whose suitability for agricultural use was in doubt.¹⁴ This applied, for example, to the dry soils of Turkestan or to the taiga regions of both Western and Eastern Siberia, which remained frozen all year round.¹⁵ The Altai Mountains, where differences in altitude resulted in a particularly "colorful" pattern of different soil types, each with its own set of prospects for utilization, also became a focus of interest in the effort to sound out the "geographical limits of agriculture."¹⁶

Within the Resettlement Administration, soil science expertise was regarded as an important input for a series of decisions that needed to be taken in the state-controlled efforts at colonization through the creation of new settlements. Soil scientists were called in wherever there was a need to plan out where settlements should be placed and how large they should be. Soil scien-

---

11 Vasilii V. Dokuchaev, "Kakie obshchie mery mogli by sposobstvovat' podniatiiu kraine nizkogo urovnia pochvovedeniia v Rossii?," in *Sochineniia*, vol. 7, ed. Akademiia nauk SSSR (Moscow: Akademiia nauk SSSR, 1953), 23–38, here 31.
12 Ibid. The argument that knowledge was lacking in Russia on the country's natural resources was a recurring theme of science policy discussions in the dimming empire. Cf. Alexander Vucinich, *Science in Russian Culture*, vol. 2, *1861–1917* (Stanford: Stanford University Press, 1970), 487.
13 Jonathan Oldfield, Julia Lajus, and Denis J. B. Shaw, "Conceptualizing and Utilizing the Natural Environment: Critical Reflections from Imperial and Soviet Russia," *Slavonic and East European Review* 93, no. 1 (2015): 1–15, here 9.
14 Ob organizatsii pochvenno-botanicheskikh ekspeditsii v Tomskom raione, 1909–1914, Rossiiskii gosudarstvennyi istoricheskii arkhiv (Russian State Historical Archive, hereafter RGIA), fond (f.) 391, opis' (op.) 3 1839 7.
15 Cf. Konstantin Glinka, "Geographische Resultate der Bodenuntersuchung im asiatischen Russland," *Pochvovedenie* 14, no. 1 (1912): 43–63, here 48–49.
16 Ibid., 54–55.

tists often also had a say in another question that depended heavily on the positioning of those settlements: that is, on the direction in which new railway lines should be laid.[17]

Thus Russian soil science, as it developed under Vasilii Dokuchaev and his students, entered as a whole into an alliance with the expansionism of the Tsarist Empire. It profited from the institutional support it received from the Resettlement Administration and managed to extend the empirical base of its studies by undertaking expeditions in the Asiatic regions of the empire. Conversely, the Russian system of colonization also profited from the methodologies and knowledge of Russian soil scientists.

International Scientific Exchange: Konstantin Glinka and Hermann Stremme

When Vasilii Dokuchaev died in November 1903, he left behind a group of students who were to continue to develop his ideas and—a fact of particular importance in the context of this story—transmit those ideas beyond Russia's frontiers and across the world. In contrast to their teacher, who was strongly oriented towards the development of the discipline as a Russian science, Dokuchaev's pupils became involved in a developing culture of international scientific exchange.

The process of internationalization in soil research began at the beginning of the twentieth century.[18] In a similar pattern to what had been going on earlier in geology and botany, researchers began to see the standardization of methods and terminologies as an important goal for international exchange efforts to aim at.[19] International conferences were organized for this purpose from 1909 on. The appearances of Russian soil scientists at these meetings impressed their international audiences, amongst whom a large number of

---

[17] Ministerstvo Gosudarstvennykh Imushchestv, Sooruzhenie Sibirskoi zheleznoi dorogi, 1894, RGIA, f. 1273, op. 1 175 1–6. In 1907 Sergei Neustruev researched the soil conditions along the planned Semipalatinsk-Vernyi railway line. Sergei Neustruev, Curriculum Vitae, Tsentral'nyi gosudarstvennyi arkhiv nauchno-tekhnicheskoi dokumentatsii Sankt-Peterburga, f. 179, op. 1–2 527 1.

[18] Elisabeth Crawford, Terry Shinn, and Sverker Sörlin, eds., *Denationalizing Science: The Contexts of International Scientific Practice* (London: Kluwer Academic Publishers, 1993); Mitchell G. Ash, "Internationalisierung und Entinternationalisierung der Wissenschaften im 19. und 20. Jahrhundert: Thesen," in *zeitgeschichte.at: Österreichischer Zeithistorikertag 1999*, ed. Manfred Lechner and Dietmar Seiler (Innsbruck: StudienVerlag, 2000), 4–12.

[19] Elisabeth Crawford, "The Universe of International Science, 1880–1939," in *Solomon's House Revisited: The Organization and Institutionalization of Science*, ed. Tore Frängsmyr (Canton, MA: Science History Publications, 1990), 251–269, here 257–258; Kaat Schulte-Fischedick and Terry Shinn, "International Phytogeographical Excursions, 1911–1923: Intellectual Convergence in Vegetation Science," in *Denationalizing Science*, ed. Crawford, Shinn, and Sörlin, 107–132.

European countries were represented. Many of them agreed with the view of German soil scientist Emil Ramann, who asserted in 1911 that soil research in Russia had developed independently in such a way that "other countries cannot offer an equivalent standard of performance."[20] The independence attributed to Russian soil science was traced to various circumstances. Among them were the large expanses available to Russian soil researchers for research; the fact that many soils in Russia had not yet been used for agricultural purposes, which meant they could be examined in a "state of nature;" and finally also the above-described encouragement the specialists received from a state that had an interest in colonial settlement and the development of new lands.

Konstantin Glinka, who had participated in international meetings since the first soil science conference in Budapest in 1909, was the most important figure in the international dissemination of Russian approaches to soil science. Glinka belonged to the inner circle of Dokuchaev's pupils. After Dokuchaev's death, Glinka became head of major soil research projects in the Tsarist Empire, and in particular, projects of the Resettlement Administration, where he served as the principal officer responsible for soil matters in colonization management in Asiatic Russia. Based on these experiences, Glinka was regarded as one of the central figures in Russian soil science from the beginning of the twentieth century on.[21] Like no one else, Glinka represented the imperial vision of a Russian soil science that worked in close step with the tasks of settlement and colonization.

At one of the first international conferences on soil science, held in Stockholm in the summer of 1910, Glinka met a young Berlin lecturer by the name of Hermann Stremme, who showed great interest in Russian soil science. At the time of his meeting with Glinka, Stremme's interest in the soil and in soil mapping had been awakened only recently.[22] He quickly turned his attention to the Russian works on the topic, to which he had found many references in the work of his older colleague Emil Ramann.[23] It appeared to him that the

---

[20] Emil Ramann, *Bodenkunde*, 3ed ed. (Berlin: Julius Springer, 1911), 4.
[21] L. V. Zykina, "Glinka, Konstantin Dmitrievich," in *Biologiia v Sankt-Peterburge, 1703–2008: Entsiklopedicheskii slovar'*, ed. Eduard I. Kol'chinskii (St. Petersburg: Nestor-Istoriia, 2011), 138. For a biography of Glinka, cf. Sergei V. Zonn, *Konstantin Dmitrievich Glinka, 1867–1927* (Moscow: Nauka, 1993).
[22] Hermann Stremme, "Bodenkarte des Deutschen Reiches," *Forschungen und Fortschritte* 13, no. 12 (1937): 149–150, here 149.
[23] Ibid. As an absorber and mediator of Russian ideas in soil science, Ramann probably had a significance comparable to that of Stremme himself. Emil Ramann, "Pochvenno-klimaticheskiia zony Evropy," *Pochvovedenie* 3, no. 2 (1901): 5–18; Gavriil I. Tanfil'ev, "Po povodu stat'i prof. Ramanna: 'Pochvenno-klimaticheskiia zony Evropy,'" *Pochvovedenie* 3, no. 2 (1901): 179–182. Cf. also Ramann's correspondence with Pavel Ototskii in early 1901, Arkhiv Rossiiskoi akademii nauk (Archive of the Russian Academy of Sciences, hereafter ARAN) St. Petersburg, f. 185 (Pavel Ototskii), op. 2 153.

ideas of Russian soil scientists were only superficially understood outside Russia. And for him the reason was obvious: not a strong Russian speaker himself, he was hardly in a position to absorb the Russians' works on the topic properly. Russian soil science had to date developed inside the bubble of the Russian language, so to speak. Its most important works were only available in Russian. Only a few short summaries had appeared in any of the standard languages of the international scientific community.

For Stremme, the encounter with Glinka and—mediated through Glinka—with Russian ideas on soil science became the defining moment of his intellectual career. For Glinka, in return, the encounter became a stepping stone on the way to his general recognition internationally as a scientist. Glinka, who was 43 years old when he met Stremme in Stockholm in 1910, had already attained a position that allowed him to operate as an international guarantor of the knowledge possessed by the Russian school of soil science. For some time now he had been working on synthetic surveys of Russian soil science, works that received broad recognition in Russia as canonical accounts of the state of knowledge in the country's soil science. For his part, Stremme—at that time 31 years old—was on the cusp of achieving his scientific independence. While he had previously primarily cropped up as a researcher in fields of knowledge in which his teacher Wilhelm Branca specialized (paleontology, geology), his independent contribution to science was to be precisely in his examination of the methods of Russian soil science and their application to Central Europe, whose soil conditions were completely different to those of Russia. Thus, in a sense, the Stockholm meeting was one between a highly competent Russian teacher and a German pupil eager to learn.

Glinka and Stremme agreed to work together on the publication of a synthesis of Russian soil science in the German language. Glinka translated a selection of his lecture notes from Russian into German, a German that was, as he put it, still quite "Russian."[24] Stremme improved on Glinka's language and used his contacts with the Borntraeger brothers' Berlin publishing house, which specialized in scientific works, to persuade them to publish the manuscript. The latter published the work in 1914,[25] making it the first presentation of Russian ideas on soil science in any of the languages of communication of the international scientific community. The presence of such a publication on the shelves was an important prerequisite for the dissemination of the Russians' ideas, which had up to then been expressed almost exclusively in Russian. Thus Glinka and Stremme's collaboration constituted a cooperative project in the transfer of knowledge from East to West.

---

[24] Konstantin Glinka, *Die Typen der Bodenbildung, ihre Klassifikation und geographische Verbreitung* (Berlin: Gebrüder Borntraeger, 1914), 1.
[25] Ibid.

Stremme remained in occasional contact with Glinka until the latter died in 1927. As Stremme himself wrote, his most important task was "to test whether the Russian research method could be applied in Germany."[26] Would the various methods and concepts developed in discussions of Russian natural surroundings and soil conditions also be fruitful in a German context?

Both through Glinka himself and through their joint work on the publication of Glinka's *Typen der Bodenbildung* (Types of Soil Formation), Stremme had become very familiar with the approaches used by the Russian school of soil science to create their typology and classification of soils. Stremme now began to search Central Europe for soils just as they had been described and typologized by Russian researchers. In retrospect, Stremme described his approach as tentative: "Knowledge of soil types seemed to me to be of great [...] significance, but the extent of that significance was not clear from the outset."[27] Stremme published his experiments for the first time in a paper published in 1914.[28] It should be noted that at the time there was simply too little data on climate and soil conditions in Germany to describe climatic soil types in the same way as Russian researchers did. In European Russia, soil scientists had begun to accumulate such data earlier and more intensively, so that by 1914 the soil conditions there had been researched with much more precision than the soils of Central Europe.

In 1914 Stremme was called to the Technical University of Danzig, a call that he followed after a period at the front during the First World War. The college was a mere ten years old when Stremme began his work there. Its foundation, as a project under the leadership of influential Prussian promoter of science policy Friedrich Althoff, received strong support from Emperor Wilhelm II. One goal of the university had been to strengthen German industry and agriculture—and thus the "German element" as a whole—in the eastern provinces of Prussia. During the interwar years, a period in which Danzig, as a free city, no longer belonged to Germany, the political attitudes of its founders nevertheless continued to prevail at the university. Stremme, who had been a member of the All-German Association (Alldeutscher Verband) since 1914 and who would later welcome the National Socialists' seizure of power, was happy to identify with the university's political line in this regard.[29]

---

[26] Hermann Stremme, *Die Arbeiten des Mineralogisch-Geologischen Instituts der Technischen Hochschule Danzig auf dem Gebiete der Bodenkartierung: Rede gehalten bei der Übergabe des Rektorats am 1. Juli 1928 von dem Rektor Professor Dr. Stremme* (Gdańsk: Technische Hochschule, 1928), 6; Stremme, "Die Verbreitung der klimatischen Bodentypen in Deutschland," in *Wilhelm Branca zum siebzigsten Geburtstage: Eine Festschrift seiner Schüler*, ed. Wilhelm Branca (Leipzig: Gebrüder Borntraeger, 1914), 16–75.

[27] Hermann Stremme, *Grundzüge der praktischen Bodenkunde* (Berlin: Gebrüder Borntraeger, 1926), 266.

[28] Stremme, "Verbreitung der klimatischen Bodentypen."

[29] NSDAP, Party file on Hermann Stremme, summary index card on the person, Bundesarchiv (Federal Archives, hereafter BArch), DS A0068 1196.

Once in possession of a university chair in Danzig, Stremme now had more freedom to implement his ideas on soil science than he had had as a private lecturer in Berlin. He took advantage of that freedom to deepen his prewar interest in Russian scholarship. When in November 1920 Danzig received the autonomous status of a Free City, its newly appointed senate entrusted Stremme with the management of soil mapping work in the Free State, a task on which work had already started before the First World War.[30] The Free State of Danzig thus became an experimental field for Stremme, in which he could put his ideas on soil science to practical test within a limited space. Stremme enjoyed taking these "new paths"—and they were very often Russian paths.[31] He understood the mapping work he did in Danzig as a "dress rehearsal" for similar work following the Russian method inside the German Reich.[32]

Stremme was particularly interested in whether black earth-type soils, to which so much significance was attached by Russian research, also existed in Germany. With some pride he reported in 1928 that he had discovered a previously unknown "steppe island" in the Rhineland whose topsoil was black earth.[33] However, in the German-speaking part of Europe, pale forest soils, which Russian soil scientists called *podzoly* were far more widespread.

Since most cereals developed from what were originally steppe plants, they tend to thrive better in steppe soils such as black earths than on cleared forest soils put under cultivation. Stremme therefore noted that *chernozem* had a higher "cultivation value" than the *podzol* forest soils, which were widespread in Germany.[34] According to Stremme, the task of human agriculture therefore basically amounted to "making podzols as chernozemic as possible."[35]

Following his Russian models, Stremme elevated the black earth to an ideal agricultural soil and recommended that German farmers work their soils in the direction of this ideal. In the mid-1930s, on the topic of the National Socialist prestige project that was the construction of *Reichsautobahnen*, Stremme complained that, according to the plans for Central Germany, *Autobahns* would cut through the black earth of "our largest areas of steppe soils" in sev-

---

[30] Hermann Stremme, "Die geologische und bodenkundliche Landesaufnahme der Freien Stadt Danzig als Beispiel einer Spezialkartierung mit Auswertungskarten," *Zeitschrift der Deutschen Geologischen Gesellschaft* 89, no. 6 (1937): 343–357, here 343.

[31] Hermann Stremme, "Neue Wege der geologischen und bodenkundlichen Landesaufnahme," *Forschungen und Fortschritte* 8, no. 4 (1932): 46–47; Hermann Stremme, "Die Bodenschätzung für Steuerzwecke," *Forschungen und Fortschritte* 9, no. 16 (1933): 235–236, here 235.

[32] Stremme, *Arbeiten des Mineralogisch-Geologischen Instituts*, 11.

[33] Hermann Stremme, "Über Steppenböden des Rheinlandes," *Chemie der Erde*, no. 1 (1928): 28–43.

[34] Stremme, *Grundzüge der praktischen Bodenkunde*, 266.

[35] Ibid.

eral places.³⁶ He also adopted the Russian terms (chernozem, podzol) for use in his scientific language.

Also like the Russians, he attached greater importance to field research than to laboratory research. In his view, "field observation" was a neglected discipline in Germany.³⁷ To correct this deficiency, Stremme recommended a reorientation towards the example of Glinka.³⁸ Anyone who positions himself "outside on and in the ground" will be rewarded by an "immediate" impression of the soils.³⁹ There was, according to Stremme, no way around such field observations when determining soil types for purposes of mapping them, while "chemical and physical testing are in general unnecessary." He went so far as to say that work of the latter type would "unnecessarily delay work" and would not generally—in contrast to "careful observation"— provide any additional valuable insights.⁴⁰

The Imperial Dimension: Herman Stremme and German Colonization and Settlement Planning

From his base in Danzig, Stremme worked closely with scientific institutions inside the German Reich. Some of his research projects were even funded by Reich-based institutions. At the end of the 1920s, for example, the Emergency Association of German Science (Notgemeinschaft der Deutschen Wissenschaft)—later to become the German Research Foundation (Deutsche Forschungsgemeinschaft)—financed Stremme's work on a soil map of Germany. He also had an involvement in numerous other mapping projects in Germany.⁴¹

Stremme worked in the closest possible cooperation with scientists in Germany in the 1920s and 1930s in the context of agro-economic research on settlement issues and spatial planning. The focus of this work was the issue of how peasant migration could be harnessed and steered to the best advantage of the state.⁴²

---

36 Hermann Stremme, "Die Bodenkarten des Deutschen Reiches und des europäischen Kontinents," *Zentralblatt Bauverwaltung* 56, no. 8 (1936): 176–182, here 182.
37 Hermann Stremme, "Die bodenkundliche Kartierung von Feldversuchen als Mittel zur Feststellung der praktisch wichtigen Bodeneigenschaften," *Zeitschrift für Pflanzenernährung, Düngung, Bodenkunde* 6, no. 6 (1927): 11–20, here 14.
38 Stremme, *Grundzüge der praktischen Bodenkunde*, Foreword with no page number.
39 Ibid.
40 Stremme, "Die bodenkundliche Kartierung von Feldversuchen," 14.
41 In the mid-1930s Stremme was also a member of the Deutsche Gesellschaft für Bodenmechanik. Stremme's membership card, Deutsche Forschungsgesellschaft für Bodenmechanik, 988 III 490 68, Apg.
42 Mechtild Rössler, "Geography and Area Planning under National Socialism," in *Science in the Third Reich*, ed. Margit Szöllösi-Janze (Oxford: Berg, 2001), 59–78, in particular pages 67–69. On the concept of "settlement" in the sense that it formed the basis of set-

Stremme was thus moving in a field of research that had developed rapidly in Germany since the turn of the century in a close interrelation with politics. During the transition from the Weimar Republic to the Nazi state, such research work became increasingly directed from the center, racially radicalized, and oriented towards the goal of state expansion. These processes were what gave settlement research and spatial planning its imperialist dimension. Researchers working in the sector were to contribute significantly to the development and implementation of the visions of the time of acquiring "living space in the East."[43]

At the beginning of the 1930s, Stremme worked closely with the German Research Institute for Agriculture and Settlements (Deutsches Forschungsinstitut für Agrar- und Siedlungswesen), based in the Dahlem district of Berlin. The director of this institute was influential agricultural economist Max Sering (1857–1939), who pursued an agenda of "interior colonization." According to Sering, a healthier agricultural structure would emerge from a reduction in large land holdings and the establishment of family farms. Sering hoped that a settlement policy along these lines would help counteract the exodus from the land, especially in the eastern regions of Prussia.[44]

Sering turned to Stremme in the early years of the 1930s because he hoped that the latter's soil maps would provide insights into the natural assets upon which the domestic colonization in the German East could be based.[45] In 1932 Stremme was commissioned by Sering's institute to produce a "soil overview map suitable for settlement planning" covering extensive areas of Eastern Pomerania.[46] Stremme stressed that his cooperation with Sering had

tlement research, see Andreas Dornheim, "Bodenreform und Siedlung: Gemeinsamkeiten und Unterschiede, Kontinuitäten und Brüche," *Zeitschrift für Agrargeschichte und Agrarsoziologie* 51, no. 2 (2003): 79–84; Götz Aly and Susanne Heim, *Vordenker der Vernichtung: Auschwitz und die deutschen Pläne für eine neue europäische Ordnung* (Frankfurt am Main: Fischer, 2013), in particular pages 59–110 and 368–413.

[43] Wolfram Pyta, "'Menschenökonomie': Das Ineinandergreifen von ländlicher Sozialraumgestaltung und rassenbiologischer Bevölkerungspolitik im NS-Staat," *Historische Zeitschrift* 273, no. 1 (2001): 31–94; Uwe Mai, *"Rasse und Raum:" Agrarpolitik, Sozial- und Raumplanung im NS-Staat* (Munich: Schöningh, 2002).

[44] On Sering, cf. Irene Stoehr, "Von Max Sering zu Konrad Meyer: Ein 'machtergreifender' Generationenwechsel in der Agrar-und Siedlungswissenschaft," in *Autarkie und Ostexpansion: Pflanzenzucht und Agrarforschung im Nationalsozialismus*, ed. Susanne Heim (Göttingen: Wallstein-Verlag, 2002), 57–90; Robert L. Nelson, "From Manitoba to the Memel: Max Sering, Inner Colonization and the German East," *Social History* 35, no. 4 (2010): 439–457.

[45] Hermann Stremme, "Die bodenkundliche Mitarbeit an den Siedlungsplänen," *Forschungen und Fortschritte* 10, no. 15 (1934): 196–198, here 197.

[46] Ibid.; Hermann Stremme, "Die bodenkundliche Siedlungskartierung: Erläutert an der Bodenkarte des Kreises Marienburg," *Planungswissenschaftliche Arbeitsgemeinschaft. Amt des Siedlungsbeauftragten der NSDAP* 3 (1934): 18–26, here 26.

resulted in an intensification of his involvement in settlement sciences, to the extent that it had entered, as he put it, a "new phase."[47]

Stremme openly advocated Russian ideas on soil science in debates among German specialists on spatial and settlement planning. In a 1934 text, he stressed that soil maps had been used for far longer and far more frequently in settlement planning in the Soviet Union, and before that in the Tsarist Empire, than in Germany. Stremme called for the Russian practice to serve as a guide for Germany.[48]

From 1934 on, Sering began to be displaced by a younger generation of National Socialist agricultural economists. They increasingly focused their settlement research on questions of expansion and race policy.[49] But Stremme remained active in the field even to the extent that his activities actually intensified under National Socialism. In July 1934 Stremme was appointed as a consultant on soil mapping on the "Staff of the Deputy Führer."[50] There he worked under Johann Wilhelm Ludowici, "one of the most influential and powerful people within the Nazi regime on issues of settlement policy."[51] Stremme himself joined the SS in 1934 as a *förderndes Mitglied* (an honorary member who made regular contributions to the organization).[52]

In an article in the *Kölnische Volkszeitung* dated May 1937, Stremme asked: "The German soil—how many farms can it support?" The article contained copious calculations to prove that the peasants of the German Reich were suffering the "bitterest land shortage." At the same time, Stremme emphasized the fact that his remarks did not constitute a "settlement plan" but were instead "intended to serve the agricultural policy of the Führer as an objective scientific basis for its calculations."[53] In the same year, Stremme and his research assistant Eberhard Ostendorff published an extensive study on the "German Reich's capacity for rural settlement." In it, against the background of the "self-evident requirement of the time" that a country should not be dependent on grain imports, Stremme advocated a "soil-based approach" to settlement issues.[54] Using mathematical techniques, he sketched out a relationship between soil quality, human calorific requirements, and population density.[55]

---

47 Stremme, "Die bodenkundliche Mitarbeit an Siedlungsplänen," 197.
48 Stremme, "Die bodenkundliche Siedlungskartierung," 21.
49 Stoehr, "Von Max Sering zu Konrad Meyer."
50 Ludowici's letter to Forster, July 7, 1934, Deutsche Forschungsgesellschaft für Bodenmechanik, 988 III 490 57, Apg.
51 Mai, *"Rasse und Raum,"* 106.
52 NSDAP, Party file on Hermann Stremme, summary index card on the person.
53 All quotes taken from Hermann Stremme, "Der deutsche Boden: Wieviel Bauernhöfe kann er tragen?," *Kölnische Volkszeitung*, 4 May 1937, 7.
54 Hermann Stremme and Eberhard Ostendorff, "Die bäuerliche Siedlungskapazität des Deutschen Reiches," *A. Petermann's Mitteilungen aus Justus Perthes' Geographischer Anstalt: Ergänzungsheft*, no. 228 (1937): 7–37, here 13–14.
55 Ibid.

Stremme increasingly positioned his work in the context of German war plans for Eastern Europe, which were becoming more concrete in the second half of the 1930s. German war planners reckoned with Soviet grain to feed their soldiers. In the postwar Europe to be built after the German victory, the agricultural resources of Eastern Europe were expected to make a substantial contribution to feeding the Aryan *Siegervolk*.[56]

But it was not just Soviet agricultural production that the occupiers hoped to acquire: the same applied to the Soviet Union's agricultural knowledge. After the invasion of the USSR, a large number of German agricultural scientists arrived in the wake of the troops to take control of Soviet agricultural institutes.[57] The scientific knowledge of the enemy would be considered as much part of the war booty of the occupying forces as the country's grain.

And German soil scientists took part in this war for knowledge as well. This observation applies to Fritz Giesecke (1896–1958), for example, who had been a professor at the Agricultural University of Berlin (Landwirtschaftliche Hochschule Berlin) since 1934 and was one of those German agricultural experts for whom "a tear-resistant combination of technical know-how and accumulated commitment to the political right guaranteed flawless—i.e. in this case regime-compliant—performance of duties."[58] At a meeting of scientists from Germany and its allies held in Dresden at the beginning of the war, Giesecke proposed the creation of a "European Working Group for Soil Science." Its purpose would be to lay the foundation for a homogeneous soil map for the whole of "Greater Europe," which would be under German domination when the war was over. As Giesecke explained to representatives of the Reich Ministry of Science, Education and Culture (Reichsministerium für Erziehung, Wissenschaft und Volksbildung) in December 1942, he looked upon this work as being relevant to the National Socialists' plans for the resettlement of Eastern Europe:

> This [i.e., a European Working Group for Soil Science, J. A.] would be of great importance for the German side on the question of resettlement—without allowing this to become known in any way, of course. [...] In my opinion we should now take advantage of the situa-

---

[56] Aly and Heim, *Vordenker der Vernichtung*, 341–351.

[57] Olga Elina, Susanne Heim, and Nils Roll-Hansen, "Plant Breeding on the Front: Imperialism, War, and Exploitation," in *Politics and Science in Wartime: Comparative International Perspectives on the Kaiser Wilhelm Institute*, ed. Carola Sachse and Mark Walker (Chicago: University of Chicago Press, 2005), 161–179, in particular pages 166–167, 177.

[58] Willi Oberkrome, "Agrarische Selbstversorgung und bäuerliche Ordnung: Die deutsche landwirtschaftliche Forschung, 1920-1960," in *Die Deutsche Forschungsgemeinschaft 1920-1970: Forschungsförderung im Spannungsfeld von Wissenschaft und Politik*, ed. Karin Orth and Willi Oberkrome (Stuttgart: Franz Steiner Verlag, 2010), 425–432, here 428.

tion: that is, we should also allow the soil scientists of other countries to work for us, in such a way as corresponds to our goals.[59]

And indeed this cynical form of international scientific "cooperation" does seem to have characterized actual German practice in the first years of the war. For example, scientific relations once more developed between Germany and the Soviet Union in the short period between the conclusion of the Molotov–Ribbentrop Pact and the German attack on the Soviet Union. Konrad Meyer (1901–1973), of all people, was responsible for contacts with Soviet agricultural scientists. He was the most powerful representative of agricultural science in Nazi Germany in matters of science policy whose research focused on the creation of *Lebensraum im Osten* for the German people. As Johannes Dafinger writes, it is obvious at least "that Meyer's 'cooperation' with Soviet science was intended to serve geopolitical purposes—after all, Meyer managed to incorporate the knowledge he had gained through this cooperation into the Nazis' *Generalplan Ost* (General Plan East)."[60]

Against this background of German interest in knowledge of agricultural conditions in Eastern Europe, Stremme became a much sought-after expert, one who was determined to prove himself through his service in establishing National Socialist rule in the East. At the beginning of 1943, Stremme and Ostendorff were commissioned by the *Abteilung Planung Ost* under the *Reichsministerium Speer,* the body responsible for wartime economic affairs, to deal with "all questions of soil mapping and soil research that may arise in the technical Planung Ost."[61] Konrad Meyer came to the conclusion that "soil science, alongside economics and, of course, folk and racial biology, provided the inspiration for Ostplanung."[62]

For war planners like Meyer, Stremme might have provided an appropriate substitute for "cooperation" with Soviet experts, to whom he could hardly be expected to give his complete trust. In contrast to them, Stremme was a loyal German soil scientist. At the same time, there was hardly another scientist outside the Soviet Union who knew the Russian work on the Eastern European soil resources that they now wanted to take possession of.

---

[59] Internationale Bodenkundliche Gesellschaft, "Fritz Giesecke an Reichsministerium für Wissenschaft, Erziehung und Volksbildung," 10 December 1942, BArch, R 4901 3056.

[60] Johannes Dafinger, *Wissenschaft im außenpolitischen Kalkül des "Dritten Reiches": Deutsch-sowjetische Wissenschaftsbeziehungen vor und nach Abschluss des Hitler-Stalin-Paktes* (Berlin: Neofelis-Verlag, 2014).

[61] Reichsministerium Speer to the Gauhauptmann of Danzig, 21.1.1943, BArch, R3 4123; NSDAP, Party file on Eberhard Ostendorff, BArch, DS A 49 2779.

[62] Konrad Meyer, "Bodenkunde und Bodenpolitik," *Zeitschrift für Pflanzenernährung, Düngung, Bodenkunde* 29, no. 1 (1943): 2–13, here 10.

Conclusion

Knowledge undergoes a transformation when it is transferred from one context to another. Stremme interpreted what he learned from Glinka in his own particular way. Firstly, his interpretation concerned the scope of application of the knowledge he had learnt. He applied Russian approaches to the discipline to the natural spaces of Central Europe, regions that Dokuchaev and his students had not had in mind while doing their research. On the one hand, he noted that Russian methods were capable of being universalized to a certain extent.[63] At the same time, though, he also realized that soil conditions in Central Europe were more complex than those in Russia. While it was possible to identify relatively clearly delineated, horizontally running soil zones in European Russia, conditions in the German-speaking area of Central Europe more closely resembled a patchwork carpet: soil conditions there varied across small distances.

Above all, however, in the process of transfer described here, the knowledge was transformed in terms of its strategic value in a context of imperial expansion. In the communication between Glinka and Stremme, this strategic dimension of knowledge seemingly played no role. The communication between the two on the topic of soils was a conversation between natural scientists. It was only in the context of a National Socialist settlement and spatial research effort geared to the territorial expansion of the state—a purely German enterprise that did not carry on its business in international scientific forums—that Stremme began interpreting knowledge of Russian soil science in relation to its value in the effort to conquer and appropriate living space in Eastern Europe.

Translated from German by Jaime Hyland.

---

[63] Stremme, *Arbeiten des Mineralogisch-Geologischen Instituts*, 6.

David Moon[1]

SCIENTIFIC INNOVATION IN THE RUSSIAN EMPIRE: THE CASE OF GENETIC SOIL SCIENCE

In the summer of 1930, German soil scientist Friedrich Schucht took part in the Second International Congress of Soil Science in Leningrad and Moscow in the Soviet Union. It was followed by a four-week excursion around the Soviet Union's vast territory. In a speech at Kislovodsk in the North Caucasus, Schucht remarked:

> We soil scientists [...] have got to know all the soil zones of Russia from north to south. We are at the foot of a high mountain chain [the Caucasus], and here Russian soil scientists have [...] reported that the soil zones appear fully in a vertical direction. The Russian soil scientists have been the masters [i.e., teachers] of all of us, and they have prompted all soil scientists of the world to undertake this research.[2]

Schucht was acknowledging both the wide variety of soils in different regions of the Soviet Union and also the innovative work of Russian scientists in devising new ways of understanding soils in the context of the environment, including the topography, in which they had formed. The organizing committee had deliberately designed the excursion to demonstrate both of these.[3] Indeed, by 1930, "Russian" soil science was becoming the basis for the discipline around the world.[4]

This Russian scientific innovation, which they named "genetic soil science,"[5] was a result of pioneering work led by Vasilii Dokuchaev (1846–1903) in the 1870s and 1880s. It came about as a result of field work in Russia's fertile black-earth (chernozem) region. Dokuchaev explained how the black earth,

---

[1] The author acknowledges the support of the Collegium Carolinum, the Leverhulme Trust, UK Arts and Humanities Research Council, British Academy, and Santander Universities.

[2] Arkhiv Rossiiskoi akademii nauk (Archive of the Russian Academy of Sciences, hereafter ARAN), f. (fond) 487, opis' (op.) 1, delo (d.) 27, listy (ll.) 71–72; "Excursion," *Proceedings and Papers of the Second International Congress of Soil Science* 7 (1935): 87–146. See also Fritz Giesecke, "Friedrich Wilhelm Schucht," *Bodenkunde und Pflanzenernährung* 21–22, no. 1 (1940): 1–9.

[3] ARAN, f. 487, op. 1, d. 12, ll. 58–68 and ARAN, f. 487, op. 1, d. 14, l. 37.

[4] See Jan Arend, *Russlands Bodenkunde in der Welt: Eine ost-westliche Transfergeschichte, 1880–1945* (Göttingen: Vandenhoeck & Ruprecht, 2017) and the essay by Jan Arend in this volume.

[5] "Genetic"—from "genesis"—refers to the origins of soils, not the science of genetics.

and all soils, had formed in the environment in which they were located. He stated that soils were "the result of the extremely complex interaction of local climate, plant and animal organisms, the composition and structure of parent rocks, the relief of the locality, [and] finally, the age of the land."[6] Genetic soil scientists understood soils as natural and historical bodies in their own right and established soil science as an independent discipline. Before Dokuchaev, and until scientists around the world accepted the new approach, soils were studied either by agronomists, as the layer of earth used to grow crops, or by geologists, who considered them simply degraded rocks.[7] The new conception of soils informed the advice scientists gave to farmers on cultivating them and governments on policies concerning agriculture.[8]

This essay asks why this scientific innovation occurred in the Russian Empire in the late nineteenth century. Several studies have emphasized Dokuchaev's personal role.[9] Catherine Evtuhov argued that his contribution was rooted in the "very specific cultural and historical context" of Russia after the "Great Reforms" of the 1860s, including the abolition of serfdom and establishment of elected local councils (zemstva).[10] Without denying Dokuchaev's importance, this essay develops Evtuhov's point in a wider context.

Specialists on the preconditions for innovations in science and technology in general have highlighted three essential elements: a need for an innovation; competent people with relevant knowledge; and financial support.[11] The fact that the innovation in soil science took place in an empire is also significant. Historians of science have emphasized the importance of imperial powers in providing infrastructure for field science.[12] Drawing on David Livingstone's

---

[6] V. V. Dokuchaev, "Russkii chernozem," in *Izbrannye sochineniia*, 3 vols., ed. V. V. Dokuchaev (Moscow: Gos. izd-vo sel'khoz. lit., 1948–1949), vol. 1, 27 (first published in 1883).

[7] I. A. Krupenikov, *Istoriia pochvovedeniia: Ot vremeni ego zarozhdeniia do nashikh dnei* (Moscow: Nauka, 1981), 150–192.

[8] See David Moon, *The Plough that Broke the Steppes: Agriculture and Environment on Russia's Grasslands, 1700–1914* (Oxford: Oxford University Press, 2013), 167–172.

[9] See, for example, Krupenikov, *Istoriia pochvovedeniia*, 151–171; I. V. Ivanov, *Istoriia otechestvennogo pochvovedeniia*, 2 vols. (Moscow: Nauka, 2003), vol. 1, 46–79; G. V. Dobrovol'skii, *Lektsii po istorii pochvovedeniia* (Moscow: Izd-vo MGU, 2010), 78–97.

[10] Catherine Evtuhov, "The Roots of Dokuchaev's Scientific Contributions: Cadastral Soil Mapping and Agro-Environmental Issues," in *Footprints in the Soil: People and Ideas in Soil History*, ed. Benno P. Warkentin (Amsterdam: Elsevier, 2006), 125–148 (quotation on 125).

[11] See, for example, Joseph F. Engelberger, *Robotics in Practice: Management and Applications of Industrial Robots* (London: Kogan Page, 1980), 119; Charles Kalmanek, "The essential elements of successful innovation," *ACM SIGCOMM Computer Communication Review* 42, no. 2 (2012): 105–109; John H. Marburger III, "Science, technology and innovation in a 21st century context," *Policy Sciences* 44, no. 3 (2011): 209–213.

[12] Henrika Kuklick and Robert E. Kohler, "Introduction: Science in the Field," *Osiris* 11 (1996): 7–10.

call to "put science in its place,"¹³ the present essay adds another element: the geography of the Russian Empire. The vast territory provided a variety of natural environments and soils for the Russian scientists to study. In an important recent book, Jeremy Vetter has argued for the importance of fieldwork by scientists in the production of new knowledge about the "natural world." His book was about the American West between 1860 and 1910, but his argument has wider relevance to other places and other times.¹⁴ In locating the scientific innovation of genetic soil science in its geographical context, moreover, this study is a response to Kelly O'Neill's challenge to historians of Russia to think about "space,"¹⁵ and draws on insights from spatial history to understand the "space" of the Russian Empire as a source of creativity, as opposed to older assertions of "environmental determinism."¹⁶

Thus, this study makes a connection between "science" and "empire," and places them at the heart of the Russian innovation of genetic soil science. It analyzes in turn: the needs of the imperial Russian government for a new way of understanding soils; the institutional support, including finance, it provided for science and education; and the geography of the empire. In the final section attention turns to the Russian soil scientists. As a sub-text, this essay also asks why this innovation did not occur in and spread from another country.

The Needs of the Imperial Russian Government

The needs of the Russian government were deeply rooted in the soil. Farming provided a way of life for most of its enormous and growing population, whose livelihood and that of the state it supported through taxation, thus depended on its success.¹⁷ Exporting agricultural produce, especially grain grown in the fertile black earth, was increasingly important as a source of revenue.¹⁸ Yet,

---

13   David N. Livingstone, *Putting Science in Its Place: Geographies of Scientific Knowledge* (Chicago: University of Chicago Press, 2003). Two British geographers have also drawn on Livingstone's insight in a Russian context. Jonathan Oldfield and Denis J. B. Shaw, *The Development of Russian Environmental Thought: Scientific and Geographical Perspectives on the Natural Environment* (London: Routledge, 2016), 4–5, 48–77.

14   Jeremy Vetter, *Field Life: Science in the American West during the Railroad Era* (Pittsburgh: University of Pittsburgh Press, 2016)

15   She made this challenge at her seminar "Wrangling Space into Russian Imperial History," The Radcliffe Institute for Advanced Study, Harvard University, 24–25 February 2012.

16   Mark Bassin, Christopher Ely, and Melissa K. Stockdale, "Introduction: Russian Space," in *Space, Place, and Power in Modern Russia: Essays in the New Spatial History*, ed. Mark Bassin, Christopher Ely, and Melissa K. Stockdale (DeKalb: Northern Illinois University Press, 2010), 6–7.

17   See David Moon, "Estimating the Peasant Population of Late Imperial Russia from the 1897 Census," *Europe-Asia Studies* 48, no. 1 (1996): 141–153.

18   See Malcolm E. Falkus, "Russia and the International Wheat Trade, 1861–1914," *Economica* 33 (1966): 416–429.

Russian agriculture continued to be hit by periodic bad harvests, often caused by droughts that led to short-term and regional crises, which were particularly acute in the central black-earth region.[19] In the aftermath of the abolition of serfdom of 1861, the government's concern for the state of agriculture led it to set up a commission headed by Peter Valuev, the minister of state domains, in 1872. It reported that Russian agricultural productivity was low in comparison with other European countries and the United States, in spite of the fertility of the black earth, which Russia had in large expanses. The commission attributed this to, amongst other factors, outmoded, extensive agricultural methods and the increasing and uneven burden of taxes and other obligations levied on the population. Its recommendations included using more intensive methods of cultivation that took account of the need to conserve and augment soil fertility, expanding agricultural education, and reforming the system of taxation to take into account the resources, for example soils, available to taxpayers.[20] The commission's recommendations were not fully implemented, but the report is evidence for the government's sense of its needs.

The government's needs with regard to the soil were shaped by another reform of the 1860s: the creation of elected provincial and district councils (zemstva). They raised revenues by taxing the population and were responsible, amongst other matters, for schools and economic development. Given the importance of agriculture, *zemstva* saw the need to assess the productive capacity of the land as a basis for levying taxes. One of "the most dynamic" of the new *zemstva*, the importance of which we will return to later, was that of Nizhnii Novgorod province, east of Moscow, which straddled the contrasting forest and steppe regions with their differing soils.[21]

Thus, a new way of understanding soils was important in the Russian Empire in the late nineteenth century, not just for purely scientific reasons, but because it would meet the needs of the imperial government in the centre and authorities in the provinces. But, Russia was not unique in this period in having a largely rural population and depending heavily on agriculture. This was true of most countries outside northwestern Europe. In the United States in 1870, for example, 70 percent of the population lived in rural areas and agricultural produce comprised 55 percent of GNP.[22]

---

[19] Stephen G. Wheatcroft, "Crises and the Condition of the Peasantry in Late Imperial Russia," in *Peasant Economy, Culture, and Politics of European Russia, 1800–1921*, ed. Esther Kingston-Mann and Timothy Mixter (Princeton: Princeton University Press, 1991), 128–172.

[20] *Doklad vysochaishe uchrezhdennoi komissii dlia issledovaniia nyneshnego polozheniia sel'skogo khoziaistva i sel'skoi proizvoditel'nosti v Rossii* (St. Petersburg, 1873).

[21] Evtuhov, "The Roots," 132–138; Evtuhov, *Portrait of a Russian Province: Economy, Society, and Civilization in Nineteenth-Century Nizhnii Novgorod* (Pittsburgh, PA: University of Pittsburgh Press, 2011), 13 and 24.

[22] Ronald Seavoy, *An Economic History of the United States from 1607 to the Present* (Abingdon: Routledge, 2007), 221.

Institutional Support for Science and Education in the Russian Empire

The Russian government recognized that its needs would be served by promoting science. Over the eighteenth and nineteenth centuries, it provided increasing institutional and financial support for research. It had an especial interest in agricultural sciences.[23] Foundations had been laid by Peter the Great (r. 1682–1725), who established the Russian Academy of Sciences in 1724.[24] The academy took a keen interest in collecting data and locating resources, including fertile soil, for exploitation. Over the eighteenth century, it sponsored lengthy expeditions in which teams of naturalists, mostly German or German-trained, explored large parts of the vast empire.[25] The infrastructure to support scientific research was developed over the eighteenth and nineteenth centuries. Catherine the Great (r. 1762–1796) established the Free Economic Society for the Encouragement of Agriculture and Husbandry in 1765. It produced a journal to discuss the latest ideas on farming and published information on agriculture and natural resources. In the nineteenth century, the society sponsored scientific research.[26] Another institution that sponsored scientific research, especially expeditions to outlying parts of the empire, was the Russian Geographical Society, founded in 1845. Its interests included cartography, climatology, botany, zoology, and geology.[27] Over the nineteenth century, agricultural societies were established with official support in many provinces and regions to promote agriculture. Some carried out research on experimental farms.[28]

State support for agriculture was enhanced in 1837 with the creation of the Ministry of State Domains to administer its extensive land holdings and peasants. The ministry was set up in part in response to a drought, crop failure, and famine in the early 1830s as well as with an aim to better manage the state's resources. It sponsored scientific research, a growing network of exper-

---

[23] See Ol'ga Elina, *Ot tsarskikh sadov do sovetskikh polei: Istoriia sel'sko-khoziaistvennykh opytnykh uchrezhdenii XVIII-20-e gody XX veka*, 2 vols. (Moscow: Rossiiskaia Akademiia Nauk, 2008).

[24] See Alexander Vucinich, *Science in Russian Culture: A History to 1860* (London: Peter Owen, 1965), 38–74.

[25] V. F. Gnucheva, "Geograficheskii departament akademii nauk XVIII veka," *Trudy Arkhiva Akademii Nauk SSSR* 6 (1946): 1–445; David Moon, "The Russian Academy of Sciences Expeditions to the Steppes in the late-Eighteenth Century," *Slavonic and East European Review* 88 (2010): 204–236.

[26] See A. I. Khodnev, *Istoriia Imperatorskogo Vol'nogo Ekonomicheskogo Obshchestva s 1765 do 1865 goda* (St. Petersburg: Obshchestvennaia pol'za, 1865); Colum Leckey, *Patrons of Enlightenment: The Free Economic Society in Eighteenth-Century Russia* (Newark: University of Delaware Press, 2011).

[27] See P. P. Semenov-Tian-Shanskii, *Istoriia poluvekovoi deiatel'nosti Imperatorskogo Russkogo obshchestva, 1845–1895*, 3 vols. (St. Petersburg, 1896).

[28] See S. A. Kozlov, *Agrarnye traditsii i novatsii v doreformennoi Rossii* (Moscow: Rosspen, 2002); S. A. Kozlov, *Agrarnaia ratsionalizatsiia v Tsentral'no-Nechernozemnoi Rossii v poreformennyi period* (Moscow: Rossiiskaia Akademiia Nauk, 2008).

imental farms and forestry plantations, and educational establishments. The reforms of the 1860s, in particular the abolition of serfdom, increased the ministry's role in promoting agriculture, agricultural sciences, and education. The ministry published a journal with articles on scientific agriculture, the state of farming, and the natural environment, including the soil, in different regions of the empire.[29] The establishment of *zemstva* in the 1860s led to more support for agricultural sciences and education in the provinces. More enterprising *zemstva* funded research into the environment and resources, including soils.[30]

Parallel to the development of state support for science, and essential to its success, was the creation of a system of secular education. The Russian government came to realize that its need for trained personnel, including scientists, outweighed concerns that educating people would be detrimental to social stability in a hierarchical society. Again, Peter the Great laid the basis.[31] In fits and starts over the eighteenth and nineteenth centuries, universities were established in Moscow, St. Petersburg, and some major provincial centres, which took students from a growing network of secondary and primary schools.[32] The state educational system supplemented that of the Russian Orthodox Church. In the nineteenth century the church schools' curriculum was reformed to include more secular subjects.[33]

Thus, by the second half of the nineteenth century, the educational systems of the Russian Empire had trained competent people with relevant knowledge who could engage in scientific exploration and research at high levels. The government had also created institutions, including universities, that supported and carried out scientific work on Russia's natural environment and resources. The Russian scientists who devised the new soil science in the late nineteenth century, therefore, were an outcome of almost two centuries of official support for and investment in science and education. Loren Graham concluded: "By 1900 Russia had produced a number of scientists known throughout the international science community." He cited soil science as an example.[34]

---

[29] *Istoricheskoe obozrenie piatidesiatiletnei deiatel'nosti ministerstva gosudarstvennykh imushchestv, 1837–1887*, 5 vols. (St. Petersburg, 1888); Elina, *Ot tsarskikh sadov do sovetskikh polei*, 26–56.

[30] See G. P. Sazonov, *Obzor deiatel'nosti zemstv po sel'skomu khoziaistvu, 1865–1895 gg.*, 3 vols. (St. Petersburg, 1896).

[31] See, for example, Max Okenfuss, "Technical Training in Russia under Peter the Great," *History of Education Quarterly* 13 (1973): 325–345.

[32] On universities, see the essay by Andrej Andreev in this volume.

[33] See Gregory L. Freeze, *The Parish Clergy in Nineteenth-Century Russia* (Princeton, NJ: Princeton University Press, 1983), 102–143, 319–329, 354–363, 433–440.

[34] Loren R. Graham, *Science in Russia and the Soviet Union* (Cambridge: Cambridge University Press, 1993), 53 and 230.

Nevertheless, Russia was a relative newcomer to the "international scientific community." Other European countries had older and more extensive systems of education and research. Across the Atlantic, by the late nineteenth century, the US federal government had established an extensive system of "land-grant" colleges to provide higher education in agriculture, sciences, and engineering as well as the liberal arts. They were supplemented by state agricultural and mechanical colleges.[35] The US Department of Agriculture and individual states, moreover, set up a network of agricultural experiment stations that was the envy of Russian scientists, including Dokuchaev and his colleagues.[36]

The Geography of the Russian Empire

The vast Russian Empire, which encompassed much of Eastern Europe and northern and central Asia, offered a rich diversity of environments for its scientists to study. The range of natural regions and soil types and, as will be explained, their precise configuration in space, proved crucial to the Russian innovation in soil science. Russian efforts to describe and map their environment can be traced back in time. From the late fifteenth to the seventeenth centuries, the grand dukes and tsars commissioned cadastres (pistsovye knigi) of the Russian land to facilitate levying obligations.[37] In 1627 Tsar Michael Romanov ordered a description of his lands to assist in organizing their defence.[38] Over the seventeenth century, Russian explorers charted Siberia and other, mostly outlying, regions.[39] However, as Denis Shaw has pointed out, the origins of the "Russian geographical tradition" can be traced back to the initiative of Peter the Great. The scientific and governmental institutions he and his successors established engaged in detailed study of the empire's geography, natural resources, and the economic activities of its population.[40] Catherine the Great commissioned a "General Land Survey" in 1765, which

---

[35] Mary Jean Bowman, "The Land-Grant Colleges and Universities in Human-Resource Development," *Journal of Economic History* 22 (1962): 523–546.

[36] See Sankt-Peterburgskii filial Arkhiva Rossiiskoi akademii nauk (hereafter PF ARAN), f. 184, dd. 137, 138, 144.

[37] See A. A. Frolov, "Struktura pistsovykh knig derevenskoi piatiny 40-kh godov XVI v. po dannym kompleksa istochnikov kontsa XV-serediny XVI v.," *Drevniaia Rus': Voprosy medievistiki* 4 (2007): 69–79.

[38] K. N. Serbina, ed., *Kniga Bol'shomu chertezhu* (Moscow: AN SSSR, 1950).

[39] See V. A. Esakov, *Ocherki istorii geografii v Rossii XVIII-nachalo XX veka* (Moscow: Editorial URSS, 1999), 14–15.

[40] Denis J. B. Shaw, "Geographical Practice and Its Significance in Peter the Great's Russia," *Journal of Historical Geography* 22 (1996): 160–176; Oldfield and Shaw, *Development*, 19–47.

gathered important data.⁴¹ The education of future tsar Alexander II (r. 1855–1881) included a tour of Russia in company of the geographer, historian, and statistician Konstantin Arsen'ev in 1837. A decade later, Arsen'ev produced a compendium of knowledge of the empire's geography that surveyed the climate, topography, soils, lakes and rivers, forests, forms of land use, and economic activities, paying attention to regional differences.⁴² Detailed geographical information was presented in two series of provincial studies published by the imperial general staff between 1848 and 1868.⁴³ The Russian Geographical Society published a five-volume *Geographical and Statistical Dictionary of the Russian Empire*, compiled by explorer and geographer Petr Semenov, between 1863 and 1885.⁴⁴ Thus, by the late nineteenth century, Russian geographers and scientists were well aware that the Russian Empire contained a range of environmental regions, including tundra, forests, steppe grasslands, desert, and mountains. These regions had contrasting climates, flora and fauna, geologies, topographies, and soils.

The size and diversity of the Russian Empire contrasted with its European neighbours to the west. Those such as Britain and the Netherlands, which had extensive empires, ruled lands with different natural conditions, but they were separated by thousands of kilometres of ocean. Beyond Europe, another large state that, like the Russian Empire, had diverse and contiguous natural regions was the United States. The United States also needed an understanding of soils to support its agricultural economy, and the federal government and states provided significant support for education and science. The preconditions for an innovation in soil science outlined in this essay's introduction thus existed in two countries in the late nineteenth century: the Russian Empire and the United States. But, it was in the first that a new soil science was devised and went on to form the basis for a new discipline around the globe.

The Russian Innovation of Genetic Soil Science

Scientists in the Russian Empire paid growing attention to the soil. The Ministry of State Domains funded scientists to study the empire's soils and climate, and published the first detailed maps of these in 1851. Scientific interest

---

⁴¹ L. V. Milov, *Issledovanie ob "Ekonomicheskikh primechaniiakh" k general'nomu mezhevaniiu* (Moscow: izd-vo Moskovskogo universiteta, 1965).
⁴² Konstantin Arsen'ev, *Statisticheskie ocherki Rossii* (St. Petersburg, 1848). On the author, see "Arsen'ev, Konstantin Ivanovich," in *Entsiklopedicheskii slovar'* (hereafter ES), vol. 2 (St. Petersburg, 1890), 174–175.
⁴³ *Voenno-statisticheskoe obozrenie Rossiiskoi imperii*, 17 vols. (St. Petersburg, 1848–1858); *Materialy dlia geografii i statistiki Rossii, sobrannye ofitserami General'nogo Shtaba*, 39 vols. (St. Petersburg, 1859–1868).
⁴⁴ P. P. Semenov, *Geografichesko-statisticheskii slovar' Rossiiskoi imperii*, 5 vols. (St. Petersburg, 1863–1885).

in the soil had developed over the eighteenth and nineteenth centuries. Attention focussed on the black earth of the steppe grasslands, because of its fertility in contrast with the soils of the forest heartland around Moscow. Scientists speculated whether the black earth owed its origins and productive capacity to a former sea or a past forest that may once have covered the region. Or, and this was long a minority opinion, perhaps the wild grasses that covered the steppe before it was ploughed up were the key. In the 1860s microscopic analysis by Franz Joseph Ruprecht, the director of the Academy of Sciences' botanical museum in St. Petersburg, showed that the organic matter in the black earth comprised decomposed grasses.[45]

Developments in the Russian Empire in the wake of the reforms of the 1860s prompted closer attention to the soil. Droughts in the black-earth region in 1873 and 1875 undermined hopes for more successful farming in a countryside now rid of serfdom. The droughts alarmed the Free Economic Society into funding a scientific expedition to study the region's soil. In the 1880s, *zemstva*, including Nizhnii Novgorod, began to fund teams of scientists to evaluate the land for purposes of tax assessment.[46] There were competent people with appropriate training to take on these assignments as a result of official support for science and education over the preceding almost two centuries. The scientists had been educated at Russian universities and had expertise in various disciplines, including climatology, botany, zoology, geology, and geography. They had the funding and resources to carry out fundamental research. Travelling around the empire for fieldwork was facilitated by recent developments in transport, such as railways and river steamers, which were also a result of government investment. The late nineteenth-century scientists benefitted also from an intellectual milieu which encouraged specialists in various disciplines to exchange ideas in academic societies. They were familiar with the work of Alexander von Humboldt, who had emphasized the interconnectedness of natural phenomena, indicating the importance of studying what we would now term "the environment" in its entirety, rather than component parts in isolation. The Russian scientists sought to work for the benefit of society, addressing burning questions of the day, as well as advancing science.[47]

---

[45] Anastasia A. Fedotova, "The Origins of the Russian Chernozem Soil (Black Earth): Franz Joseph Ruprecht's 'Geo-Botanical Researches into the Chernozem' of 1866," *Environment and History* 16 (2010): 271–294; M. V. Loskutova and A. A. Fedotova, *Stanovlenie prikladnykh biologicheskikh issledovanii v Rossii: Vzaimodeistvie nauki i praktiki v XIX – nachale XX vv.* (St. Petersburg: Nestor-Istoriia, 2014), 40–78.

[46] Krupenikov, *Istoriia pochvovedeniia*, 155, 162, 163; Evtuhov, "The Roots," 134–135.

[47] See Evtuhov, "The Roots," 131; Oldfield and Shaw, *Development*, 37–40, 49–55, 58–62; Elizabeth A. Hachten, "In Service to Science and Society: Scientists and the Public in Late-Nineteenth-Century Russia," *Osiris* 17 (2002): 171–209; Loskutova and Fedotova, *Stanovlenie*, 40–78. Humboldt's work was published in Russian: A. fon Gumbol'dt, *Kosmos: Opyt fizicheskogo miroopisaniia*, 3 vols. (Moscow, 1862–1863).

From the 1870s, groups of scientists engaged in soil research in different parts of the Russian Empire. During the expeditions in the black-earth region funded by the Free Economic Society and in Nizhnii Novgorod province, supported by the *zemstvo*, they collected samples of soils in cross-sections to a depth of over a metre. The scientists came to call the cross sections "profiles," and divided them into three layers or "horizons:" the upper or A horizon was the top soil; the middle or B horizon was the subsoil or transition layer; and the bottom or C horizon—the parent rock. The scientists subjected the profiles, and each horizon, to detailed analysis of their physical properties, chemical composition, organic matter, and moisture content. The scientists also gathered data on the local environments where they collected the samples: climatologists analysed the climate; botanists—the flora; zoologists—the fauna; geologists—the underlying rocks; geographers—the topography; and they pooled expertise to calculate the age of the soil. They came to realize—and this was crucial to developing the new theory—that similarities and differences in soils collected in different localities depended not on one factor, such as the geology, but, and in keeping with Humboldt's holistic approach, on the combined influences of all the component parts of the environment. In line with Darwinist ideas, moreover, they understood that soils evolved over many centuries. Thus, the parent rock in the C horizon was broken down under the combined influences of the climate (precipitation and temperature), the organic matter (plants and animals) on top of it, the topography (level or hilly), over time. Analysing the soil profiles from the bottom to the top, from the C to the A horizons, moreover, allowed them to see the process of soil formation in action.[48]

While Russian soil scientists were most concerned with the fertile black earth and were funded to undertake further applied work to address agricultural problems in the region,[49] they were also interested in other types of soil elsewhere in the empire. They carried out fieldwork, for example, along the border between forest and steppe, and in the Central Asian deserts and the Caucasus Mountains. Their research, informed by the new genetic soil science, showed that different types of soils formed in different conditions. The black earth had formed in the semi-arid climate of the grassland plains. Further south and east, in regions with more arid climates and sparser vegetation, the scientists classified the less fertile and lighter-coloured earth as chestnut (kashtanovye) soils. In the heartland, different soils, classified as

---

[48] Dokuchaev, "Russkii chernozem;" Dokuchaev, "K voprosu o pereotsenke zemel' evropeiskoi i aziatskoi Rossii s klassifikatsii pochv," in *Sochineniia*, 9 vols. (Moscow: Izd-vo Akademii Nauk SSSR, 1949–1961), vol. 6, 256–343 (first published 1898). See also Moon, *The Plough*, 75–86.

[49] David Moon, "The Environmental History of the Russian Steppes: Vasilii Dokuchaev and the Harvest Failure of 1891," *Transactions of the Royal Historical Society* 15 (2005): 149–174.

podzoly, had formed in the more humid climate under the forest. In the Caucasus Mountains—as Schucht noted in 1930—the Russian scientists observed that soils changed in a vertical direction, from the foothills to the peaks, which they considered analogous to the changes in soils from north to south on the Russian plain. By the 1890s Russian soil scientists had produced an entire system of classification of the different soils throughout large parts of the empire, from the tundra in the north, through forests and steppes, to deserts, mountains, and sub-tropical regions in the south.[50]

As has been hinted earlier, the very configuration of the natural regions on the vast Russian plain was fortuitous, creating "an ideal experimental situation" for Russian soil scientists to achieve their innovation. The absence of high mountains means that regions with similar temperature and humidity form belts that run roughly from east to west. Since the types of plants that grow depend to a large extent on climate, the vegetation zones, in particular, forest and grassland, also form belts running from east to west. In North America, in contrast, the presence of the Rocky Mountains extending from north to south in the west of the continent means that while belts with similar temperatures run, as in Russia, from east to west, humidity belts run from north to south in consequence of the mountains' rain shadow, greatly complicating the situation for scientists seeking to relate soils to the environment. "It is not surprising," as American botanist Homer Leroy Shantz noted, "that the Russians first glimpsed the significance of the soil profile." The Russian scientists were further assisted by the large expanses of virgin soils they were able to study. This was in contrast to their counterparts in Western Europe, where most soils had been changed appreciably by many centuries of farming. American soil scientist Curtis F. Marbut commented the Russian workers were able to see, with remarkable clearness, relationships that in western Europe were either obscurely expressed or not expressed at all.[51]

The importance of the geography of the Russian Empire in the origins of the new soil science was explained by Russian-born American forestry scientist Raphael Zon in 1930. In trying to pursuade the director of research of the

---

[50] See PF ARAN, f. 184, op. 1, d. 112, ll. 1–21; V. V. Dokuchaev, "K uchenie o zonakh prirody: Gorizontal'nye i vertikal'nye pochvennye zony," in *Sochineniia*, vol. 6, 398–414 (first published 1899); Dokuchaev, "Noveishaia klassifikatsiia pochv," *Pochvovednie* 2 (1900): appendix; N. M. Sibirtsev, "Ob osnovaniiakh geneticheskoi klassifikatsii pochv," in *Izbrannye sochineniia*, 2 vols. (Moscow: Gos. Izdat-vo Sel-khoz. lit., 1953), vol. 2, 271–293 (first published 1895); P. Ototskii, "Pochvy," *ES*, vol. 54 (St. Petersburg, 1899), 54–57, including N. Sibirtsev, "Skhematicheskaia pochvennaia karta Evropeiskoi Rossii," facing 54. See also Ivanov, *Istoriia*, vol. 1, 46–111.

[51] Homer Leroy Shantz, "A Memoir of Curtis Fletcher Marbut," *Annals of the Association of American Geographers* 26, no. 2 (1936): 113; Curtis F. Marbut, "Introduction," in *Pedology*, ed. Jacob Joffe, 2nd ed. (New Brunswick, NJ: Pedology Publications, 1949), ix; Joffe, ed., *Pedology*, 18–19; N. Sibirtsev, *Chernozem v raznykh stranakh* (Warsaw, 1896), 41–45.

US Forest Service to authorize him to visit the Soviet Union to attend the Second International Congress of Soil Science, he wrote:

> Russia [...] presents something entirely different [from Europe] [...] as far as the vastness of the territory and variety of climate, soils, and forest types are concerned [...] Because of the vastness [...], the Russians succeeded better than anyone else in understanding the genetics of soil complexes, and because they dealt with virgin soils, they, more than anyone else, correlated forests and soils.[52]

Returning to the late nineteenth-century Russian scientists who devised the new way of understanding soils: the scientist who was awarded the funding, put together and led the teams of scientists, and articulated and defended the new theory was Vasilii Dokuchaev. The reason for leaving his role to last is not to undermine his importance, which has been emphasized by other scholars, but to underline the significance of the wider contexts, in particular the preconditions for scientific innovations outlined earlier, in which he was working. Dokuchaev's contribution, however, was singular. His background as the son of a village priest who was educated at church schools before switching to natural sciences at St. Petersburg University was not unusual among his contemporaries. Other aspects of his training and career trajectory did differ from the norm. Most young Russian scientists spent time at a university outside Russia, typically in Germany, but Dokuchaev preferred to carry out fieldwork in Russia. In contrast to most Russian scientists, who read Western European languages and kept up to date with the international scientific literature, Dokuchaev was poor at languages and relied on his colleagues to tell him about the latest work abroad. As his graduate research progressed, he worked independently as his interests transcended conventional disciplinary boundaries. What could have been handicaps for another person seem to have aided Dokuchaev. Unencumbered by received wisdom, his imagination, ability to see connections, and clarity of expression enabled him to devise and promote an innovation in science.[53]

Dokuchaev had his critics. His chief rival was Pavel Kostychev. At Dokuchaev's defence of his doctoral thesis—his landmark monograph on the Rus-

---

[52] Zon to E. H. Clapp, 21 April 1930, Raphael Zon papers, Box 7, Folder 1, Minnesota Historical Society, St Paul. (Zon did not attend the congress as the federal government would not allow its employees in their official capacities to travel to the Soviet Union, which it did not recognize, in their official capacities.)

[53] For recent discussions, see B. F. Aprarin, "Dokuchaev Vasilii Vasil'evich," in *Pochvovedenie v Sankt-Peterburge. XIX–XXI vv.: Biograficheskie ocherki*, ed. N. N. Matinian (St. Petersburg: Serebrianyi vek, 2013), 98–103; E. S. Kul'pin-Gubaidullin, "Vasilii Dokuchaev kak predtecha biosferno-kosmicheskogo istorizma: Sud'ba uchenogo i sud'by Rossii," *Obshchestvennye nauki i sovremennost'* 2 (2010): 103–113; Oldfield and Shaw, *Development*, 48–77. On his limited knowledge of German and French, see, Aleksandr Ivanovich Voeikov to Eugene W. Hilgard, 6/18 January 1892, Hilgard, E. W., Incoming Letters, Box 23, File: Voeikov, Aleksandr Ivanovich, The Hilgard family papers [Hilgard Papers], Bancroft Library, Berkeley, California.

sian black earth—Kostychev argued that he had overstated the role of climate in soil formation and had not gathered sufficient soil samples to support a new theory. Kostychev, whose background was in agronomy, was a distinguished soil scientist in his own right. He argued for the primacy of vegetation in the formation of the black earth and carried out original work on the organic matter (humus) in soil. Kostychev perhaps had a deeper knowledge of the scientific literature and may have been more methodical in his research. He did not, however, propose a rival theory or impede the progress of a new discipline based on Dokuchaev's ideas.[54]

Dokuchaev also—and very importantly—institutionalized the new soil science in Russia and trained a new generation of scientists to continue the research. He persuaded the Free Economic Society to set up a soil commission in 1888. It started publication of the world's first journal in the field, *Pochvovedenie* (Soil Science) in 1899. In 1894 Dokuchaev created the first academic department of genetic soil science at the Novoaleksandriia Institute of Agriculture, of which he was director, in Russian Poland. The teams of scientists Dokuchaev led on his expeditions included younger scholars, for example Nikolai Sibirtsev and Konstantin Glinka, whom he trained in the new approach. They both went on to hold the chair at the Novoaleksandriia Institute. Another of his students, Pavel Ototskii, was the first editor of the journal *Pochvovedenie*. Dokuchaev's students, some of whom idolized their professor, devoted their professional lives to promoting the new discipline and training the subsequent generation of genetic soil scientists.[55]

Emboldened by their new understanding of soils and perhaps also by the convenient correlation of climatic, vegetation, and soil zones in the Russian Empire, the Russian soil scientists laid the foundations for the global application of their new discipline. They extended their system of natural zones and soil types around the northern hemisphere, for example, classifying the black earth of the Great Plains of North America as *chernozem* like that of the steppes.[56] Glinka wrote a book explaining the principles behind genetic soil science in German, published in 1914, to facilitate wider international recep-

---

[54] P. A. Kostychev, "Po voprosu o proiskhozhdenii chernozema," *Sel'skoe khoziaistvo i lesovodstvo* 147 (1884): 259–282; Kostychev, *Pochvy chernozemnoi oblasti Rossii, ikh proiskhozhdenie, sostav i svoistva* (St. Petersburg, 1886). See also Dobrovol'skii, *Lektsii*, 82–88.

[55] See Dobrovol'skii, *Lektsii*, 88–109; Ivanov, *Istoriia*, vol. 1, 112–183; Oldfield and Shaw, *Development*, 52–54.

[56] A. N. Krasnov, "Travianye stepi Severnogo polushariia," *Izvestiia Obshchestva liubitelei estestvoznaniia, antropologii i etnografii* 81, no. 7 (1894): 1–294; Sibirtsev, *Chernozem*; V. V. Dokuchaev, "Klassifikatsiia pochv prof. V. V. Dokuchaev (severnoe polusharie)," in *Sochineniia*, vol. 6, 526–531. See also Oldfield and Shaw, *Development*, 61–64.

tion of the new discipline.⁵⁷ The internationalization of genetic soil science was well under way by the time of the Second International Congress in the Soviet Union in 1930. Nevertheless, several of the delegates commented on witnessing the Russian soil scientists demonstrating their techniques for analysing soil profiles in the field. Towards the end of the excursion, in Khar'kov, Professor Hermann Stremme of the Danzig Free State (today's Gdańsk) commented to his hosts:

> Even those of us who were already familiar with this method, all the same were struck how it enabled you in a short time to determine the most difficult soil profiles and you mobilized all conceivable biological, chemical, and meteorological data in order to make everything comprehensible to us, and it was all done magnificently.⁵⁸

Conclusion

This essay has argued for the importance of the wider contexts in the Russian Empire in the late nineteenth century to explain why the innovation in soil science occurred in this place and at this time. Indeed, the argument consciously mirrors that of the new genetic soil science itself, which emphasized the wider environmental contexts in which the soils had formed. Thus, in the Russian Empire in the late nineteenth century, the preconditions for a scientific innovation in this field were all present: the Russian government needed such an innovation; it had provided significant institutional and financial support for science and education; finally, the geography of the vast Russian Empire itself, with its convenient configuration of natural regions, provided ideal raw material for its scientists while engaging in fieldwork to uncover the connections between climate, organic matter, parent rock, and topography, over time in the formation of soils. The contribution of Dokuchaev and his colleagues was considerable, but the wider contexts in the Russian Empire in this period facilitated their achievement.

Why did the innovation in soil science not take place and become institutionalized elsewhere, for example in the United States, where the preconditions—need, institutional support, and a vast and diverse geography—were all in place? In fact, it almost did. Independently and starting before Dokuchaev, the German-born scientist Eugene W. Hilgard (1833–1916) came up with similar ideas. While the climatic and vegetation zones in the United States were not as conveniently laid out as in the Russian Empire to reveal the connections, it is significant that Hilgard discovered a relationship between climate, specifically precipitation, and soils while working successively in hu-

---

57   Konstantin D. Glinka, *Die Typen der Bodenbildung: Ihre Klassifikation und geographische Verbreitung* (Berlin: Verlag von Gebrüder Borntraeger, 1914). See also the essay by Jan Arend in this volume.
58   ARAN, f. 487, op. 1, d. 27, l. 107.

mid Mississippi and arid California.[59] Hilgard became aware of Dokuchaev's work (but could not read it in the original Russian) only after he developed his own ideas. In 1894 he wrote to French scientist Jean Vilbouchevitch that he found Dokuchaev's field work "as closely parallel to my own as I could wish," but had disagreements over his chemical analysis.[60] It took a scientist of Hilgard's ability to realize the relationship between climate and soils in the United States, but he did not articulate his ideas as clearly in a bold theory as Dokuchaev. Nor did Hilgard's innovative ideas fall on such receptive ground as Dokuchaev's. Instead, he encountered stiff resistance from Milton Whitney, the first chief of the Division of Agricultural Soil at the US Department of Agriculture. Whitney insisted on older, geological understandings of soils, and persisted in this error until his death in 1927.[61] Hilgard expressed his frustration with Whitney's ignorance, for example in a letter to the Russian scientist Aleksandr Voeikov in 1908.[62] Paradoxically, American soil scientists, including Marbut, learned about the new way of understanding soils from their Russian counterparts, rather than Hilgard.[63] On the other hand, several Russian soil scientists have acknowledged the importance of Hilgard's work.[64]

When the First International Congress of Soil Science was held in the United States in 1927, the Soviet soil scientists comprised the largest foreign delegation; their leader, Glinka, was treated as guest of honour; Marbut's translation of his 1914 book in German was displayed in a prominent place; and Glinka was elected president of the society and organizer of the next congress. On the excursion that followed, Marbut showed the delegates *podzols* and *chernozems* in North America.[65] The importance of the Russian black

---

[59] Eugene Woldemar Hilgard, "A Report on the Relations of Soil to Climate," *US Department of Agriculture: Weather Bureau. Bulletin* 3 (1892); Hans Jenny, *E. W. Hilgard and the Birth of Modern Soil Science* (Pisa: Collana Della Rivista "Agrochimica," 1961). Jenny argued that Hilgard should be considered the co-founder of the new soil science.

[60] Hilgard to Vilbouchevitch, 24 May 1894, Hilgard, E. W.: Outgoing Letters, Letterpress copy books, vol. 20, 57–58, Hilgard Papers, Bancroft Library, Berkeley, California.

[61] See Ronald Amundson, "Philosophical Developments in Pedology in the United States: Eugene Hilgard and Milton Whitney," in *Footprints*, ed. Warkentin, 149–165.

[62] Hilgard to Woeikof [sic], 2 Jan 1908, Hilgard, E. W.: Outgoing Letters, Letterpress copy books, vol. 26, 108–110, Hilgard Papers, Bancroft Library, Berkeley, California.

[63] See Curtis Fletcher Marbut, *Soils: Their Genesis and Classification* (n.p.: Soil Science Society of America, 1951), 17–18, 25. For a fuller discussion of the adoption, under Marbut's leadership, of Russian soil science by the US Soil Survey, see David Moon, *The American Steppes: The Unexpected Russian Roots of Great Plains Agriculture, 1870s-1930s* (Cambridge: Cambridge University Press, 2020), 188–276.

[64] For example, N. M. Tulaikov, "K voprosu o vliianii klimata na kharakter pochv: Po povodu rabot prof. E. W. Gilgard'a," *Pochvovedenie* 9, no. 3 (1907): 315–328; Dobrovol'skii, *Lektsii*, 129–130; Krupenikov, *Istoriia pochvovedeniia*, 153, 172, 179. In contrast to Jenny, however, they did not consider Hilgard a co-founder of the new soil science.

[65] See A. A. Iarilov, "Na kongresse i o kongresse," *Biulleten' pochvoveda* 5–8 (1927): 120–148; Curtis F. Marbut, "The Transcontinental Excursion," *Ist International Congress of*

earth (chernozem) and the need for a deeper understanding of soils were acknowledged by Soviet soil scientists when they persuaded the Soviet government to authorize them to accept the invitation to take part in the congress in the United States. Arsenii Iarilov wrote:

> Our "<u>chernozem</u>" is the most formed, developed and most [...] "qualified" representative of an absolutely <u>special</u>, independent formation of nature, which is distinguished from all its other bodies, <u>soil</u>. Therefore, it was precisely in studying the chernozem that it was most easy to create a <u>new</u> natural-historical science – <u>Soil science</u>. Life helped by demanding soil maps for land evaluation.[66]

Soil Science: Proceedings 5 (1927): 40–88. Glinka died later in 1927. Marbut adopted into English the Russian terms "podzol" and "chernozem" for these types of soils.

[66] ARAN, f. 487, op. 1, d. 5, l. 11. Underlinings in the original.

LIST OF ABBREVIATIONS AND ACRONYMS

ARAN	Arkhiv Rossiiskoi Akademii Nauk (Archive of the Russian Academy of Sciences)
ASTRA	Asociația Transilvană Pentru Literatura Română și Cultura Poporului Român (Transylvanian Association for Romanian Literature and the Culture of the Romanian People)
BArch	Bundesarchiv (Federal Archive)
BHStA	Bayerisches Hauptstaatsarchiv (Bavarian State Archives)
BSOI	Bestand Südost-Institut
ČAVU	Česká akademie císaře Františka Josefa pro vědy, slovesnost a umění (Emperor Franz Josef Czech Academy for Sciences, Literature, and Arts)
GBA	Archiv der Geologischen Bundesanstalt (Archive of the Geological Survey)
GIM	Gosudarstvennyi Istoricheskii muzei (State Historical Museum)
KČSN	Královská česká společnost nauk (Royal Bohemian Society of Sciences)
KEPS	Kommissiia po izucheniiu estestvennykh proizvoditel'nykh sil strany (Commission for the Exploration of Russia's Natural Resources)
MGH	Monumenta Germaniae Historica
NEP	Novaia ekonomicheskaia politika (New Economic Policy)
OLEAE	Imperatorskoe obshchestvo liubitelei estestvoznaniia, antropologii i etnografii (Society of Devotees of Natural Science, Anthropology, and Ethnography at the University of Moscow)
PAU	Polska Akademia Umiejętności (Polish Academy of Learning)
PSZ	Polnoe Sobranie Zakonov (Complete Collection of Laws)
RGIA	Rossiiskii gosudarstvennyi istoricheskii arkhiv (Russian State Historical Archive)
Supplex	Supplex Libellus Valachorum
Theresianum	Theresianisch-Leopoldinische Akademie

INDEX OF PERSONAL NAMES

Ablonczy, Balázs 213
Albert, Eduard 38, 41
Áldásy, Antal 92
Alexander I (Emperor of Russia) 56, 102, 105, 107, 117
Alexander II (Emperor of Russia) 45, 49, 135, 138 f., 312
Alexics, György 93
Alth, Alojzy 39
Althoff, Friedrich 297
Alton, Johann 91
Antal, Gyula 89
Apponyi, Albert 219
Arany, János 93
Arsen'ev, Konstantin 312
Ásbóth, Oszkár 93
Aschbach, Joseph 89
Ash, Mitchell 275

Bach, Alexander von 27
Bagster, Gerard George 91
Baintner, János 89
Balbi, Adriano 211
Ballagi, Aladár 92
Ballif, Philipp 122–124
Balling, Karl Joseph Napoleon 23
Ballmann, Johann Michael 198
Baranskii, Nikolaj M. 244–246, 248
Baric, Daniel 10 f., 13, 136, 147, 149
Bartol'd, Vasilii V. 174–176, 184
Bartoszewicz, Joachim 222
Bassin, Mark 239

Baudouin de Courtenay, Jan 183 f.
Becker, Philipp August 91
Beer, Hieronymus 87
Behr, Rudolf 91
Benndorf, Otto 131
Beöthy, Zsolt 93
Beran, Jiří 28, 39
Berger, Ev. János 94
Berzeviczy, Albert 213
Bethlen, István 224
Beurle, Carl 264
Bezard, Lucien 93
Bittner, Maximilian 91
Blache, Paul Vidal de la 208, 211
Blanc, Eduard 146
Bodnár, Zsigmond 93
Bogusławski, Wilhelm 52
Bolla, Martin 198
Boller, Anton 90, 154
Bormann, Eugen 90
Bowman, Isaiah 218
Bráf, Albín 38 f.
Branca, Wilhelm 296
Brandl, Vincenc 37
Brotanek, Rudolf 91
Brunhes, Jean de 211
Buchen, Tim 59
Budenz, József 93
Büdinger, Max 90
Bühler, Georg 91, 154 f.
Bujak, Franciszek 216
Burenin, Viktor Petrovich 142

Catherine II ("the Great," Empress of Russia)   109, 309, 311
Čelakovský, František Ladislav   30
Čelakovský, Ladislav Josef   30
Chertkov, Aleksandr Dmitrievich   230 f.
Cholnoky, Jenő   213
Christian, Viktor   167 f.
Ciocan, János   93
Clusius, Carolus   279
Collaud, Károly   92
Copernicus, Nikolaus   44, 50
Cousin, Victor   56, 71
Cserép, Jószef   92
Csokor, Johann Nepomuk   94
Cvijić, Jovan   208 f.
Czartoryski, Adam   105
Czekanowski, Jan   221
Czirbusz, Géza   211
Czoernig, Carl von   193

Dafinger, Johannes   303
Darányi, Ignác   213
Darwin, Charles   207
Den, Vladimir Ėduardovič   243–245, 248 f.
Depkat, Volker   137
Dézsi, Lajos   93
Dietl, Joseph (Józef)   56
Dmowski, Roman   218
Dobrovský, Josef   34
Dokuchaev, Vasilii   290–295, 304, 311, 316–319
Dondukov-Korsakov, Mikhail Aleksandrovich   114
Dopsch, Alfons   89
Dril', Dmitrii   257

Drucki-Lubecki, Franciszek Ksawery   60
Druzhinin, Sergei I.   179
Dryden, John   96
Durdík, Josef   40
Durége, Heinrich Jacob Karl   35
Dzhura-Bek   144

Eder, Josef Karl   198
Egan, James   92 f.
Ermakov, Nikolai Andreevich   140 f., 143 f.
Evtuhov, Catherine   306
Eybesfeld, Sigmund Conrad von   33

Ferdinand I (Holy Roman Emperor)   153
Ferdinand I (Emperor of Austria)   57
Ferdinand Maximilian (Archduke)   156
Ferenc, József   92
Ferri, Enrico   255
Ficker, Heinrich von   16 f.
Filat'ev, Vladimir Ivanovich   108
Frähn, Christian Martin   181 f.
Frank, Susi K.   240
Franz I (Emperor of Austria, as Holy Roman Emperor Franz II)   57, 280
Franz Joseph (Emperor of Austria)   23, 42, 156, 276, 287
Franzos, Karl Emil   163
Frauenstädt, Paul   263
Frič, Antonín   30, 32
Frič, Josef František   30

Index of Personal Names

Gasprinskii, Ismail 176 f.
Gasser, Rudolf von 145
Gautsch von Frankenthurn, Paul 37, 42, 159
Gebauer, Jan 34
Geim, Ivan Andreevich 108
Geraci, Robert 269
Geyer, Rudolf 91, 165, 168
Giammaria, Cattaneo 90
Giesecke, Fritz 302
Girardin, Saint-Marc 62, 69, 75
Gischig, Joseph 90
Glaser, Eduard 155
Glinka, Konstantin 290, 293–297, 299, 304, 317, 319 f.
Glowacki, Johann 91
Goethe, Johann Wolfgang von 106
Goldenthal, Jakob 90, 154
Goldziher, Ignaz 93, 158
Golitsyn, Alexander Nikolayevich 107
Golovkin, Yurii Alexandrovich 114 f.
Gorchakov, Aleksandr Mikhailovich 136
Graham, Loren 310
Gratacap, Marc 91
Grech, Ivan Mikhailovich 230
Gretsch, Johann-Ernst von 230
Grigor'ev, Vasilii V. 172
Grohmann, Adolf 167
Gross, Hans 256, 264
Grum-Grzhimailo, Grigorii Efimovich 146

Haberlandt, Michael 160
Haenke, Thaddäus 280
Haffner, August 91, 154
Haidinger, Wilhelm von 285
Halecki, Oskar 216
Hamann, Günther 156
Hamerling, Robert 154
Hammer-Purgstall, Joseph von 72, 155, 157 f., 163
Hampel, József 92
Hanslik, Erwin 164
Hanusz, Johann 91
Hatala, Péter 93
Hauer, Franz 286
Hauler, Edmund 90
Hausmann, Guido 20
Hausmann zu Stetten, Franz von 283
Haverfield, Francis John 133
Hée, Nadin 77 f.
Heinrich, Gusztáv 93
Helfert, Alexander von 162 f.
Hellbeck, Jochen 137
Heller, Klaus 137
Hermann, Tomáš 24
Herz, Hugo 264 f.
Hettner, Alfred 234, 242, 245 f., 249
Heufler, Ludwig von 284
Heyne, Christian Gottlob 110
Hilgard, Eugene W. 318 f.
Hlávka, Josef 14, 23–25, 36–42
Hodinka, Antal 92
Hoernes, Moriz (Moritz) 124, 126 f.
Hoffmann, Pál 88
Höfler, Konstantin von 23, 35
Högel, Hugo 264
Högl, Johann 90

Hörmann, Kosta 127
Horvát, Árpád 92
Horváth, Cyrill 93
Hrozny, Friedrich (Hrozný, Bedřich) 91, 165, 167
Hrushevs'kyi, Mychailo 247
Huber, Alfons 90
Hultzsch, Eugen 91
Humboldt, Alexander von 106, 210, 313 f.
Humboldt, Wilhelm von 106
Hunfalvy, Pál 201

Iakovkin, Ilya Fedorovich 109
Ianovskii, Kirill P. 177, 180
Iarilov, Arsenii, 320
Il'minskii, Nikolai I. 181, 184
Illés, József 88

Jäger, Albert 89
Janko, Jan 24, 28, 30
Jankowski, Czesław 211
Jedlicki, Jerzy 51 f.
Jennel, Engelbert 265
Jireček, Hermenegild 23
Jireček, Josef 23 f., 33, 37
Jireček, Josef Konstantin 91
Jobst, Kerstin 173
Joseph II (Holy Roman Emperor) 198
Judson, Pieter M. 78, 273
Jung, Julius 122, 201

Kachenovskii, Mikhail Trofimovich 109
Kállay, Benjamin von 123 f., 126 f., 129, 131
Kappeler, Andreas 171

Karabacek, Josef 91, 154 f.
Karácsonyi, János 215
Karl Ludwig (Archduke) 42
Karneev, Egor Vasil'evich 108, 111
Karneev, Zakharii Iakovlevich 107
Károlyi, Mihály 217
Katanov, Nikolai F. 181, 183 f.
Kaufman, Konstantin Petrovich von 135
Kaufmann, Stefan 213
Kaunitz, Anton von 157
Kautz, Gyula 89
Kayurszky, György 94
Kazem-Bek, Aleksandr K. 172, 181
Kégl, Sándor 93
Kerékgyártó, Árpád Alajos 92
Kieniewicz, Stefan 45
Kirste, Johann 151, 154
Kjellén, Rudolf 207, 243 f., 249
Klinger, Friedrich Maximilian 103, 112
Kliuchevskii, Vasilii O. 19, 240
Klun, Vincenz 89
Kmety, Károly 89
Kobeko, Dmitrii Fomich 143–145
Koch, Wilhelm Daniel 15, 279, 281 f.
Komáromy, András 92
Komlosy, Andrea 192 f.
Konek, Sándor 89
Korányi, Frigyes 94
Korsh, Fëdor E. 174
Kostychev, Pavel 316 f.
Kovachich, György 198
Kovalevskii, Maksim Maksimovich 266

Kraelitz von Greifenhorst, Friedrich 157, 166
Krall, Jacob 91
Kraus, Alfred 37
Krcsmárik, János 88, 92
Kremer, Alfred von 158–160
Kremer-Auenrode, Hugo von 88
Kretschmayr, Heinrich 89
Kropotkin, Petr A. 235 f., 248
Kubitschek, Wilhelm 89 f.
Kühnert, Franz 91
Kúnos, Ignác 93
Kurella, Hans 258, 261–263
Kutrzeba, Stanisław 216
Kutschera, Hugo 123
Kuzsinszky, Bálint 92

Lamanskii, Vladimir I. 174
Lambert, David 59
Lánczy, Gyula 92
Laube, Gustav Karl 35
Laurian, Treboniu 200
Laurin, Franz 86, 94
Leciejewski, Johann 91
Lega, Georg 90
Lenhossék, Mihály Ignác 74
Lenin, Vladimir 244 f.
Leopold II (Holy Roman Emperor) 198
Lewis, Lajos 92 f.
Lieve, Karl Andreevich 112
Liszt, Franz von 256 f., 259, 267
Livingstone, David 306 f.
Lóczy, Lajos 213
Lombroso, Cesare 255–258, 261, 269 f., 272
Lord, Robert H. 218

Lorenz, Joseph 89
Lorenz, Ottokar 89
Lotheissen, Ferdinand 91
Ludowici, Johann Wilhelm 301
Ludwig, Alfred 35

Mach, Ernst 35
Mackinder, Halford 207, 241, 248
Maddalena, Edgardo 91
Magnitskii, Mikhail Leont'evich 108, 180
Mahler, Ede 92
Makanec, Julije 124
Málek, Ivan 31
Maly, Joseph Karl 278 f.
Manteuffel, Gotthard Andreas 103
Marbut, Curtis F. 315, 319 f.
Marczali, Hernik 92
Margalits, Ede 93
Maria Theresia (Holy Roman Empress) 153, 157
Markovits, Johann 90
Marr, Nicolai 172, 174, 178
Martonne, Emmanuel de 208, 217 f., 224
Masaryk, Tomáš Garrique (T. G.) 34 f., 39–41, 167
Matasci, Damiano 56 f.
Mayr, Aurél 93
Medveczky, Frigyes 93
Mehringer, Rudolf 91
Menčík, Ferdinand 91
Mesgnien Meninski, Franz von 157
Messi, Antal 92 f.
Metternich, Klemens von 56 f., 62, 65, 72

Meyer, Konrad 303
Meyer-Lübke, Wilhelm 91
Mianowski, Józef 45
Michael Romanov (Tsar of Russia) 311
Micińska, Magdalena 46, 52
Mickiewicz, Adam 44
Mika, Sándor 92
Miller, Vsevolod F. 174
Milosich, Franz Xaver 90
Mogilner, Marina 251, 271
Moldovan, Gheorghe 200, 202
Molnár, Géza 93
Molostvov, Vladimir P. 183
Mommsen, Theodor 124 f., 133
Moon, David 16
Morczko, Marian 209
Morel, Bénédict A. 269
Much, Rudolf 91
Müller, David Heinrich 91, 154 f., 168
Müller, Friedrich 90 f., 154 f.
Müller, Heinrich 91
Münchhausen, Gerlach Adolph von 107
Murav'ëv, Mikhail Nikitich 105 f., 108
Musil, Alois 166 f.
Musin-Pushkin, Mikhail N. 111, 113 f., 181 f.
Mussafia, Adolph 90 f.
Mutschenbacher, Alajos 92
Mžik, Hans von 151, 163

Näcke, Paul 258 f.
Nawroczyński, Bogdan 47
Neilreich, August 279, 281 f.
Némethy, Géza 93

Neruda, Jan 41
Neugeboren, Daniel 197
Neumann, Leopold 88
Neumann, Wilhelm Anton 153
Nicholas I (Emperor of Russia) 110, 113, 115, 138
Nietzsche, Friedrich 92
Nikolai, Aleksandr 177
Novosil'tsev, Nikolai Nikolaevich 105
Nuić, Anđeo 129

O'Neill, Kelly 307
Offmański, Mieczysław 47
Olay, Ferenc 225
Ol'denburg, Sergei Fëdorovich 138, 146 f., 175, 180
Orłowicz, Mieczysław 49 f.
Ostendorff, Eberhard 301, 303
Osten-Saken, Fëdor Romanovich 138, 140–146
Ototskii, Pavel 317
Ottenhals, Emil von 90

Palacký, František 23, 27, 32
Parrot, Georg Friedrich 102
Patrubány, Lukács 93
Patsch, Carl 10, 13, 16, 19, 119–123, 125–133, 136, 139 f., 147 f.
Pázmány, Zoltán 89
Pecz, Vilmos 93
Penck, Albrecht 208 f., 216, 222, 246
Perovskii, Alexey Alexeyevich 108, 112
Perthes, Justus 222
Peter I ("the Great," Emperor of Russia) 309–311

Index of Personal Names

Petőfi, Sándor 93
Petrovskii, Nikolai Fëdorovich 7–9, 12, 135 f., 138–149
Petz, Gedeon 93
Phillips, Georg 87 f.
Piłsudski, Józef 218
Pisarev, Aleksandr Aleksandrovich 108, 110 f.
Pol, Wincenty 210
Polak, Jakob 91
Polner, Ödon 89
Popoviciu, József 93
Posnikov, Aleksandr S. 178
Potocki, Józef 51
Potocki, Seweryn 105 f., 108, 110
Prażmowski, Adam 50
Precht, Johann Joseph 62
Przheval'skii, Nikolai Mikhailovich 145 f.
Pughe, Francis H. 91
Pulszky, Ágost 88
Purkyně, Jan Evangelista 14, 23–32, 39, 41 f.
Puşcariu, Sextil 91
Puschmann, Theodor 87

Radlov, Vasilii V. 184
Rákosi, Sándor 93
Ramann, Emil 295
Ratzel, Friedrich 207, 233, 238, 242–244, 246, 249
Razumovskii, Aleksei Kirillovich 105 f., 108 f.
Razumovskii, Kirill Grigorjewitsch 105
Reclus, Elisée 234 f.
Redlich, Oswald 90
Reinisch, Leo 91, 154 f.

Reméle, Johann 90
Rešetar, Milan Ritter von 91
Rey, Armand 91
Rezek, Antonín 36
Richthofen, Ferdinand von 234
Riedl, Frigyes 93
Riedl, Szende 92
Rieger, František Ladislav 38
Ritter, Carl (Karl) 210, 233, 240
Rolf, Malte 59, 147
Román, Sándor 92 f.
Romer, Eugeniusz 209 f., 212, 216, 218, 222 f.
Rómer, Flóris 91
Rosen, Viktor 170, 174 f., 184
Rößler, Robert 201
Rozen, Viktor Romanovich 146 f.
Rudnyc'kyi, Stepan 209, 215, 246–249
Rudolf (Crown Prince of Austria) 162
Rumovskii, Stepan Iakovlevich 104, 108 f., 181
Runich, Dmitrii Pavlovich 108
Ruprecht, Fanz Joseph 313
Ruzsicska, János 94

Sachau, Eduard 154
Sacher-Masoch, Leopold von 27
Sachs, Julius 30
Šafařík, Pavel Josef 23, 29
Sághy, Gyula 89
Said, Edward 161, 169
Sakhnovskaia, Sofia Alekseevna 139
Sandl, Markus 207
Savitskii, Petr N. 241
Scala, Arthur von 162

Schimmelpenninck van der Oye, David 176
Schipper, Jacob 91
Schlagintweit, Adolf 145, 149
Schreber, Johann Christian von 106
Schroeder, Leopold 91
Schucht, Friedrich 305, 315
Schultes, Joseph August 280 f.
Schwarz, Gusztáv 88
Šebek, František 25
Šembera, Alois 90
Semenov-Tian-Shanskii, Petr P. 233, 312
Semenov-Tian-Shanskii, Veniamin 235–246, 248 f.
Sering, Max 300 f.
Seydler, August 35
Shantz, Homer Leroy 315
Shaw, Denis 311
Sibirtsev, Nikolai 317
Sickel, Theodor 90
Siegel, Heinrich 88
Simonyi, Zsigmond 93
Sklenař, Joseph 91
Smirnov, Nikolaj 231
Smith, Neil 218
Sokolov, A. T. 233
Solov'ev, Sergei M. 240
Somarides, Eugen 91
Somogyi, Manó 89
Sonnenfels, Josef von 109
Sprenger, Aloys 145, 158
Srbik, Heinrich von 90
Stalin, Joseph 245 f., 248
Stanislas II. Augustus Poniatowski (King of Poland) 60

Stasiulevich, Mikhail Matveevich 143
Staszic, Stanisław 43 f., 47
Stephan, Anke 148
Stockinger, Julius 91
Stöwesand, Walther 262, 268
Strakhov, Peter Ivanovich 108
Stremme, Hermann 289 f., 294–304, 318
Stremoukhov, Pëtr Nikolaevich 141 f.
Stroganov, Sergei Grigoryevich 99, 114–116
Strossmayer, Josip 23, 30
Sujkowski, Antoni 221
Superville, Daniel de 107
Suvorin, Aleksei Sergeevich 142
Szádeczky, Lajos 92
Szentmiklósi, Marton 88
Szinnyei, József 93

Takács, Lajos 88
Tamul, Villu 102
Tarnovskaia, Praskov'ia Nikolaevna 270
Tarnovskii, Evgenii N. 267
Tęgoborski, Ludwik 15, 57, 59–66, 68–76
Tęgoborski, Walerian 60
Teleki, Géza 213
Teleki, Pál 212–214, 217 f., 220, 224
Thewrewk, Emil 93
Tigranov, Grigorii F. 179
Tisza, István 213 f.
Toldy, Ferenc 92
Tolz, Vera 17, 149, 171 f.
Tomaschek, Johann Adolph 87 f.

Index of Personal Names

Tomek, Václav Vladivoj 56
Torma, Károly 92
Truhelka, Ćiro 124, 131
Tumanov, Georgii M. 178 f.
Turba, Gusta 89

Uebersberger, Hans 89
Unger, Franz 282
Uvarov, Sergei Semënovich 106, 113–117

Vaijayantî, Yâdavaprakâśa's 155
Valuev, Peter 308
Vámbéry, Ármin (Arminius) 92 f., 143
Vasenev, Aleksei Danilovich 140
Vécsey, Tamás 88
Veghy, Anton 88
Vernadskii, Vladimir I. 237
Vetter, Jeremy 307
Vil'd, Genrikh Ivanovich 144
Vilbouchevitch, Jean 319
Vincenti, Carl Ferdinand von 152
Vocel, Jan Erazim 23
Voeikov, Aleksandr 319
Volkonskii, Grigory Petrovich 114
Vondrák, Wenzel 91
Vörösmarty, Mihály 93

Wahrmund, Adolph 90 f.
Waldner, Victor 88
Waltenhofen zu Eglofsheimb, Adalbert Carl von 33
Warschauer, Adolf 49
Weinberg, Richard 259, 268
Whitney, Milton 319

Widmanstetter, Johann Albrecht 153
Wiesner, Adolf 61
Wilde, Oscar 72
Wilde, William 59, 72–76
Wilhelm II (German Emperor) 297
Willkomm, Moritz 35
Winkelbauer, Thomas 78, 80
Wulfen, Franz Xaver 278

Yolland, Arthur Battishill 93

Zakrevskii, Ignatii Platonovich 257
Zambra, Péter 93
Zhukovskii, Valentin A. 174
Zingerle, Wolfram 91
Zippe, Franz Xaver Maxmilian 23
Zolnai, Gyula 93
Zon, Raphael 315 f.
Zuev, Nikolaj I. 231 f.

CONTRIBUTORS

Prof. Andrej Andreev, Professor, Faculty of History, Lomonosov Moscow State University and Faculty of History, St. Tikhon's Orthodox University Moscow (Russian Federation)

Dr. Jan Arend, Researcher, Institute for Eastern European History and Area Studies, Universität Tübingen (Germany)

Dr. Daniel Baric, Associate Professor, Department of Slavic Studies, Sorbonne University, Paris (France)

Dr. Johannes Feichtinger, Senior Research Associate, Institute of Culture Studies and Theatre History, Austrian Academy of Sciences, Vienna (Austria)

Doc. PhDr. Martin Franc, Ph.D., Head of the Department for the History of the Academy of Sciences, Masarykův ústav a Archiv Akademie věd České republiky, Prague (Czech Republic)

Matthias Golbeck, M.A., Research Fellow, Institute of History, Rheinische Friedrich-Wilhelms-Universität Bonn (Germany)

Prof. Dr. Peter Haslinger, Director of the Herder Institute for Historical Research on East Central Europe, Marburg; Professor, Historical Institute, Justus-Liebig-Universität Gießen (Germany)

Prof. Dr. Guido Hausmann, Head of History Division, Leibniz-Institut für Ost- und Südosteuropaforschung; Professor, History Department, Universität Regensburg (Germany)

Prof. Dr. Mark Sven Hengerer, Professor for Early Modern History, History Department, Ludwig-Maximilians-Universität Munich (Germany)

Prof. Dr. Maciej Janowski, Professor, Institute of History, Polska Akademia Nauk, Warsaw (Poland)

Prof. Mag. Dr. Marianne Klemun, Professor, Department of History, University of Vienna (Austria)

Dr. Arpine Maniero, Researcher, Collegium Carolinum, Munich (Germany)

Prof. David Moon, PhD., Professor, Department of History, University of York (UK); Visiting Professor, Nazarbayev University, Nur-Sultan (Kazakhstan)

Sabrina Rospert, M.A., PhD student LMU München/Université Paris I Panthéon-Sorbonne, Research Fellow at the German Historical Institute Paris (France)

Dr. Jan Surman, Research Fellow, Poletayev Institute for Theoretical and Historical Studies in the Humanities, National Research University Higher School of Economics (Russian Federation)

Dr. Borbála Zsuzsanna Török, University Assistant at the Institute of Austrian Historical Research, University of Vienna (Austria)

Prof. Dr. Volker Zimmermann, Research Fellow, Collegium Carolinum, Munich (Germany); Adjunct Professor, Department of Historical Studies, Heinrich-Heine-Universität Düsseldorf (Germany)

NEUES STANDARDWERK ZUR JÜDISCHEN GESCHICHTE DER BÖHMISCHEN LÄNDER

Kateřina Čapková | Hillel J. Kieval (Hg.)
Zwischen Prag und Nikolsburg
Jüdisches Leben in den böhmischen Ländern

Veröffentlichungen des Collegium Carolinum, Band 140

2020. 428 Seiten, mit 76 Abb., 23 Tab. und 14 Karten, gebunden
€ 70,00 D
ISBN 978-3-525-36427-7

Seit rund zwei Jahrzehnten erfreuen sich die jüdische Geschichte und Kultur der böhmischen Länder eines wachsenden Interesses. Damit rückt der historisch multiethnische Charakter der Region verstärkt ins Zentrum der Aufmerksamkeit. Vor diesem Hintergrund ist es umso erstaunlicher, dass bislang noch keine innovative Synthese dieser Forschung vorlag. Vorliegendes Buch aus der Feder eines internationalen Autorenteams nimmt sich daher erstmals der Herausforderung an, die jüdische Erfahrung in den böhmischen Ländern als integralen und untrennbaren Bestandteil der Entwicklung Mitteleuropas vom 16. Jahrhundert bis heute zu erzählen und zu analysieren. Dabei geht es ebenso um Kontakte der jüdischen Bevölkerung mit ihren nichtjüdischen Nachbarn wie um den Blick in die Provinz, das heißt in die ländlichen Regionen und Gemeinden abseits der großen städtischen Zentren Prag, Brno und Ostrava.

Vandenhoeck & Ruprecht Verlage
www.vandenhoeck-ruprecht-verlage.com

1968: CZECHOSLOVAKIA'S REFORM EXPERIMENT

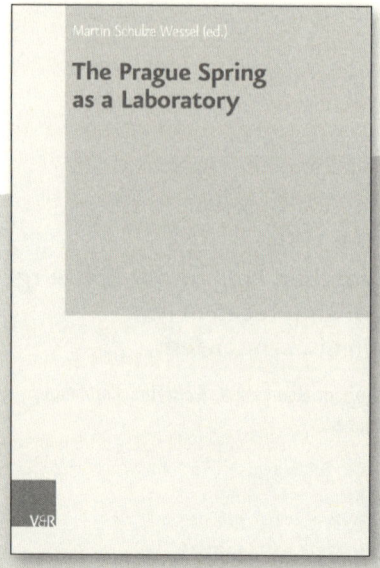

Martin Schulze Wessel (ed.)
The Prague Spring as a Laboratory

Bad Wiesseer Tagungen des Collegium Carolinum, Vol. 40

2019. 312 pages, hardcover
€ 50,00 D
ISBN 978-3-525-35598-5

Retrospectively, the Prague Spring is generally interpreted as a coherent but unsuccessful experiment in finding a synthesis of Western democracy and socialism. However, this perspective ignores that different groups and individuals participated in these developments and shaped the reform experiment in Czechoslovakia with their completely varying professional, generational, national, and gender-specific experiences. What appears as a goal-oriented reform movement or as an "interrupted revolution" in hindsight looked in the eyes of the protagonists rather like the situation in a laboratory. The volume focuses on the protagonists' ideas of politics, society, and their reform plans. Of particular interest is the question which new thoughts about the interrelation of politics, science, economics, and arts were developed in Czechoslovakia.

Vandenhoeck & Ruprecht Verlage
www.vandenhoeck-ruprecht-verlage.com

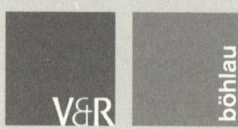

HUMANITÄRE MASSNAHME ODER NATIONALSOZIALISTISCHE KRIEGSPOLITIK?

Martin Zückert / Michal Schvarc / Martina Fiamová
Die Evakuierung der Deutschen aus der Slowakei 1944/45
Verlauf, Kontexte, Folgen

Veröffentlichungen des Collegium Carolinum, Band 139

2019. 349 Seiten, gebunden
€ 50,00 D
ISBN 978-3-525-31075-5

Im Herbst 1944 ordnete Heinrich Himmler angesichts des Vorrückens der sowjetischen Truppen nach Westen die Evakuierung der deutschen Bevölkerung aus Ostmittel- und Südosteuropa an. Lässt sich diese Evakuierung, wie nach 1945 in der deutschen Erinnerung überwiegend geschehen, ausschließlich als humanitäre Maßnahme im Interesse der von Kriegshandlungen bedrohten Zivilbevölkerung verstehen? Oder standen dahinter nicht auch Zielsetzungen der nationalsozialistischen Kriegspolitik? Am Beispiel der Deutschen in der Slowakei untersuchen die Autoren Hintergründe, Abläufe und Zusammenhänge der Evakuierung in den Jahren 1944–1945.

Vandenhoeck & Ruprecht Verlage
www.vandenhoeck-ruprecht-verlage.com

LOYALITÄT ALS ANSATZ FÜR DEN IMPERIENVERGLEICH: DAS RUSSLÄNDISCHE REICH UND DIE HABSBURGERMONARCHIE

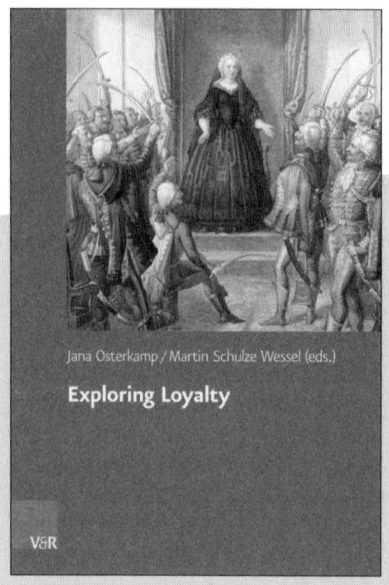

Jana Osterkamp /
Martin Schulze Wessel (eds.)
Exploring Loyalty

Veröffentlichungen des Collegium Carolinum, Band 136

2017. 240 Seiten, gebunden
€ 50,– D
ISBN 978-3-525-37317-0

Das Konzept »Loyalität« ermöglicht einen neuen und frischen Blick auf politische Kulturen in Geschichte und Gegenwart sowie politischen Wandel. Loyalität steht neben neueren Ansätzen aus der Geschichtsschreibung zu Emotionen und politischer Kultur wie Vertrauen, Treue, Solidarität, Patriotismus oder Identität. Zugleich bildet der Begriff besser als andere die Mehrschichtigkeit sozialer Beziehungen ab und erfasst deren vertikale und horizontale Ausprägungen. Dieser Band untersucht Loyalitäten in der Geschichte der Region Ost- und Ostmitteleuropa im 19. und 20. Jahrhundert. Der Vergleich umfasst sowohl das Russländische und das Habsburger Reich als auch die dem Zerfall beider Imperien folgende Zeit nach dem Ersten Weltkrieg. Die Autorinnen und Autoren gehen dabei nationalen, konfessionellen, politischen und militärischen Loyalitäten nach.

Vandenhoeck & Ruprecht Verlage

www.vandenhoeck-ruprecht-verlage.com